Interior Design and Decoration

Interior
Design and
Decoration

Premavathy Seetharaman PhD
Reader, Department of Resource Management
Institute of Home Economics (University of Delhi)

Parveen Pannu PhD
Reader, Department of Resource Management
Institute of Home Economics (University of Delhi)

CBSPD

CBS Publishers & Distributors Pvt Ltd

New Delhi • Bengaluru • Chennai • Kochi • Kolkata • Lucknow • Mumbai
Hyderabad • Jharkhand • Nagpur • Patna • Pune • Uttarakhand

Interior Design and Decoration

ISBN: 978-81-239-1192-2 (PB)
      978-81-239-1212-7 (HB)

Copyright © Authors

First Edition: 2005
Reprint: 2007, 2009, 2010, 2012, 2013, 2014, 2015, 2017, 2019, **2025**

Published by **Satish Kumar Jain** and produced by **Varun Jain** for

CBS Publishers & Distributors Pvt Ltd

4819/XI Prahlad Street, 24 Ansari Road, Daryaganj, New Delhi 110 002, India.
Ph: 011-23266838, 23289259   Website: www.cbspd.com
                                e-mail: delhi@cbspd.com

Corporate Office: 204 FIE, Industrial Area, Patparganj, Delhi 110 092
Ph: 011-4934 4934   Fax: 011-4934 4935
                           e-mail: publishing@cbspd.com; publicity@cbspd.com

Branches

- **Bengaluru:** Seema House 2975, 17th Cross, KR Road, Banasankari 2nd Stage, Bengaluru 560 070, Karnataka, India
  Ph: +91-80-26771678/79   Fax: +91-80-26771680   e-mail: bangalore@cbspd.com
- **Chennai:** 7, Subbaraya Street, Shenoy Nagar, Chennai 600 030, Tamil Nadu, India
  Ph: +91-44-26680620, 26681266   Fax: +91-44-42032115   e-mail: chennai@cbspd.com
- **Kochi:** 42/1325, 1326, Power House Road, Opp KSEB, Power House, Ernakulum Kochi 682 018, Kerala, India
  Ph: +91-484-4059061-65,67   Fax: +91-484-4059065   e-mail: kochi@cbspd.com
- **Kolkata:** 147, Hind Ceramics Compound, 1st Floor, Nilgunj Road, Belghoria, Kolkata-700056, West Bengal, India
  Ph: +033-25633055, 033-25633056   e-mail: kolkata@cbspd.com
- **Lucknow:** Basement, Khushnuma Complex, 7 Meerabai Marg (Behind Jawahar Bhawan), Lucknow-226001, UP, India
  Ph: +0522-4000032   e-mail: tiwari.lucknow@cbspd.com
- **Mumbai:** PWD Shed, Gala no 25/26, Ramchandra Bhatt Marg, Next to JJ Hospital Gate no. 2, Opp. Union Bank of India, Noorbaug, Mumbai-400009, Maharashtra, India
  Ph: 022-66661880/89   e-mail: mumbai@cbspd.com

Representatives

- Hyderabad   0-9885175004
- Patna   0-9334159340
- Jharkhand   0-9811541605
- Pune   0-9664372571
- Nagpur   0-8692091830
- Uttarakhand 0-9716462459

Printed at SRK Graphics, Delhi (India)

PREFACE

Values in home making have changed over the years. More housewives are now employed outside the house and thus run two jobs. This has enabled them to increase their purchasing power and spend more on expensive items. They like to live not only in comfortable houses but also in attractive and aesthetically decorated homes. No one can, however, achieve this objective without proper planning. One needs to devote a good number of hours in the home to decorate it to one's satisfaction. Regardless of how one achieves this goal, there is a need to do some ground work even in the purchase of a simple item like a table lamp or a cushion cover, so that it is a welcome and useful addition to the home.

Whether a person knows it or not, every one develops his or her own tastes from early years and this individuality is indelibly stamped in the environment from the very beginning. A home that speaks warmly and welcomingly of the people who live there, reflects their tastes and needs through their successful decorative endeavours.

Essentials of good decorating are easy to learn and enjoyable to practice. It is true that all are not born with the gift of good taste but any one can learn to develop and acquire the skill to appreciate good art and designing by applying the principles of beauty deliberately. Learning to appreciate the fundamental elements of art and design would help a person to create, acquire or arrange objects of art and beauty aesthetically.

Many books have been brought out in the recent years that instruct and inform the reader on the specific issues on interior decoration. However the present book aims to extend its scope to include not only the basics on interior decoration but also offers the reader a complete manual on the subject taking into account the place occupied by it in the making of a home which defines the life style of its occupants. The book intends to give practical ideas on interior decoration to a student learner or an inexperienced housewife in a way that is not only most satisfying but one which will also suit her and her family's tastes and above all, the family's budget.

A good deal of effort has gone into the collection of available literature on the subject, the extent of availability of art objects, contemporary and traditional designs and materials. A number of illustrations have been included to supplement the ideas in words. The subject matter has been split into four major units—Aesthetics and Design Basics, Interior Designing, Interior Furnishings and Interior Decoration and each Unit has four chapters each devoted to deal with the various aspects of Interior Decoration in detail.

Unit I dealing with Aesthetics and Design basics would enable the reader to become familiar with good taste in design concepts and the elements of art and design, while Unit II focuses on creating awareness on the various aspects of Interior decoration. Colour being one of the most important aspects of Interior Decoration, has been dealt with extensively in two different chapters. Lights and Lighting, walls and wall finishes and ceilings which provide the major aesthetic background for the interiors have also been dealt with exclusively in the other chapters of the book.

Floors and floor coverings, soft furnishings and furniture and their selection have been discussed in detail in Unit III. Selection and use of art objects to enhance the beauty and decor of the interiors has also been discussed in this unit in separate chapters.

The subject of Interior Decoration will be incomplete without a discussion on accessories popularly known to enhance the beauty and the look of the indoors. The last unit is, therefore, devoted to the aesthetic decoration of the interior by use of flowers, and flower arrangements, antiques and objects of art, traditional and modern paintings, attractive Rangoli patterns etc. This unit also contains a chapter that traces the historical development as well as trends in Interior Decoration in India over the years.

Having been members of the teaching profession for over more than two decades, the authors have kept in mind the needs of the present day students to provide a comprehensive but inexpensive book on the subject which they can even retain and use long after they have left the portals of learning. However, the authors do not claim that everything that everyone wanted to know on the subject has been included in the book. The authors would, therefore, welcome any suggestions in this regard.

The authors gratefully acknowledge the valuable suggestions offered by many friends and colleagues in the making of this book. These have been given due consideration and a good many of them were also adopted.

Thanks in particular are due to—Miss Monica Murali for her help in the preparation of the drafts in the initial stages of the book; Miss S. Gayatri for help in organising the information in appropriate formats and for some intricate drawings; Mr. N. Seetharaman who assisted in the proof-reading of the book at a critical stage, to meet the deadlines of the Publishers.

Authors

CONTENTS

AESTHETICS AND DESIGN BASICS

> *Aesthetics and Good Taste*
> *Design: Concepts and Types*
> *Elements of Art*
> *Principles of Design*

In home planning and interior designing, it is extremely important to keep in mind continually the idea of the home as a setting for a happy family life. A *home-like effect* is one of the rarest qualities a house can possess and depends entirely upon the basic idea of enjoying living in a home. A good architectural style will accomplish its purpose only when it is fully exploited and utilised properly while designing and furnishing its interiors.

When we choose objects and design our homes by using them, we are doing two things. In the first instance, we gratify some need or desire, and secondly, we are, through the qualities possessed by these objects, unconsciously stating our personality to everyone with the power and insight to interpret the meaning behind our choice. As we surround ourselves with beauty, art actually becomes a part of our life and personality.

Few people are born with the rare gift of good taste. However, it is possible for others to develop and acquire good taste in art and designing by applying the principles of beauty deliberately. Through constant effort and practice, a person can easily display good taste in all his creations unconsciously.

Keeping these factors in mind, unit I of this book is devoted to the various aspects of good designing. It begins with a chapter on developing good taste and its significance in our day-to-day life. The second chapter deals with the details on the types of design and also their requirement, since interest in beauty is not merely confined to the people who create artistic objects; it concerns everyone of us who select and arrange such items like furniture, furnishings, pictures, accessories etc. for decorating the interiors.

For any decoration to be successful, it is necessary to understand what constitutes a good design. The entire purpose of spending the scarce resources like time, effort and money on decoration is to make it attractive and pleasing to the eye. A complete knowledge on the various fundamental elements of art and design would help a person to create and arrange objects in an aesthetic manner. The third chapter is devoted to the details on the elements of art.

To create beautiful objects consistently, one need to follow certain guidelines in selecting and arranging art elements. These guidelines, otherwise known as the principles of art, are common to all forms of art like painting, sculpture, architecture, handicrafts etc. These principles are not only the formulae for creating beautiful objects, but also help in judging whether an object is aesthetically good or poor. The fourth chapter in this unit, therefore, deals with these principles and their application in home interiors.

Chapter 1

AESTHETICS AND GOOD TASTE

Man has a strong urge to express his creative and aesthetic instincts in visual form. Since his home is the centre of all his activities, and his way of life and thoughts spring from within this protective shelter, man from the earliest times appears to have begun at first to paint and draw simple pictures on the walls and floors of his dwellings. This has helped him to satisfy his inner creative artistic impulse – the love of the beautiful. The inherent instinct for decoration, therefore, seems to have closely bound up with the structure of the home or other objects he produced or came into contact with, in all forms and shapes.

Thus, at first, the particular architecture of the age seems to have lent the form, the place and the basic material for the artistic creations of man to find fulfillment. Within each structure, he then gave of his fullest to supply the objects of his needs through his special mental and physical abilities and his particular sensibilities. His home, his place of worship, his implements, the receptacles for his food and drink and ritualistic offerings, his apparel and all other commodities of utility and dedication were thus fashioned with deep thought according to his way of life. Inspired by the surroundings of nature, he decorated these with an instinctive refinement to produce things of beauty.

Form became pre-eminent and got rooted in function. It became an important factor in all the things man came across. Then ornamentation came to highlight the beauty of the basic material. Thus the intrinsic beauty of a particular object was never lost and its grace of shape or texture was highlighted by its surface decoration.

HISTORICAL PERSPECTIVE OF ART AND DECORATION IN INDIA

Indian art is an immediate expression of Indian civilization as a whole. It represents beliefs and philosophies, ideals and outlooks, the materialized vitality of the society and its spiritual endeavour in varying stages of development. From the dawn of history, India has been the meeting place of peoples and cultures, interacting upon one another – coalescing, absorbing or conquering. The historical process of social intermingling in the national pattern began with the Aryans displacing the established Dravidian inhabitants which was possible partly by the passive acceptance by the latter many of the former's beliefs. The other participants in this process of intermingling were the

Parthians, the Greeks, the Sakas, the Huns, the Turks, the Afghans, and the Mongols. They all had brought their own intellectual and emotional experiences and during their reign as sovereigns of the land sought to impose them on the people they had conquered.

However, these things did not hinder the Indians in their quest for art and beauty. Indians were culturally strong and these invasions did not stop them from creating beautiful things in pottery, sculpture, fabrics, handicrafts, etc. Indians utilized the imposed cultures to improve their aesthetic knowledge in their pursuit of art and culture.

The Indians have always been conscious of decorative ornaments. But a very striking feature of the Indian life, has been the 'socialisation of the arts'. Thus, the Indians have responded to their instinct for artistic embellishment in their day-to-day life with so much vigour that there is no fast, festival or social event in the family which is not celebrated with all the traditional decorative ritualism, such as the birth of shoots at the close of winter, the ripening of the paddy or the golden wheat or such family events as the naming ceremony of the child, the weddings, the first entry into a new house or even the dawning of a new day. All these have inspired people to indulge in a luxurious ornamentation of the floors and walls of houses and courtyards with patterns of singular charm and graceful variety. The women especially pour their very gleeful souls into the work and transform the cold bare surfaces of the walls and floors into a glowing mass of decorative designs with mere powders and paste, and also with flowers

SIGNIFICANCE OF GOOD TASTE

A thing of beauty is a joy for ever (Fig. 1.1 – see clour plate 1). All of us enjoy beautiful things be it our possessions or surroundings. So, art becomes a part of our life and influences our personality. In fact, the impression created by us on anyone with whom we interact is due to our mannerisms, clothes and all other things we exhibit which is a reflection of our taste. Beauty is not something which is for occasional enjoyment, but it is present in everything we do, choose and use, and therefore, it leads to the experience of enjoyment and pleasure.

We, as human beings have an instinct to accumulate and to possess and for that reason, every time one goes to the market, one chooses the things that are pleasing to him or appear to be good. One is always using consciously or unconsciously, the power to choose, which comes from the principles fundamental to good taste. Now the question arises as to what a good taste means. Is it something inborn, can it be acquired or are all of us blessed with that gift? Do we all have an opportunity to experience all the good things in our life? Can we generalize it for everyone?

Good taste in the field of art is the application of principles of design to the problems in life where appearance as well as utility are together considered. These principles of design are the guides for producing a desired result. They are dynamic and flexible but are never static. We can take the example of fashion which keeps changing with the passage of time. This is because, the things we have enjoyed for some period of time become monotonous and no longer please us. So, we try to change, and arrive at a new form which is pleasing to our eyes and it comes to stay. However, after some period, it takes on another new form. Similarly, with the passage of time, we have changed from brass utensils to stainless steel utensils and more recently plastics have taken their place. Still recent ones are the earthernwares that are used in the modern microwave cooking. One cannot say that the brass utensils, plates, kettles, pots etc. were not beautiful, but they only became monotonous. In fact traditional brassware is now the in-thing to be used for decorating the big drawing rooms, hotel lobbies, stage settings etc. and thus have become the most sought after decorative pieces. All these reflect our quest for good taste which goes on changing with time.

According to Goldstein, good taste is 'doing unconsciously the right thing at the right time, in the right way'. Good taste gives a bias to intellectuals. It is a quality of inner soul and is possessed by a very few of us from birth. However, experience has proved that good taste in art can be acquired too. One can acquire good taste by practising the principles of design deliberately, until the time is reached when the right thing is done unconsciously.

The taste of an individual might be an inborn ability or the result of that person's educational and cultural experiences, values and attitudes. It is purely subjective and the emotional response created by the socialization process and is affected by the environment. According to Roach and Eicher, "taste is exhibited within a social context and is judged in relation to standards for taste that have grown out of the behaviour patterns of the social group".

In this context, a few things are important to be understood. Most people who enjoy the effect of richness, need to differentiate between the rich and gaudy, while others who like simple things, need to know the difference between plain and monotonous. Similarly, most of us come to think of art as a synonym for decoration and an object must be ornamented if it has to have an 'art quality'. This is not true. A simple pot of clay created by a potter, may be beautiful because of its lines and curves. On the other hand, a fully embroidered silk saree may not look attractive because of poor colour combination and disharmony between the cloth and threads used for embroidery. Generally, it is seen that over-decoration is one of the major faults which is committed while creating beautiful things. A little training can guide a person, as to how a mere variation in proportion or texture or an addition of matching or contrasting colour can enhance the quality of an art object.

Taste, to a large extent, is moulded by the things that surround us. Family taste is developed by the homemaker through the objects she selects. Therefore it becomes all the more important for a homemaker to be trained and given the knowledge about the principles of art. If she chooses things of beauty, it enables the other members of the family to develop good taste and appreciate beautiful things. By living in a beautiful surrounding, they get an opportunity to modify and form their own tastes.

Most of us face certain limitations while we are choosing objects for our use. We have to compromise when a thing is not available at the time of purchase, or the prices are very high. Under such circumstances, it is important that one does not make a poor choice and compromise on their standards.

One of the misconceptions about art is that it is confined to refer to painting or sculpture. Art is, however, not limited to these only. Its scope is much larger and deeper. Painting or sculpture is just a form of an artistic expression. Art is not only limited to those who create things of art, but to all those who indulge in the appreciation of art objects and are engaged in a creative experience. Active appreciation and enjoyment is a form of participation in it. Those who enjoy beautiful things and help others to enjoy and modify their taste are also working for art in the same manner as those who are creating things of art.

A woman who is choosing beautiful furnishings for her home or a man who is choosing beautiful clothes for his children, or children who are selecting beautiful toys for themselves are also practising art.

DESIGN AND GOOD TASTE

Design is selecting and arranging materials with order and beauty as the two chief aims. One can use just ordinary paints and canvas to create a painting which is enjoyed and appreciated by many.

Great artists do create things of art with beauty and order. Many of us have the same access to paint and canvass, but may not be able to create something which can please even one's own self, because of lack of imagination. However, the ability and skill to create beautiful objects and surroundings can be acquired through a scientific training in the knowledge of art elements and its principles.

One can acquire a good taste from reading and make improvements by observing and perceiving objects of a good design. Nostalgia also plays an important role in the taste of many. It can be seen from the fact that the traditional or classical decoration seen in early hand-crafted and antique furniture are currently popular and are in great demand in the present day.

The choice of people varies in the selection of a particular style. Some people favour style of a traditional period, others who want to break away from the tradition, find interest in contemporary designs and still others prefer eclectics, a mixture of different styles. The personal taste is developed by the selection and appraisal of something appealing to the eye, i.e. accepting what is suitable to that person's life style and rejecting what is not. A good taste gets reflected in the possessions that one acquires. The result of a good taste becomes so rewarding that it even improves the personality of that person.

In a broad sense, an item is said to be in good taste if it is well proportioned, integrated, beautiful, original, fine, sincere, appropriate, logical, direct and efficient with its form defining its function, while on the other hand, an item is considered to be in bad taste, if the observer feels that it is shoddy, fake, cheap, insincere, vulgar, etc. However, people may disagree with a person's individual judgement and classification in all these things.

Developing good taste is a continuous process. A person who is developing finer discrimination, keener perception and mature appreciation, might discover that his previous taste has become inadequate, and perhaps may even abandon it, and accept a new one.

OBJECTIVES OF AESTHETIC PLANNING

Houses often have a kind of aura, sensed when one is inside. There is an atmosphere attributable to the personality of someone who has built the house on a chosen spot and dwelt in it with the family for some time. The way a house looks affects not only the purpose of shelter, but also the promotion of spiritual, intellectual, physical and social growth of the family and its members. Beautiful home surroundings constitute the most important factor in the development of visually good taste, since daily contact with beauty leaves a lasting impact (Fig. 1.2 – see clour plate 2).

Every person has a natural ability to recognize beauty and functional ability in houses and their furnishings. This forms the basis of planning good interiors and exteriors. Beauty can also be enhanced by studying and observing various designs, and by experiencing them while they are being created. Thus, one can seek to establish beautiful things around him.

While beauty is an important aspect in planning interiors, it is not, however, the only one. A house and its furnishing plan should not only incorporate the elements of expression of the personality of the owner, it should also function effectively. The ultimate goal of planning a good interior is the successful integration of the three objectives of aesthetic planning. They are:

- Beauty,
- Expressiveness, and,
- functionalism

For instance, a sofa in a house is expected to serve the purpose of sitting (i.e. functionalism). The sofa is designed in a particular way, giving a style that the owner has selected to express his own personality ie. expressiveness. Then the sofa has to look beautiful so that it can draw appreciation from family members, as well as the visitors (ie. beauty).

BEAUTY

The definition of beauty is different for different people. For some, beauty is used 'to describe well designed and pleasing things'; to some others 'beauty is a quality that is rarely obtainable'. One simple definition of beauty is 'a combination of qualities that is pleasing to the trained eye'. Some say, that 'beauty is the thread of simplicity that runs through the whole design'. According to an oriental proverb 'One man's beauty is another man's ugliness'.

Aesthetics is the philosophy of beauty. Aestheticians have studied nature and the objects made by men and have come to the conclusion that there are certain guiding premises that determine the factors to appreciate designs. These factors are the elements and principles of design. The elements of design include line, form, colour, texture, pattern, light and space, while the principles of art and design include proportion, balance, rhythm, repetition, harmony and emphasis.

The study of elements and principles of art develops the ability of an individual to judge the appearance of all manmade objects. This knowledge enables a consumer-buyer to distinguish between beauty and mere fashion-value asset where new things appear and ideas change rapidly.

EXPRESSIVENESS

One way to approach the subject of selecting, decorating and furnishing a home is to seek to express some definite idea or theme in it. The term 'expressiveness' implies the power to excite emotional response that is lacking in the word 'character' and it avoids the suggestion of human attributes which are contained in the word 'personality'. Talbot F. Hamlin says, "All good architecture should have the gift of expressiveness. Every building, every well-designed room should carry in itself at-least one message of cheer or rest".

Some of the ideas that are expressed consciously or unconsciously in the homes are repose, nature, warmth, coolness, delicacy, intimacy, sophistication, animation, strength, freshness, antiquity, honesty and sincerity. The dishonest things that can be avoided are things like imitation of fire-places, fuels, coals/logs, wood, stone, flame and candles, and the plastics.

A home that expresses *formality* usually expresses dignity, strength, reserve and impressiveness also. The features that contribute to this effect in a house are unbroken lines, large spaces and a symmetrical arrangement, like the front of a house in which the two vertical walls are alike. In an interior, formality results from symmetry and also from the use of conservative colours. The furniture is usually, though not necessarily, traditional in style and with formal balance. The family that creates a home of this type, generally lives a conventional, dignified orderly life made possible by efficient service. A house which expresses dignity is not a mere representation of that quality, but an active assertion influencing the emotions and behaviour of all those who also enter it.

Informality, unpretentious, friendly, hospitality and intimate charm are expressed through various means. Bright, warm colours and simple, comfortable furniture have these characteristics. However, houses can also express informality through modest size, asymmetrical balance and broken lines.

Another type of informality is that which stresses *naturalness* or primitiveness. A house of this character may express the following themes – simplicity, handmade quality, sincerity, thrift, playfulness, rugged force, unpretentious, originality or a protest against artificiality. Among the factors that contribute to the attainment of the natural effect are the use of native material or the styles, handwork showing natural irregularities in structure, direct treatments, inexpensive materials and pleasant or primitive colours. Laboured effects, fine finish and imitations are avoided.

Primitive means simple, crude, old fashioned and characterized by the style of early times. It means merely a natural state, unrefined, unpolished, unfinished, showing lack of skills in workmanship. The words primitive and crude are used in describing the quality of sincerity that is prized today in many forms of art, from primitive sculpture to pleasant wall paintings.

Persons with highly trained taste often prefer articles of primitive or pleasant construction because such products usually have satisfying realness whereas the products of more highly organized society are often artificial.

The *modern* home expresses the spirit of this modern age. Most of the definitions for a modern home indicate the importance of functionalism. The families that choose modern furnishings, are usually young, courageous, experimental and logical. They are interested in a style that is expressive of their own time.

The *personality* of the family determines the idea to be expressed in a home. Personality of the family is the reflection of the family's qualities and characteristics. It is, therefore, important to keep these things as the basis for decorating the interiors of the home in which the family is going to live. The family interest, which has a permanent significance and not a mere passing fad, should provide the inspiration for a plan of decorating and finishing.

If a family likes to do things in a *formal* way with careful reward for the conventions, that attitude should be reflected in its choice of architecture and in home furnishings. If a family, on the other hand, has an *informal*, domestic, stay-at-home attitude, it should select a more picturesque, but simple type of house, garden and furnishings.

FUNCTIONALISM

The homes of today should be as functional as the machines of today. They should give the maximum service and comfort. Such houses also demand maximum care and maintenance. To achieve this goal, every phase of home planning and furnishing should be based on function. The number of rooms and their arrangement depend upon what will best serve the family. Outdoor areas too, are divided according to the function, with places to eat, play and exercise. The layout of the garden will also be governed by this rule of being functional.

The selection of individual articles should be governed by a critical judgement of how well they fulfill the function. Some common mistakes in their selection are the ones like the lamps that focus direct light in the reader's eyes, vases that tip easily and pitchers with spouts that do not pour well. All these mistakes in their function are due to their incorrect form. The right shape of any article is the one that will give the best function and usually, it is the shape that makes it to look the best.

As form follows function, materials too should follow function. Garden furniture should be water-proof and upholstery fabrics should be durable. The home does not permit its occupants to find peace, comfort and relaxation, if it is not functional. Function should necessarily be modified for the sake of appearance. The integration of beauty and function makes the house an ideal home.

HOW TO DEVELOP A GOOD TASTE

Taste is the orientation of an individual that results in his judgement while selecting things. Though

PLATE 1

Figure 1.1. A thing of beauty is a joy for everyone. Flowers are the Nature's best creations, appreciated by everyone for their colour, pattern, texture and appeal

PLATE 2

Figure 1.2. In the two settings, picture (a) gives a simple, calm and balanced look ......

...... whereas in picture, (b) though the furniture and the background are beautifully set they result in gaudiness because of over-decoration

some of us are born with this rare gift, it is also possible to develop a good taste by cultivating abilities to

- Observe
- Analyze
- Collect ideas
- Make wise purchases
- Be critical about cost of products i.e. be economical

Observe

For developing a good personal taste, the first step is to start observing things in the surroundings. Visits to museums, historical homes, local furniture show rooms, exhibitions and other displays would also help a person to see things and compare them for good and poor features. Reading newspapers and magazines would give some more ideas about the things that are available in the market. With careful observation, a person will be able to notice colour, pattern, line, shape, texture, balance, emphasis, rhythm, scale and proportion in all these objects. An analysis of these things would reveal that some of them exhibit good taste, whereas some others do not exhibit such qualities. Such an exercise would help a person to develop a habit of looking for aesthetically appealing objects. Thus, one will learn to recognize objects with good design and discard poorly designed objects.

Analyze

When a person comes across certain items, it will be of help to examine them in detail. If they are appealing, analyze the qualities that exhibit good design or vice versa. Before deciding to possess an item, study each one of them as an achievement or failure in design. Thus a complete analysis of the items would enable a person to look for the appealing features or their defects. In due course of time one will learn to develop an ability to analyze all the objects for good taste before finalizing the choice.

Collect Ideas

While going through newspapers and magazines it is also important to collect and compile the articles that are appealing. It might be a picture of a room, a piece of furniture or an accessory item. Once a number of articles are collected, one should try to compare and eliminate some ideas and develop preference for one item over another. Only through observing many objects and comparing them, one can develop a personal taste to analyze and make a better choice. For this purpose, one has to have a record of all the ideas and this can be made possible by collecting and compiling articles of interest.

Make wise purchases

It is not always necessary that the mass acceptability of an item is a good criterion of a design. Just as fashionable clothing become outdated, so do home furnishings. It is, therefore, wiser to start discriminating purchases however small they may be. There is no point in regretting later on about purchases which are not according to one's taste.

Be economical

It is important to understand that cost is not always a good criterion of wise purchase of an article of a good design. There are expensive items that are in poor taste while there are inexpensive, well-designed items that exhibit good taste. Therefore it is good to learn to be critical and analyze the various aspects of a good design rather than be carried away by the cost of the products. Cost is not a major factor for judging or deciding the design as good or poor. Good workmanship, attractive colour combinations and innovative ideas are the factors that should lead to a final choice.

The development of good taste or good design sense has a practical aspect. A person who usually goes through all the above steps will discover the kinds of decoration that will be most suitable to his needs and desires which applies even to a person with a limited budget. Gradually, one will begin to eliminate some items and prefer others that are of good design. The owner of an object that follows the principles of good design will receive more pleasure, learn to appreciate it and enjoy it permanently.

Thus it can be said that the development of good taste or good design sense is a continuous process in an individual's life. The taste has a practical aspect which was recognized even by the primitive man while designing and decorating his dwelling and persists in the contemporary world where a designer or a home maker leaves no stone unturned to project oneself in the interior of her residence. A home calls for energy, acquisition, involvement and improvement. Awareness on the part of the designer is required to look for things that are in good taste. The family members too, would learn to appreciate objects of good taste when surrounded by things of beauty.

The strength of a design lies in its ability to be rooted in the particular cultural environment. India has a rich heritage to define this as an everyday culture. The challenge lies in a successful interpretation of all these cultural inputs.

Creativity by definition cannot allow a rational route. However, interior design is an applied art form. There is an inspiration based on the space and the mood it evokes. This then becomes the foundation for a good taste upon which we build the structure bearing in mind the practical necessities and the profile we want.

Chapter 2

DESIGN: CONCEPTS AND TYPES

For a long time, individuals, groups and the entire nation have been concerned with countless design problems. The problem starts with the planning of their homes, their furnishings, and their apparel, and extends upto their community areas. Throughout history, the taste of designers has been undergoing alterations constantly and has resulted in the succession of various forms of designs.

CONCEPT OF DESIGN

In the field of art, the term 'design' is interpreted in a number of ways. One of these definitions refers to a particular organisation of elements for a special object of art. Another definition of design refers to the act of designing, selecting and arranging of these elements for a particular purpose.

Thus, we might speak of the design of a salad bowl at micro level or the design of a city at macro level. Each attempt while making a design is a definite act which should represent creative thinking. When we create a design, we make use of elements like lines, shapes, forms, textures, and colours, either as an object in space or as a drawing on a piece of paper. These are, therefore, referred to as elements of design. These are the tools and materials with which every one makes a design. These elements are discussed in more detail in a separate chapter later in this book, while the design aspect is taken care of in this chapter.

In the last century, the attitude towards design has gradually changed. Formerly it was more frequently thought of in relation to an applied decoration or pattern, such as a wall paper or a printed fabric. Today we refer more to the art of designing which encompasses every aspect of selection and arrangement. These may include the problem of selecting lines and colours for painting a picture or design on a drawing board or paper or for the furnishings of a room. Thus creating a design is a problem for everyone and not just for students alone with their so called artistic ability. Consequently, each one of us needs a thorough background to appreciate, to select and to arrange the elements to create designs.

While evaluating and arranging the elements, one has to keep in mind the various factors of organisation or principles of design such as balance, proportion, emphasis, harmony and rhythm. More information about these principles have been given in a separate chapter.

The term "*design*" also refers to several meanings like purpose, aim, intention, plan, scheme, selection and organisation. When these terms are put together, they describe the total design procedure of deciding on one's aim, developing a plan or approach and selecting and organising the form and materials best suited to the purpose. In the visual art, designer works with the elements of art and design, by selecting and arranging them to suit a purpose, which might be to create something that is purely functional or purely ornamental or both.

In simple words, design is the selecting and arranging of materials, with two aims – order and beauty, the order denoting organisation or structure, and beauty that shows character through the interpretation of an idea by an individual. In doing so, the designer expresses various ideas, moods and values, and such expressions evoke some response in the minds of the viewers. In a nutshell, "*design can be defined as the selection and arrangement of lines, form, colour and texture of an object in space, or a drawing of it on a piece of paper*".

PURPOSE OF DESIGN

Design, as ever, is directed by the use and combination of tools and materials, methods of construction and use as well as the designer's personality and creativity.

The purpose of this chapter is to help the students of design to develop a philosophy and a creative approach to design as *a way of life*, and to learn to analyse a design and make a wise choice, considering both utility and beauty. By learning to use and live with such selections of artistic objects, design finds a definite place in one's daily life. No text or a course by itself can guarantee that a student would automatically develop the knowledge and understanding of designs. It takes the eye, the hand, the brain and the heart to make a good design. Therefore, the students must therefore study, see, feel and evaluate their reactions to explore and experiment new designs. With increased understanding of designs, one can not only create a functional environment, but can also enjoy a more satisfying life in a beautiful environment.

TYPES OF DESIGN

There is no meaningful classification that would be useful to evaluate all designs. We must learn to judge the expressiveness of any design, by the way all the elements are blended together to create the total effect. It is not possible to study line without colour and texture. Neither is it possible to ignore the effects that colour and texture have on each other. Therefore the over-all design effect results from the intermingling of these elements, and we must learn to understand how these elements work with one another in producing something that is pleasing to the eye. In this manner, a number of designs can be created. In all, designs can broadly be divided into two categories. They are

- Structural design
- Decorative design

While structural design is basic to all objects, decorative design refers to the surface enrichment of these objects. Thus designing the structure of an object is a "*must*" while decorative designs could be 'optional'. Structural design is "*essential*" to all objects, and decoration is the "*luxury*" to the object.

In structural design the structure determines the form, and enrichment comes from the materials used. It is the design made by the size, form, color and texture of the object whether it be the object itself in space or a drawing of that object worked out on paper. Any line, colour/material that have been applied to structural design for the purpose of adding a richer quality and to constitute its decorative design. Since structural design is essential to every object, it is more important than the decorative design. Much of our modern design is of this type, with form dictated by function. There is very little ornamentation as far as such designs are concerned.

Requirements of a good structural design*

The phrase "form follows function" is correctly applied to all structural designs. The real meaning is that form and function are complementary to each other. In practice, we find a large number of objects with different design, meant for different uses. It is not necessary that all these objects are of good designs. Some may be good and others, poor. A few guidelines are followed to judge whether a design is good or poor. If the designs conforms to these guidelines or requirements, the design can be called a good one, and vice versa. Goldstein lists four requirements for a good structural design, when the object is intended for use. These are the pre-requisites for a good structural design. While meeting these requirements, the designer considers whether the form, colour and texture have given enough interest to the object or, if a sense of bareness remains, it needs to be relieved. In some structural designs, the forms, colours and textures are already so beautiful that one does not feel the necessity for decoration.

The requirements for a good structural design are
- In addition to be being beautiful, it should be suited to its purpose
- It should be simple
- It should be well-proportioned
- It should be suited to the material of which it is made and to the processes which will be followed in making it.

The design also has to do with the function of an object or the purposes for which a form was created. Becoming familiar with the function of an object may lead to a greater appreciation of the resulting solution. This function can always be looked at as both aesthetic and utilitarian.

It is important that the design should suit the material of which it is made and to the processes followed in making it. While purchasing any article, one must make a close scrutiny of the article, as to what is the purpose, and how it was made. The honesty of the designer and the technique involved in making an object should get reflected at the mere appearance or look of it.

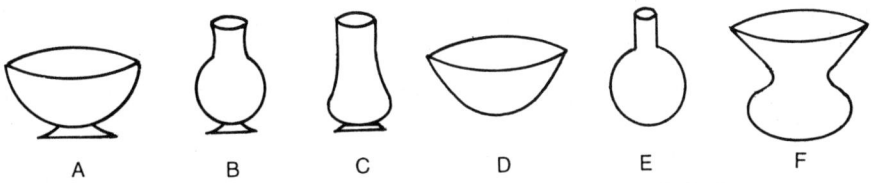

A   B   C   D   E   F

Figure 2.1. Bowls for use as Flower Vases in different designs

The bowls in figure 2.1 are given as the examples to indicate good or poor structural designs to serve as flower vases. You will find that the structural designs of bowls A, B, and C are not only simple, but also beautiful and proportionate. The structural designs are such that they would be suited to the purpose. When they are made of good china clay and baked well, they make perfect designs for use as flower vases or even simply as decorative pieces. The form and texture of these

---

* Goldstein and Goldstein, Art in Everyday Life Prentice Hall Inc. NJ.

bowls are so beautiful that one does not feel the necessity for any decoration. However the designs of bowls D, E, and F are not only poor, but are also not suitable for the purpose because of either wide or narrow top part or an unstable base indicating bad proportions. The wide or a narrow top of a bowl makes it unbalanced and unstable. In addition, if these are made of a material like light weight plastic, their material might also be not suitable to hold water and flowers. These bowls are likely to fall and therefore, are unsuitable for the purpose. With unbalanced proportions and appearance, they may not even serve simply as accessories in a room.

Additional characteristics of form are the materials an object is made from and the techniques used to produce it. Any work in leather, glass, metal, plastic or wood suggests a feel, a surface that is characteristic of that material. Objects may look bright or dull, opaque or transparent, they may feel coarse or soft, rigid or flexible, to the touch on account of their base materials. Thus the material used in making the object, indicates not only its feel, but also its suitability for use, since the feel determines the appropriate surface for the use of any object.

Techniques of construction such as hammering, stitching, casting or spinning may create different surfaces and surface impressions. How the materials and surface qualities for a work area are organized, creates a structure of formal relationships in a work. Features such as symmetry, balance, contrast, complexity, simplicity, harmony, rhythm and unity can be seen as structural qualities, as is evident from the bowls given in Figure 2.1.

Decorative design

Decorative design is the surface enrichment of a structural design. By *surface enrichment*, it is meant the applied decoration, such as a printed or painted design, one that is etched, carved, appliqued, or otherwise executed to decorate the surface of the area. Any lines, colours or materials that have been applied to a structural design for the purpose of adding a richer quality to it constitute its decorative design. Decoration in general, is the luxury of design. The elements are simplified, exaggerated, rearranged and even distorted by these decorative designs. Therefore it can be said that decorative design can be created for meeting various purposes and ideas.

Requirements of a good decorative design

Once the designer has decided that the object will be enhanced by decoration and that its structural design is simple and beautiful as well as functional, the designer can plan his decoration.

Whether it is the decoration of a vase, chair, table or a costume, the design should fulfill the following conditions:

- The decoration should strengthen the shape of the object by emphasising or harmonizing with its proportions.
- The decoration should be used in moderation.
- It should reinforce the function of the object.
- The decoration should be suitable for the material and the service it must give.
- There should be enough background space to give an effect of simplicity and dignity to the design.
- Surface patterns should cover the surface quietly.
- The background shapes should be as carefully studied and as beautiful as the patterns placed against them.
- It should express individuality and creative thinking.

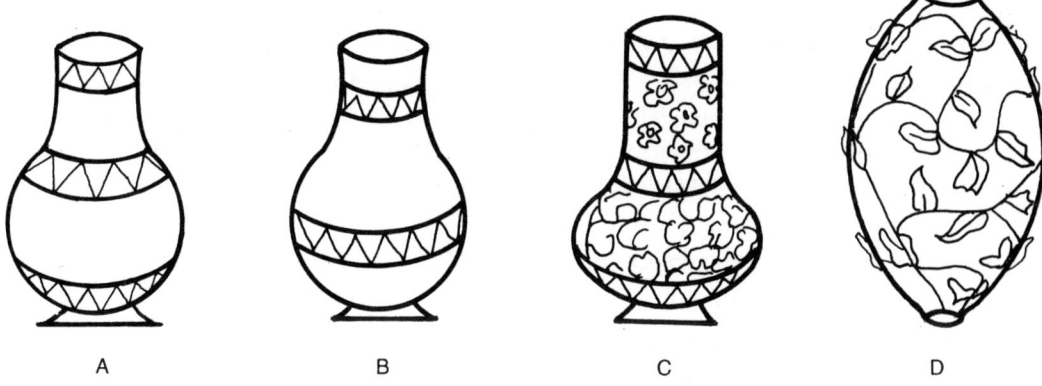

Figure 2.2. Some Bowls with Decorative designs

A close scrutiny of the bowls given in Figure 2.2 would show that bowls A, B, and C have a good and consistent design whereas bowl D lacks these, with a narrow top and an unstable base. The base of the other bowls are sturdy and even. The size of the top of these bowls also seem to be well balanced and can hold flowers, if used as flower vases. The design done on bowl A to decorate its surface, seem to supplement and strengthen the shape of the object, because they are placed at the structural points. The design is also used in moderation to enhance the beautiful structural design. In the bowl B which is similar to A structurally, the decorative design is not placed at the structural points. The decorative design, therefore, has failed to strengthen the shape of the object, and the design pattern looks out of place.

The decorative pattern on bowls C and D covers the whole area of the pot. An analysis of the design on bowls A and B shows that they are better seen against the visible background space which enhances the beauty of these objects. Thus the background space plays a significant role in making the design more attractive and emphatic. The decorative design on bowl C is attractive and elaborate, and seems to cover the background space evenly and quietly. However, the decorative design on bowl D does not seem to have the same effect. It is, therefore, necessary that any decorative design placed on a surface should become a part of the structure of the object, and should not seem to go away from the background space. Thus a good decorative design shows fine quality, and serves the purpose of being a functional one, when adopted to conventionalise, or stylise the basic design.

While designing a decorative object, it is important to see that the design is suitable to the material of which it is being made. If care is taken to keep this point in mind, the design is made suitable to its purpose or use. When the decorative design is placed all over the space on the bowl, the bowl itself becomes beautiful and can be used exclusively as a decorative piece on a mantel place. When the decorative design is used in moderation and placed at structural points, with enough background space, the decorative design will not only be highlighted, but also makes the object most suitable for its purpose.

Finally all decorative designs can be categorised as unique creations, when it expresses individuality. Creative thinking plays a significant role in preparing beautiful and exemplary objects. Designs are created carefully to express their functional quality, i.e. their exclusive use and the purpose, as a decorative object or otherwise. Some objects are created for the sake of beauty only and are not connected in the designer's mind with practical use. When an object is chosen for their appearance alone, they should be judged critically, because our taste must grow up toward beauty and we must learn to discard objects that are inferior aesthetically.

The designer whose work shows fine quality, adapts and conventionalizes or stylizes his design to suit the material he is using. He does not attempt to deceive by imitating real objects, such as flowers and fruits, done in materials like wood, clay, plastics or threads, but having decided to take flower/leaf idea as the theme of his decoration, he alters it to suit:

- the shape of the object
- the purpose for which the object will be used
- the limitation of the material
- the tools and processes he must use while making the object.

Thus, a lot of effort is put in to create objects either for beauty or for functional use.

JUDGEMENT OF STRUCTURAL AND DECORATIVE DESIGNS

A good judgment of structural and decorative design is essential for any person while selecting an object. The flower vases as shown in figure 2.1 and 2.2 can be taken as examples. All the flower vases are intended to hold flowers, all have structural designs, but there the similarity ends. The figure 1 reveals that the design can be appreciated as the purpose for flower vase. It will be used as a background for flowers and thus, must be less conspicuous than the flowers placed in them. Then the designer should consider the structural points which would strengthen the design. He believed that decoration should be placed at the structural points of the design. With suitable tools he worked out good patterns, as can be seen in figure 2.2. When the proportions and shapes in the design were pleasing and the background shapes agreeable, it can lead to the creation of beautiful objects, and great satisfaction to all.

If a design is to give satisfaction, it cannot stop at being merely correct, but should go beyond that for the person to get maximum amount of satisfaction from the object he has purchased. It must have character or style and individuality. A distinctive feature of a designed object is style. When a design possesses a positive and dramatic appearance, it is said to be 'decorative' i.e. it has a 'decorative quality'. These two terms denote characters in a design and are different from decorative design. Decorative design means any decoration added to the structural design. 'Decorative quality' is never found in a design that is 'pretty' or 'sentimental' and it does not appear in imitative designs. Simplicity is an attribute of 'decorative quality', while fussy elaboration is never associated with a good decorative design.

Generally, style may refer to a tradition or culture, such as Indian, European, Italian, Danish or Egyptian. Style may relate to a school of art or movement such as cubistic, abstract, classical, romantic, Bauhaus, futuristic etc. Design characteristics in any given period are generally the result of exchange of ideas among the artists of that period. Designs in fashion may also relate to design in painting, furnishings, homes etc.

Structural and Decorative design in architecture

A building's structural system is formed according to the composition of its materials and how they react to the forces applied to them. The structural form and material composition, in turn, influence the dimensions, proportion and arrangement of the interior spaces within the building volume.

PLATE 3

Figure 2.3. Some structurally good designs in furniture

Figure 2.4. Decorative designs in furniture styles

PLATE 4

(a)

(b)

Figure 2.5 (a) & (b). Utility in structural & decorative designs in homes

Unity must be a basic factor in any consideration of design. The relationship of form, colour and texture produces the design. We look at the design of a house in terms of its relation to the terrain. We also look at each room in relation to the whole house. But, we also see the design of the room itself as a unit. In no other architecture has there been more emphasis upon structural design than in the type that is commonly called 'organic' or functional. Although the work varies according to the interpretation of different architects, the essential character remains the same. Decorative design is infrequently used, and when it is used, it shows unmistakable reserve. It cannot be assumed that simplicity alone will make a design good. The design emphasis should be placed upon the skillful proportions in the mass of the house and in the relationship of the separate part. Necessary elements have become handsome through the sheer quality of the spacing.

The Greeks knew the principle of making structural design more important than decoration, and they applied it to everything they did. The ancient Greek building, Parthemon, is a typical example of dignity and restraint in design decorations. When one looks at this building, the first impression is that of a beautiful structure and after that the decoration comes to one's attention. This decoration is not put on a lavish hand, but with a great deal of reserve, and it has been placed only on those spaces which have grown out of the construction of the building. Appreciation of the design of the Parthemon should lead one to enjoy fine structural design and a reserve in decoration in any building.

A simple and well proportioned design in house building should so lead the shapes of all parts of the building to grow easily and naturally out of the principal mass of the building that the first impression should be that of unity. A thoughtful scrutiny should show that the lines and textures are so pleasantly related to the land and the landscape behind it, that the house should seem a natural part of its setting. The simple structural design should be marked by the continuity of the roof line without any obstruction. The window openings should be arranged in an orderly fashion, and their shapes are consistent and well related to the wall spaces. The treatments of the walls and wooden panels of doors and windows should give an impression of unaffected sincerity. The landscape and plantations should be informal and their lines should echo the structural lines of the house. Decorative accents in the architectural designs can be limited to the use of a contrasting colour on the door and window panels/frames.

DESIGN IN FURNITURE

It is generally seen that the modern furniture design meets the demand for comfort and practicality as well as the need for quality. The result is the furniture that is simple, easy to take care of, and comfortable to live with, and reflecting the way people want to live today.

There is a steady trend towards better quality in furniture design. It is becoming generally recognized that furniture must pass the tests of good structural and decorative design. Furniture, therefore, must be useful and comfortable and soundly constructed if it is to serve its purpose well; and it should depend for its beauty more upon good structural design than upon decoration. Furniture should be so fine in its proportions and in lines that it needs very little, if any, decoration.

Today, no one style clearly predominates. There is a desire for simplification and function. Rooms frequently include a few important objects, plus stream-lined furniture and lasting materials that are easy to care for. There are fewer surfaces to clean and dust, fewer pieces to rearrange. Storage spaces are often built into the furniture or the home architecture, so that clutter is avoided. The modern furniture demonstrates the dignity and satisfaction inherent in simple and good structural designs which give a distinct impression of belonging together (Figures 2.3 and 2.4 – see colour plate 3).

More top designers than ever before are decorating products for the home that feature variety, good design and for varied purposes. In the past only the rich could afford home furnishings created by the artisans. But the modern times furnitures are equally beautiful besides being inexpensive to suit every one's pocket. The design in furniture would be deliberated in more details later in the chapter on *Furniture*.

Making and Judging Designs

The making of a design is to establish a method by which a person may learn a basic way to think in terms of the design of an object. The designing process is a form of organisation in which the elements are a collection of units. These units, in the art field, happen to be sizes, shapes, patterns and colours. Therefore, a design is built up as logically as any other type of organisation and may be compared to the plan of a building at the unit level and to the plan of a city at the unification of a large number of units put together.

Significance

Selecting and owning objects with beautiful structural or, and decorative designs has become a way of life at present. Everyone wants to possess objects for beauty and with a purpose. Therefore it has become important to learn about making and judging design for their appearance and inner qualities. Such an exercise involves the following experiences.

- For self expression, a good design is used to convey a message.
- It encourages thinking in terms of design and art.
- It provides a creative experience.
- It shows an easy and consistent method of working.
- Design is of interest to the consumers and a matter of intelligent appreciation of the things he uses.
- It is a way to develop imagination, good judgement and fine standards of taste through thoughtful study of good designs.
- The knowledge of principles of art, as a conscious or unconscious part of one's thinking, may be used as a standard for measuring quality.
- Mere adherence to the principles of design will not ensure beauty. Rather the character and style and having a strong feeling of a creative mind, may make a design significant with a realistic feeling or experience of beauty.

While making a design, the designer has to look for a number of factors. For some, it may be a natural experience, and for another, it may be a conscious effort. The making of a design comes spontaneously to a person who has the native ability as he has an unconscious feeling for organisation. Designing for him is a complete creative experience. However, the following guidelines are helpful to others who do not have this ability to create designs:

- A designer is urged to invite – to play with his pencil until something interesting appears on the paper. Uninfluenced ideas are the best if one really has something to say in terms of design.

- However, if ideas do not come, the next suggestion is to look at many excellent designs in order to store myriad impressions. This will usually make it possible for him to create something out of his own inventive or well-furnished mind.

- The designer should go to various sources to develop varied tastes and expressions.

- There are two considerations in every design; first, the shape of the object itself, called the structural design, second, the enrichment of that structure, or the decorative design. The structural design is of the greatest importance and should have the designer's first consideration, and only when necessary decorative ideas should be worked out.

- Although the different uses for designs will impose different limitations for working them out, the same general method may be followed in making designs for any purpose, such as a design for a woven linen, a room, a costume, etc.

To have a design coherence, and a relative value for the object, the worker has to follow the following steps:

Step-1. Prepare the plan for the structural design

Step-2. Divide the entire area into blocks and then decide the decoration for each of the blocks.

Step-3. Work out the details for individual blocks and then connect the design of each block with that of the other blocks, and then to the entire structure.

It is important to form a habit of looking at any design for its effect, and then to work out the details, otherwise the design effect is likely to be unrelated without any reference to the whole structure. Thus creating a design involves a systematic thoughtful experience.

Outline of Procedure in the making of a Good Design:

It is necessary to make an outline of procedure in the making of a good design, whether structural or decorative.

Structural Design:

It is the selection and arrangement of the size, form, colour and texture of the object. Since the structural design is the fundamental or essential part of an object, it must be beautiful in itself. Structural Design is conditioned by –

- The use to which the object will be put
- The person for whom the object is planned
- The surroundings in which it will be used
- The appropriate standards for a good design (Principles of design).

Decorative Design:

It refers to the added enrichment to the object. Decorative design may be optional. If it is included, it should be used to enhance the structural design. Decorative design is conditioned by –

- The structural design of the of the object
- The use to which the object will be put
- The person for whom the object is planned

- The surroundings in which it will be used
- The personality of the designer (which determines the character or quality of the design)
- Standards for good design
- The amount of decoration desirable
- The nature of the design – motif or pattern
  - (i) Sources – abstract or geometric or natural forms
  - (ii) Treatment – conventionalization, or adaptations to the materials used and the use to which the object will be put, (pictorial or naturalistic).

METHOD OF WORKING OUT A DESIGN

While working out a design, a systematic method would yield better results. The method for working out a structural and a decorative design is as follows:

Structural Design

- (i) Plan the size, shape, and colour of the object, according to the guidelines for designing, namely, proportion, harmony, texture, ideas (suitability to purpose), and colour, rhythm, balance and emphasis.
- (ii) Execute the structural design, using appropriate material according to their use or purpose.

Table 2.1: Method of Working out a Design

Structural Design	Decorative Design
• Keep in mind its use and purpose	• Study the structure of the object
• Plan the size, shape and colour of the object	• Prepare a layout of the design
• Follow the guidelines (principles) of design	• Try sketching of decorative design-sizes and shapes in their positions.
• Select the appropriate materials	• Judge the design according to art principles
• Execute the design.	• Work out the final decorative design
	• Arrange and execute the design.

Decorative Design

- (i) Initially, to prepare a decorative design, a *layout* is required. The aim in the layout is to secure and orderly arrangement in a geometric plan or layout of the structural design.
- (ii) Within the structural design the principal masses of form and colour of the decoration are indicated by sketching or *blocking in* the sizes and shapes in their positions. Here the design is in an experimental state. One should play freely with shapes and colours, for, here one has to invent new designs. After something has been put down, the design may then be judged according to the art principles i.e. emphasis, proportion, harmony, rhythm, and balance.
- (iii) Once the layout is ready, and the experimentation of shapes and colours is over, workout for the details of the decorative design begins. The objective at this step is to secure beauty, character

and order. For meeting this objective of the decorative design, it becomes easy to arrange and judge forms and colours according to the art principles as mentioned earlier, and look for individuality.

INTERIOR DESIGN METHODOLOGY*

A scientific method can be adopted for designing interiors. This methodology can be broadly divided into four stages and thirteen steps. Every interior design scheme should go through these four stages for a comprehensive understanding of its usefulness.

Stage I

It may be divided into four major steps. They are Interior design purpose, Activity performed, Activity details and Activity diagram.

1. *Interior design purpose:* There should be a basic reason to design an interior. The reason could be to organise the commercial needs, official needs, residential needs etc., The real purpose of organising an interior should be clearly defined in the beginning. This basic purpose should form the core of interior design and all related actions should revolve around it.

2. *Activity performed:* Once the purpose of an interior is decided, then the activity to be performed should be clearly discussed with the client. For example: if the client wants to have an interior design for an office, then how that activity is normally performed should be discussed. The entry, exit, various activity units, tables, chairs and other furniture requirement should be discussed in detail.

3. *Activity details:* Then, the details of an activity to be performed in the proposed interior should be worked out. For example the number of tables, chairs, storage units, lights, fans, AC units and other minor details should be enumerated. These details should be prepared in close consultation with the client.

4. *Activity diagram:* Based on the above discussions with the client, an Interior Designer should prepare an 'Interior Activity Diagram' showing the flow and organisation of space into various units. This is the crucial diagram based on which the entire interior design should be prepared. An Interior Designer can be innovative and creative in organising these spatial units.

Stage II

It may be divided into four major steps. They are Design process, Concept plan, Revised concept plan and Final plan.

5. *Design process:* Based on the Stage—I details, an Interior Designer should start the interior design process. In this process, an Interior Designer can be given full freedom to his thinking and creative abilities; there are no limitations and restrictions. The real ability and creativity of an Interior Designer is revealed in this process.

6. *Concept plan:* Concept plan is the basic idea of an Interior Designer. An Interior Designer should present this idea in a effective way to convince the client. It can be presented in a pictorial form or written form or both. It would be appropriate to have a combination of

---

* *Source:* Interior Design Principles and Practices.

pictorial and written forms of presentation. The drawings and language should be simple and clear, so that everybody can understand easily. Before awarding the work, the client always tries to perceive the conviction and confidence of an interior Designer in the presentation of the ideas. This concept plan should be thoroughly discussed with the client.

7. *Revised concept plan:* After thorough discussions with the client, an Interior Designer should make the necessary changes to accommodate the clients' views. Thus the revised concept plan may be prepared. The concept plan can be revised any number of times until the client is totally convinced that the requirement is satisfactorily met in the concept plan and gives the approval. The final revised concept plan becomes the basis for detailed drawings.

8. *Final plan:* Based on the revised concept plan, the detailed drawings should be prepared to a scale. Here the Computer Aided Drafting (CAD) packages are very useful for easy and quick drafting, and also to incorporate future changes. These details may include all working drawings, estimates, bill of quantities, specifications, tenders etc., All the details on interior design should be prepared, keeping in mind easy and efficient execution of work.

Stage III

It may be divided into three major steps. They are Execution & supervision, Incorporating changes and Completion of work.

9. *Execution & supervision:* Once execution of work starts, an Interior Designer should make periodic visits to supervise and see that the work is carried out as per the plan. During the execution, the client may discuss whether the requirements are properly met or not because many people cannot visualise things in the drawings. Periodic site visits by an Interior Designer and the client is necessary for satisfactory execution of the work.

10. *Incorporating changes:* An Interior Designer and the client can improve the interior during the site visits, if they find something is lacking. These on-site improvements may enhance the outlook of an interior. These changes should be incorporated during the course of work, but they should not go beyond the stipulated time and budget unless under exceptional circumstances.

11. *Completion of work:* The closing stages of the work are as important as the beginning. An Interior Designer tend to ignore the work while it is nearing completion, which is a bad practice. An Interior Designer should ensure that the work is executed precisely as per the plan with special care for safety measures, emergency provisions and compliance of local building regulations. A check-list may be prepared to know whether all the works are carried out satisfactorily or not. A successfully completed interior project is a credit to all those involved.

Stage IV

It may be divided into two major steps. They are Evaluation and Feedback.

12. *Evaluation:* Evaluation studies are the most neglected part of interior design because most of the interiors are designed, executed and forgotten. Nobody takes the interest to do a scientific evaluation study after completing the project. The results of these evaluation studies serve as feedback to similar projects in future. Once an interior is completed and put to use, a primary survey should be conducted at least after one year to know whether the essential purpose of the interior design is served or not. This survey should be scientific and based on users needs.

Table 2.2. Interior design methodology

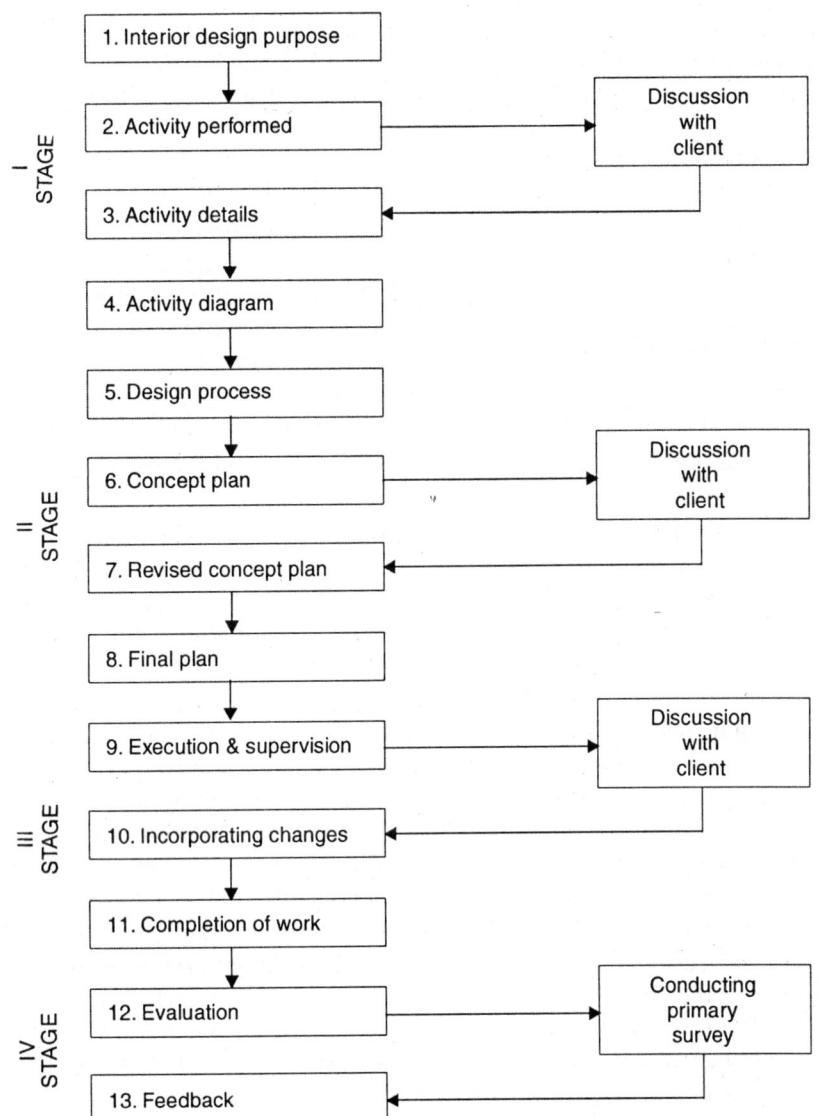

13. *Feedback:* The survey results should be scientifically analysed and the results should be recorded for future use. Many of the creative interior design concepts may satisfy the ego of respective interior designers, but ultimately it is the users who should appreciate and accept. These feedback results of various interiors form a reservoir of knowledge for other young interior designers to learn and improve their design.

Factors in Evaluation of design

This chapter is designed to help you, as a consumer and/or designer, to develop design knowledge and sensitivity. Throughout life one is constantly faced with design decisions in one way or the other. So it is important to look at the criteria that are used in evaluating a design.

1. **Honesty:** Honesty in use of materials is one evaluation point. Each material has certain natural qualities. The same is true of an individual. As individuals, we function best when we are what we are – not pretending to be something different from what we are. Honesty in the use of materials enhances the over-all quality of each piece we design. The honest use of materials make the design of the objects more appropriate and functional. Some of the examples in this direction is the use of wood like materials in the construction of furniture.

2. **Expression of today:** Another evaluation point is to judge how well the design reflects our present knowledge and the use of modern tools, equipment and materials. Whichever design we select or choose, we may ask the question, *"Is this a reflection of today?"*. It is always difficult to evaluate what exactly reflects today and it is much easier to evaluate the art of fifty years ago. Our mass production, our crowded cities with the slums, housing developments, and the freeways are some of the things that make up the everyday lives of millions of people and today's design has to reflect these, and are to be considered in the designing of the city.

    Thus the design is most likely to meet the requirements and style of today's living, and will not appear to be *"out of date."*

3. **Individuality of design:** While evaluating a design, one should look for individuality of the designer. This may be referred to as the originality of the design and not an imitation. Without this unique feature, the design loses much of its quality. At the same time, it is not necessary that all objects will have similar design quality. The individuality of the designer in creating unique designs ensures that we do improve or enrich our earlier designs and grow or add more quality to our environment.

4. **Arrangement of design elements:** How well the various design elements are organized into a pleasing object is obviously an important point to evaluate. For example, one might first eva-luate how well the various textures of the materials relate to each other. One may have the right combination of textures, but now it must be evaluated by finding out the answers for the following questions—do the colours affect the textures? What about the effect of various shapes on the texture? The multiple relationship between the different elements becomes very com-plex. There is no single mechanical way of arriving at the most pleasing combination.

    It is the most difficult task to evaluate a design that is almost right on the basis of design principles. It becomes more difficult while judging a design which lacks character or says nothing, and shows little imagination or vitality. However, when the best arrangement is achieved in connection with honesty of materials, expression of today, and individuality of the design, an overall aesthetic quality is usually observed.

Utility in Structural and Decorative designs

The relation of utility to beauty has been emphasized throughout this chapter, because the perma-nent enjoyment of the objects one purchases depends upon this relationship. A moment's thought will serve to call to mind many familiar cases of the oversight of this factor in design. We may have a handsome, but an uncomfortable chair. Its curves are graceful, and its decoration is pleasing to the eye, but the structure is so designed that it does not provide a good support to the back, and a line of decoration, which comes just across the shoulders, causes much discomfort while seated. Unlike a vase or a bowl, a chair may not exist as a purely decorative object, and so it is obvious that there is no place for an elaborately carved chair back rather than a functional one.

Among the other objects, besides uncomfortable furniture pieces, one can recall a few pitchers, teapots or cups or coffee mugs that look well and attractive on the shelf, but may have uncomfortable handles and are so designed that they drip when liquid is poured from them. Similarly a bed cover or a sheet may look attractive, but their coarse texture would hurt one's back, when used on a bed while sleeping. There might be a beautifully designed lamp shade for a table lamp, but it will throw light on the eyes or on the face of the reader rather than on the books or the table tops (Fig. 2.5 (a) and (b) – see colour plate 4).

From these examples, it can be seen that utility is a factor involved in design in general, for, the shape of an object should first of all conform to its purpose, and the decoration should not interfere with its function. It is good to remember, therefore, what exactly makes a design so good that it outlasts the swing of fashions. Two advantages may emerge or be gained by recognizing what is good in design: first, an understanding of what makes for beauty, and second, the ability to make our own choices so that they would have a lasting quality. These are the practical gains in securing objects with good design, and would lead us in the direction of economy with lasting quality and functionalism.

Chapter 3

ELEMENTS OF ART

A rt is a man-made expression of something beautiful, but beauty seems to be different for different people. To a home-maker, a beautiful home is an expression of her feelings, hopes and ways of living. While an artist might consider his studio as the most beautiful and inspiring area, some others may still consider the natural things as the most beautiful and the sources for their inspiration.

An understanding of what constitutes a good design is essential if any decoration has to be successful. The whole purpose of spending time, thoughts, effort and money on decoration is to make it attractive and pleasing to the eye. It is true that comfort and convenience are equally important, but these are more specific and easier to come by. The things a home must have, to be comfortable, are quite concrete, whereas the things that make it attractive are harder to define. This is where good designing comes into picture. To put it simply whatever may be said of design, a good design attracts and pleases the eye whereas a bad design repels it.

Design is made up of certain constituents such as line, form, colour, texture etc. These are the basic elements and form the infrastructure for creating a design. They are fundamental to all visual objects. However, pattern, light and space have also taken an increased importance as elements of art in the contemporary world. Pattern is not so distinct an element as others, but it is an important component and an essential term in interior designing and decoration. Light and space are not usually included among the elements of art, but a general appreciation of them as additional elements in art is important because of their wider applications.

Each of the basic elements is a well defined and an unique feature of every art object. Each can be manipulated by the designer, who may focus his attention on different elements in turn as his work progresses. However, the effect of each element is considered only in connection with the other elements and in relation to the organic unit which is being constructed, whether it is a picture at micro-level or a whole building, at macro-level.

The entire class of elements i.e. line, form, shape, space, light, pattern and texture, should be studied together because they function inter-dependantly. An understanding of the role of each of these elements helps to predict the impact of design or style to be created. These design elements combine with colour to produce an overall effect. These are important in all areas of creativity including home design and decoration, clothing selection and personal appearance, transportation and traffic designing, urban design and environmental enhancement, and thus, become universal in all their applications.

By recognizing and realising the significance and beauty of these elements of art, one can understand their value in approaching and developing design knowledge and sensitivity. By concentrating on the elements, one can be motivated to create and develop satisfying designs based on better perception of art elements. These are the elements that serve as tools in conveying the fundamental ideas in painting and sculpture. They also play a significant role in the creation of other art objects like handicrafts, architecture, home furnishing, household gadgets, furniture, machinery etc. To ensure the attainment of beauty while creating or designing these objects, art elements serve as an aid along with the principles of design.

To realise the objectives of beauty, expressiveness and functionalism in every area of designing, the elements of art are to be used according to the art principles – proportion, balance, harmony, rhythm and emphasis. The inter-relationship of these three are depicted in Table 3.1.

Table 3.1: Guidelines for Creating Designs

Objectives (Why)	Elements (What)	Principles (How)
Beauty	Line	Proportion
Expressiveness	Form	Balance
Functionalism	Colour	Harmony
	Texture	Rhythm
	Pattern	Emphasis
	Light	
	Space	

LINE

Line is a basic element of design and art, and contributes greatly towards the overall mood created in any arrangement. The literal meaning of the word line is a series of points joined together. A point extended becomes a line.

POINT.. LINE

Conceptually a line has only one dimension, i.e. length. In reality, a line's length visually dominates whatever thickness it must have to be made visible. The width or thickness of line may introduce a second dimension. However, it is debatable as to how thick a line may be before it becomes or identified as a shape. Unlike a point, which is static and directionless, a line is capable of expressing movement, direction and growth. Therefore, a line can portray emotion and excitement, rhythm and strength, decoration and unity.

As visible forms, lines may vary in weight and character. Whether bold or delicate, taut or limp, graceful or jagged, a line's visual character is done to our perception of its length-to-width ratio, its colour and its degree of continuity.

Line is the beginning point for designing, because it establishes shape or form and is a valuable element of composition. The human eye involuntarily travels the length of a line and different lines give different impressions. Man has associated definite ideas with certain lines because lines have been associated with the positions of the human body. When a person is lying down, he is resting or sleeping and therefore, the horizontal line naturally suggest repose, steadi-

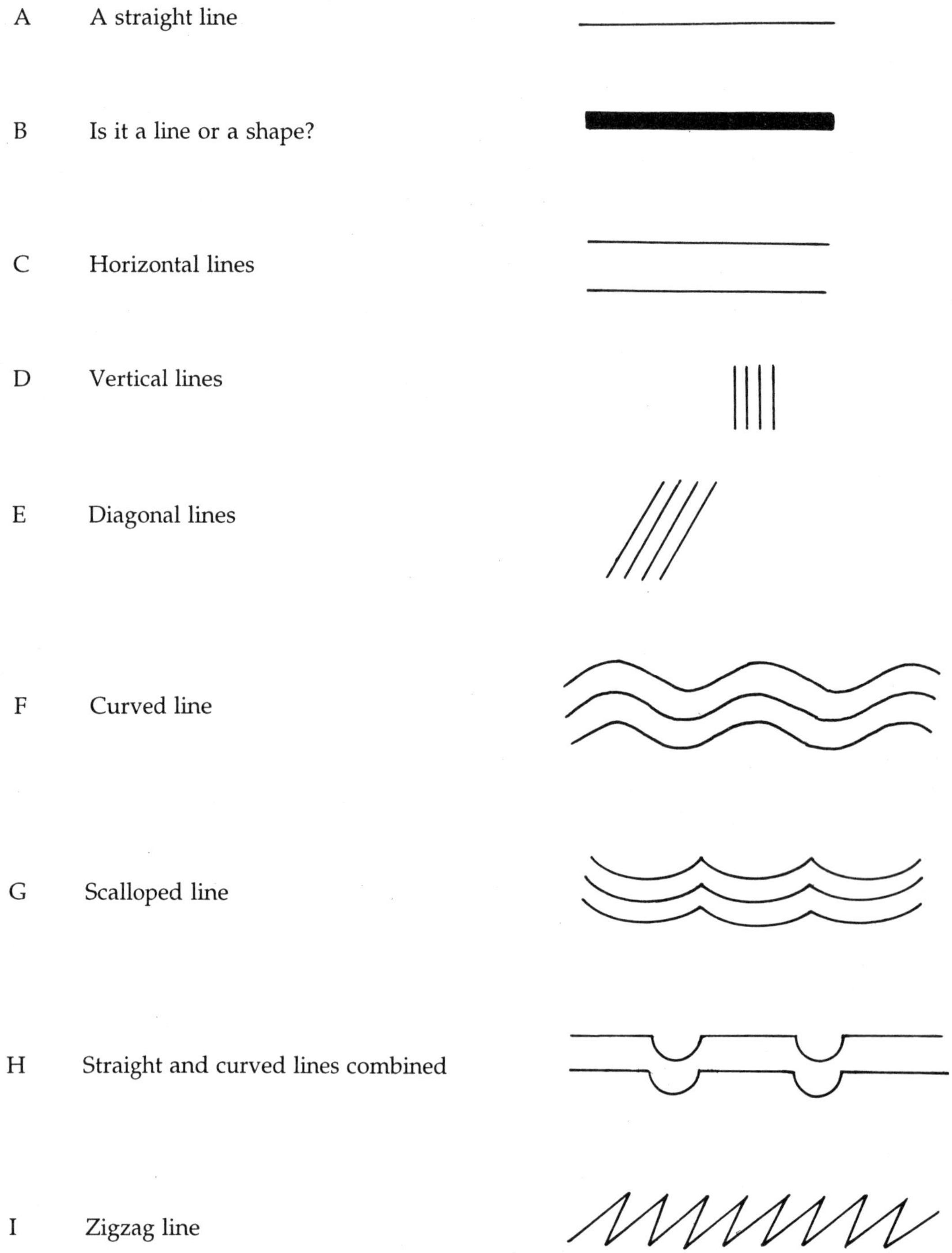

A   A straight line

B   Is it a line or a shape?

C   Horizontal lines

D   Vertical lines

E   Diagonal lines

F   Curved line

G   Scalloped line

H   Straight and curved lines combined

I   Zigzag line

Figure 3.1. What are lines?

ness and duration. When he is standing, he is at attention and ready to act in a vertical position (Vertical lines suggesting life and activity). When a man bends forward to run or to pull things, he is in a diagonal position (diagonal lines suggesting decided movement and force). In relaxation and in play, the body takes positions that are curved, thus suggesting gracious and flexible curved lines with these qualities.

Lines have positive emotional significance too, depending upon their direction and their quality. One can set the theme of a room and influence the reaction of anyone who enters it by the choice of lines. Although most designs are composed of many lines, there is often a predominance of one type that contributes to the character of design. Even simple lines evoke an emotional response, and we associate certain feelings with different types of lines.

Lines are used to create form and shape. They divide the space within a room and lead in a definite direction. They create visual impressions. For example, an impression of high ceiling can be created in a room with vertical lines. Similarly, the width of a room can be highlighted by the use of horizontal lines.

All lines fall into the category of straight, curved or a combination of these two (see figure 3.1). Straight lines can take three directions-vertical, horizontal or diagonal. Each direction of a straight line (vertical, horizontal, diagonal) creates an optical effect or illusion that must be judged by the individual to learn exactly the effect of a particular object or a room. But it is important to note that visual illusion created is not perceived by all people to the same degree. Lines, shapes, colours and textures may form illusions to produce certain effects, but we cannot be certain that the effect will be recognized in the same manner by everyone.

In interior designing, straight lines are considered intellectual rather than emotional, classic rather than romantic and sometimes severe and masculine. Curves, on the other hand, are used to achieve a more joyful, subtle and rich effect. The diagonal lines express decided restlessness.

Line should be continuous to produce an impression of unity. This does not mean that there can never be a break in the line, but with the fewer breaks, the room appears more organized. In furniture arrangement, the effect of unity is increased when equal heights are present in the room. This adds to the feeling of a continuous line around the room. When large, oversized pieces of furniture are used next to small delicate pieces, the eye is carried away from a delicate piece to a massive one, and there is no gradual movement of the eye as it passes around the room. Instead, it jumps and stops and often sees little importance. Furniture pieces should be scaled equally, so that the eye will move quietly across the room from one object to another. Lines, therefore, should be related to each other and to the surrounding factors.

I. STRAIGHT LINES

A Straight line represents the tension that exists between two points. An important characteristic of a straight line is its direction.

Vertical Lines

Vertical lines are formal and create a feeling of strength and regularity, especially when repeated. They are undoubtedly associated in our minds with an upright position of a person thus signifying alertness and conventionality. It also expresses a state of equilibrium with the force of gravity. It expresses strength and forcefulness. It is likely to be dignified and masculine, with excellence, and honourable rank. It may be severe, strong, direct and disciplined in the effect it may produce.

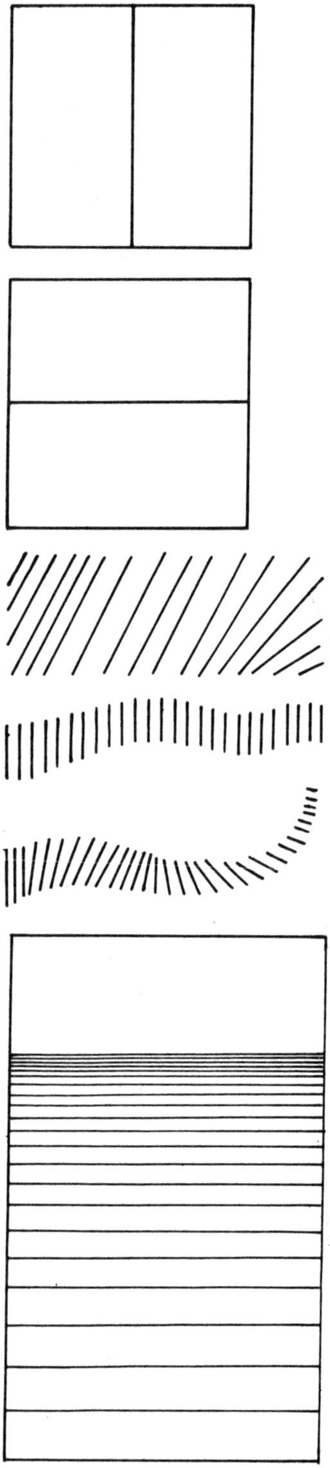

A   the vertical line tends to carry the eye up and down and lengthen the shape

B   the horizontal line tends to carry the eye across the figure and broaden the shape

C   the diagonal lines tend to make the right end appear to be wider

D   the straight lines emphasize the rhythm of a contour shape

E   the straight lines create the illusion of the railway tracks going far away

F   the horizontal lines placed progressively closer together create an illusion of steps going far away.

Figure 3.2. Lines create optical illusions

Vertical lines generally add height or length to an object and make it appear narrow. For example, a room with low ceilings and broad windows can be provided with curtains with long vertical stripes. When a vertical line is emphasized in an object, the eye of the observer measures the length of that area. But, at the same time when vertical lines are used repeatedly in quantity in an object, they add to the width. The visual effect of the vertical line is dependent on the spacing and the background colour contrast. Closely spaced, parallel vertical lines may lead an eye in upward direction, but as the space between the lines is increased, the eye may begin to measure width. Thus *widening effect* may also be produced if there is variation in the distance between the two lines, as can be seen in figure 3.2.

Horizontal Lines

Unlike vertical lines, the horizontal lines or the sideways lines are restful. They create a feeling of relaxation and informality, probably because a horizontal position is associated with sleeping or relaxing. They lend an atmosphere of tranquility, sound, stability, calmness and freedom from disturbance.

Just as vertical lines add height, horizontal lines tends to increase the width of the room. Contemporary styles in furniture and other forms of decoration are often based on low relaxed-looking horizontal lines. The horizontal line is provided by tables, benches, desks or sofas in a room.

Just as repeated vertical lines may sometimes add width, some horizontal lines spacing can produce the 'illusion of length'. This is because horizontal lines spaced closely together can create an illusion that leads the eye of the observer in an upward direction.

Diagonal Lines

The diagonal lines disturb the discipline of vertical lines and the solidity of horizontal lines. Diagonal lines suggest movement and force as it takes the shape of a body. While we are running, we bend forward, or to pull things, we place the body in a diagonal position.

Diagonal lines assume the characteristic of the vertical or horizontal line as the degree of slant approaches each extreme. The degree of slantness determines the kind of illusion created. Diagonal lines are the most difficult to use as compared to vertical or horizontal lines. They are the lines of action that seem to be pointing into space; therefore they are likely to keep the eye moving to give an uneasy feeling to a viewer and refinement to a room, thereby crossing interest. Diagonal lines are usually a disturbing element unless supported by opposing diagonal or by verticals to give them strength. Alone, they look weak and insecure which accounts for the uneasy feeling they give to the viewer.

A variety of diagonal lines used together produces a very 'busy' look which is also unsettling. This is true even if the lines are pointing in the same direction. A variety of diagonal lines going in different directions is even worse. This is because diagonal lines seem to defy gravitation. We instinctively are more comfortable with horizontal and vertical lines because they are in harmony with us and with our surroundings. That is, vertical lines repeat the lines we make ourselves; horizontal lines repeat the line of the horizon and other natural elements. The diagonal lines can be seen as rising and falling. In either case, they imply movement and are visually active and dynamic.

Zigzag lines

A *zigzag line* is a series of connected diagonal lines. A zigzag line forces the eye to shift direction alone aptly and repeatedly in an erratic and jerky movement. Because of the eye-activity caused by zigzag lines, they tend to increase the apparent mass or size of the area covered by them.

The use of diagonal lines in designing an object in home decoration can be pleasing. It is one of the best lines to incorporate when trying to camouflage poorly proportioned parts of the structure. Diagonal construction lines generally result in an informal balance. These lines can also be used successfully in supergraphics that are pointed on the wall. Such diagonals obviously need no support, so are not disturbing. They are, however, an effective way of pointing wherever you want the eye to travel, as well as being decorative.

II. CURVED LINES

Curved lines represent movement deflected by lateral forces. They tend to express gentle movement. Depending on their orientation, they can be uplifting or represent solidity and attachment to the earth. Small curves can express playfulness, energy, or the patterns of biological growth. In relaxation and play the body takes positions that are curved, so curved lines seem gracious and flexible. They achieve not only a joyful subtle and artful effect, but also a mysterious and rich effect. The curved lines can assume controlled curved lines and free-form curves.

Controlled Curved Lines

These lines have a light hearted look. Although they appeared during the baroque era of Louis XIV, they became the height of fashion during the reign of Louis XV and the Rococco style. This softly curved style is often considered the ultimate in decorative furniture and has remained popular ever since.

Free-Form Curves

The free-form curves reflect the flowing lines of nature made by waves against the shores and clouds against the sky. These irregular forms are considered contemporary, although many so far combine a suggestion of free-form curves with their tightly-controlled scrolls. Flowing forms seem to have more affinity for the new materials than they do for wood, and more free-form furnishings are appearing now that polyurethane foam and a wide *variety of plastics* have become plentiful. Just as controlled curves have a fanciful effect, free-form curves create a casual, free and easy mood desired by many homemakers today.

Yet, in order to be truly effective, curved lines—free-form or controlled, must be designed carefully. They are often seen in tied-back window curtains and in legs and corner of furniture. When the curved lines become exaggerated towards a full circle, they become very active and intense and may easily be overdone in a design. *A restrained curve is graceful, flowing and gentle.* A gradual transition in the change of direction of a curved line adds a pleasing quality to a design.

PURPOSE OF LINE:

To be beautiful and functional a line must appear to have some purpose. The line can be used to create a shape, a pattern or movement or to divide space.

(i) To Create a shape

Line is an essential element in the formation of any visual construction. Without lines, we would not be able to define shape, the characteristic by which we generally recognize things. Lines

PLATE 5

Figure 3.3. Creating line effects through the use of pictures, chairs & floor coverings

Figure 3.5. Use of forms, shapes and patterns in home interiors

PLATE 6

Figure 3.6. Textural variations in landscape designing

Figure 3.7. Use of patterns
in child's room

describes the edges of shape and separate it from the space around it. In addition, the contours of these lines imbibe the shape with their expressive qualities. In addition to describing shape, lines can articulate the edges of planes and the corners of volumes.

(ii) To Create a Pattern

Lines can also be used to create texture and patterns on the surfaces of forms. Lines and forms have traditionally been used to provide vertical support, span and express movement across space and define the edges of spatial volumes. This structural role of linear elements can be seen at the scale of both architecture and interior space and furnishings (Refer Fig. 3.3: colour plate 5).

Within the design process itself, lines are used simply as regulating devices to express relationships and to establish patterns among design elements.

(iii) To divide space and create movement

Lines may be used to divide spaces. Division of space into more than two parts might involve repetition of spaces, variation of spaces, or a combination of repetition with variety (Figure 3.3).

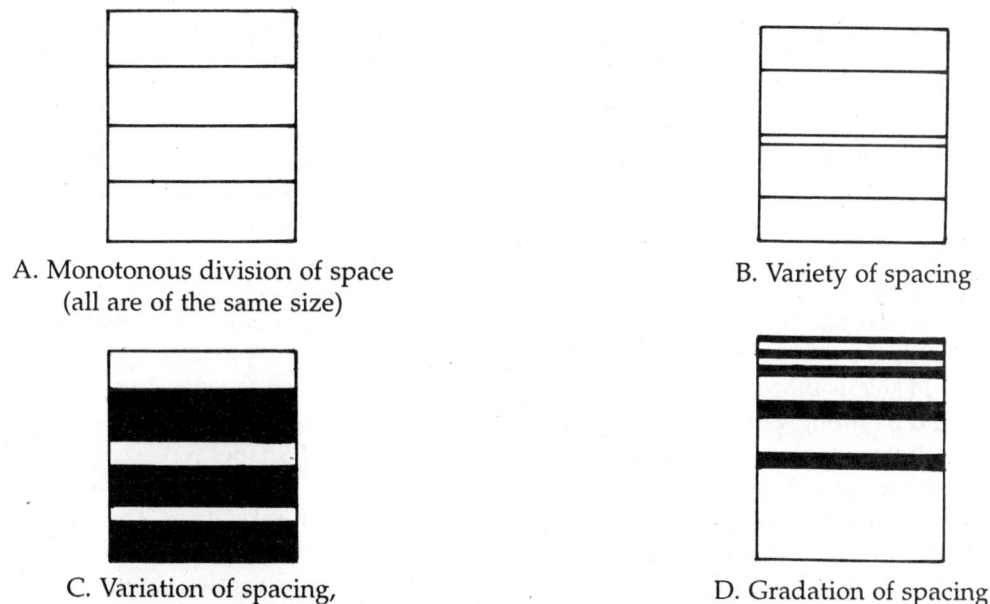

A. Monotonous division of space
(all are of the same size)

B. Variety of spacing

C. Variation of spacing,
narrow light spaces

D. Gradation of spacing

E. Space divided diagonally

Figure 3.4. Division of spaces by the use of lines for creating variety

Division of space may be done both horizontally and vertically. Diagonal lines create a dynamic effect in a composition. Pleasing proportion or divisions of space usually have some quality of strangeness or are unusual.

(iv) To create optical illusions

One may purposefully use lines in a variety of ways to make things appear different from what they seem. Horizontal lines may be used to add apparent width to a shape, and vertical lines tend to carry the eye up and down and add to the height (Figure 3.4).

FORM AND SHAPE

Shape and form are the terms that are used inter-changeably with some differences. The term 'form' is generally considered to apply to two-dimensional areas or shapes as well as to three-dimensional volumes or masses as "shape". When lines are joined to enclose space, they result in an outline, a contour, or shape. When a two dimensional shape acquires a third dimension, it becomes a form. The form of an object usually suggests its use. Form may be viewed as an enclosure of volume surrounded by limiting factors. Shape is the primary means by which we distinguish one form from another. It may refer to the contour of a line, the outline of a plane, or the boundary of a mass. In each case, shape is defined by the specific configuration of the lines or planes which separate a form from the background or surrounding space.

Form is an important element of home-planning and decoration. Without beauty of form, application of excellent colour, texture and decoration are of no use. Just imagine a beautiful tea-pot which is rich in colour and subtle in effect but when used, the liquid spills on its sides. Thus, its functional incompetence mars its beauty. A large number of such ill-formed objects can be seen around us. Similarly, in kitchen, one can see such ill-formed frying and sauce pans which trip when placed on stove because of the use of materials which make heavy handles and lighter base, though they may be beautifully finished. Therefore, it can be said that two essentials of good form of an object are that it

- should suit its function
- should be strongly influenced by the material wih which it is made.

There are three broad categories of shapes and forms. **Natural Shapes** represent the images and forms of our natural world. These shapes may be abstracted, usually through a process of simplification, and still retain the essential characteristics of their natural sources. The second type of shape and form is **abstract.** This type of shape and form is derived from objects in nature or from other things that are familiar to us; however, they have been distorted, exaggerated, and reorganized, and at times beyond recognition. The third type of shape and form is called **non-objective.** Non-objective shapes make no obvious reference to a specific object or to a particular subject matter. In this grouping, geometric forms and biographic shapes are found. Some non-objective shape may result from a process such as calligraphy and carry meaning as symbols. Others may be geometric and elicit responses based on this purely visual qualities (Figure 3.5 – see colour plate 5).

Geometric shapes dominate the built environment of both architecture and interior designing. There are three separate and distinct types of geometric shapes:

- Rectilinear   —   square or rectangle
- Angular   —   triangle or pyramid
- Curvilinear   —   circle, sphere, cone, cylinder

In their most regular form, **curvilinear shapes** are circular while rectilinear shapes include the series of polygons which can be inscribed within a circle. In all these, the most significant geometric shapes include the circle, triangle and square. Extended into the third dimension, these primary shapes generate the sphere, cylinder, cone, pyramid and the cube. Seldom can one find a room with only one form. Most interiors illustrate a combination of forms—curved lamp shades, rectilinear sofa and area rugs and angular ceiling or staircase, are some of the examples. Each of these groups of forms suggests certain distinctive characteristics.

Squares and rectangles

The **square form** represents the pure and the rational. The equality of its four sides and its four right angles contributes to its regularity and visual clarity. A square shape has no preferred or dominant direction. It is a stable, tranquil figure when resting on one of its sides, but becomes dynamic when standing on one of its corners. The square form epitomizes strength but, used exclusively, it tends to become tedious (hence the descriptive term 'square' for certain people). This solid sameness can be relieved by hanging draperies from ceiling to floor, by placing a piece of furniture at right angles to the wall, or by using round or rectangular tables to counteract the squareness.

All **rectangles** can be considered to be variations of the square with the addition of width or length. While the clarity and stability of rectangular shapes can lead to visual monotony, variety can be introduced by varying their size, proportion, colour, texture, placement or orientation.

Rectangular shapes are clearly the norm in architectural designing. The rectangular forms have more widespread acceptance as they are easily measured and handled, fit together, have a sturdy, secure relationship of exactly 90° which gives a sense of definiteness and certainty and establish unity and rhythm when repeated. Inspite of its definiteness and regularity, it can encompass great diversity. They can rest with stability on one side, insecurely on one side or precariously on one corner, each position calling forth a different emotional response.

To the ancient Greeks, the oblong or rectangle was the form most pleasing to the eye. The Greek 'golden oblong' is ideally 2 units to 3 units, or 3 units to 5. During the Classic Adam Period, architects made a point of planning rooms in this idealized proportion.

Triangles

Triangles contribute unity and balance. The triangle represents stability. Triangular shapes and patterns are often used in structural systems since their configuration cannot be altered without bending or breaking one of their sides. Forming a purely visual point of view, a triangular shape is also stable when resting on one of its sides. When tipped to stand on one of its points, however, the triangular shape becomes dynamic. It can exist in a precarious state of balance or imply motion as it tends to fall over onto one of its sides.

The dynamic quality of a triangular shape is also due to the angular relationships of its three sides. Because these angles can vary, triangles are more flexible than squares and rectangles. In addition, triangles can be conveniently combined to form any number of square, rectangular and other polygonal shapes. Pyramid and triangles differ from rectangles and squares in their pointed, dynamic character and express greater flexibility. Small repeated triangles and diamond shapes in textiles, tiles, wall-paper etc. add briskness to interiors. A room with nothing but square and rectangular furniture can be varied and brought into balance with triangular tables or a pattern that incorporates triangles in its design.

Circles

The circular forms are also useful to change a room's pace and offset the rigid right angles of square and rectangular furniture. The circle is a compact, introverted shape which has as its natural focus on its counterpoint. It represents unity, continuity and economy of form. They are man's and nature's most conservative and economical forms as they enclose the greatest area and volume with least surfacing. A circular shape is normally stable and self-centering in its environment. When associated with other lines and shapes, however, a circle can appear to have an apparent motion. Round mirrors, pictures, tables, rugs can be introduced to supply diversity, as can some curved back chairs.

Other curvilinear lines and shapes can be seen to be fragments or combinations of circular shapes. Whether regular or irregular, curvilinear shapes are capable of expressing softness of form, fluidity of movement, or the nature of biological growth. Cones and cylinders imply a dynamic directional movement not formed in circles and spheres. Cones reach a climatic terminal peak and cylinders seem to continue for ever.

Although repetition is one way to achieve rhythm, too many of the same forms can become uninteresting. For example, a rectangular mirror over a rectangular wall space, can seem monotonous because of the excessive repetition of one form. Therefore it becomes important to break this monotony by the introduction of a circle in a square or a rectangle.

While designing the interior spaces, the first area to be considered is the enclosure of the whole house or apartment i.e. at macro level. Interesting concepts of spatial relationships are evident in modern residential architecture, where there is a conscious effort to design the house in terms of the geographical terrain. The desire to relate the interior to the exterior has led to the extensive use of window walls and huge picture windows. However, it affects the privacy of its inmates. The solution for this can be the use of lacy grill patterns in facades to let in light and air. Those on the inside can see out but outsiders cannot see in.

On the microlevel, it is the room itself, the height of the ceiling, the placement of the windows, doorways and the other structural features—for example, a fireplace, will determine to a large extent how the area is to be utilized. Every form and shape will introduce new area relationships. Furniture groups, pictures on the wall, draperies at the windows, floor coverings, and accessories will all contribute to the space divisions. If a house has a long, narrow room, it can be divided into different functional spaces such as a living area and a dining area or a study area with the help of partitions. Some rooms have alcoves that will determine such area divisions. It can be done effectively by furniture placement, room dividers, area rugs, colour, pattern and texture.

COLOUR

Colour is an important art element which no one can ignore. Appreciation of colour, largely an emotional process, is felt by nearly everyone, whereas appreciation of other art elements such as line form, texture etc., a large intellectual process, is not so common. Colour is a source of universal pleasure. Colour is something which is used by everyone to delight them and also to fortify their living environment by its stimulating effect.

The interior decorator should understand the theory of coloured light as well as the theory of coloured pigments. Light is a flexible decorative medium with enormous potentialities. Similarly coloured pigments are of significance as they are widely used and appreciated by the people of all walks of life – young/old, poor/rich, rural/urban, literate/illiterate, male/female and so on. The modern theatre has shown the possibilities of coloured light in obtaining the desired effects. Changes of colour schemes in homes, clubs, hotels, shops etc. are made by the use of light rather than by pigments. It is possible to produce pictures by the use of coloured lights instead of painted pictures. This can be achieved either by stationary or by mobile coloured lights to produce natural or abstract pictures.

Like other art elements colours exhibit their qualities. The three basic qualities of colour – hue, value and intensity or chroma – can be measured with considerable exactness. These three are distinct from one another. Hue is synonymous with the word colour itself, and so, refer to the name of colour. While value refers to the amount of lightness (white) or darkness (black) in a colour, intensity refers to its brightness or dullness. Some colours like red, orange and yellow are warm and advancing colours, and the colours like blue, green and purple are cool and receding. Darker colours (shades) are heavier and lighter colours (tints) are lighter. These colours can successfully be used in home interiors to create warmth or cool feelings or to reduce or increase the size of a room.

Colours also create emotional effects. White is a symbol of innocence, purity and peace, whereas black is used for mourning, evil etc. Red colour may represent love or martyrdom, and blue, sincerity or hope. Colour, because of its emotional effect on us, is largely responsible for the atmosphere of a room. A colour scheme of room may be soothing or irritating, cheering or depressing, charming or boring, welcoming or repelling. Different colours excite different colour emotions, though it may differ in its degree on different people. For example, some persons are more sensitive and more stimulated to the colour effects, and some, at lesser levels.

There are some neutral colours like white, black and grey. Such colours are more valuable in interior decoration as background colours. Some earthy colours such as umbers, siennas, and ochres, and almost all of reds, yellows and greens, suggest natural earthy characters. The so-called acid colours such as magenta, blue-green, cyan blue etc. give an idea of artificiality. These ideas should be considered by the interior decorator in the use of colours.

Colour is such an important element that volumes and volumes of its nature and qualities can be written. Because of this unique quality of colours, it is discussed in detail in two separate chapters later in this book. This will help an interior decorator to utilize the colours successfully.

TEXTURE

When we mention the word texture, we immediately think of touching something. The term texture now generally refers to the tactile quality of the surface of any object, although originally it applied only to textiles. Texture is that specific quality of a surface which results from its three-dimensioned structure. Texture is most often used to describe the relative smoothness or roughness of a surface. It can also be used to describe the characteristic surface qualities of familiar materials, such as the roughness of stone, the smoothness of cream, the grain of wood, and the weave of a fabric. It refers also to the way the small constituent parts are combined in a substance. For example, a poor porridge may have a grainy texture and granite a granular texture. The pliability or rigidity of objects also has textural significance as it affects the quality of the surface.

There are two basic types of texture – *'tactile texture'* that is real and can be felt by touch, and *'visual texture'* that is perceived and can be seen by the eye. All tactile textures provide visual texture as well. Visual texture, on the other hand, may be real or illusory. Texture is the sensory impression that is understood by sight as well as by touch.

Our sense of sight and touch are closely intertwined. As our eyes read the visual texture of a surface, we often respond to its apparent tactile quality without actually touching it. The visual aspect of texture is perceived by the eye because of the degree of light absorption and reflection on the surface of the material. Lustrous texture are seen in satins and dull texture in fuzzy wool.

Fabric, metal, leather and straw, each has a distinctive texture. Some adjectives, used to describe textures are smooth, heavy, fine, crisp, glossy etc. We base these physical reactions to the textural qualities of surfaces on previous associations with similar materials. So the artist depicts textures on canvas as he paints a woman wearing a beautiful satin dress with glowing folds, or a bowl of flowers with lovely soft petals. The textures he portrays give character to his work and help to express his ideas.

'Hand' is the term used to refer to the tactile aspects of the fabric. Coarseness, softness or rigidity is recognized by feel. Texture is fully comprehended by touch, but it is not always necessary to feel an object to understand its tactile qualities, as sight recall the memory of tough.

More often textural qualities are felt or observed on fabrics used in the home interiors. The texture of the fabric is affected by the characteristics of the raw material used and by the production processes involved, from fabric to the final stage. Texture determines how the cloth is to be used. For example soft velvet dictates its use in sophisticated, classic interiors and coarse denims etc. for more casual and informal settings. Sometimes, the term texture is also used in other decorative objects where the finish given to an object gives an illusion of texture as in the case of marblized flower pots. Just as lines, shapes and colours convey messages, there are important modifying factors in our perception of textures and the surfaces they articulate (Figure 3.6 – see colour plate 6).

The ability to select texture that reflects the self concept of an individual is achieved when there is an understanding of the character or idea projected by the textures. The distinctive individual qualities of some textures typify particular moods and feelings. A variety of factors must be examined carefully in order to be able to identify the character they usually project. For example the response learned by feeling satin differs from that of velvet. Texture is an element of design that can be effectively used to express individuality. It is valuable in giving a particular character and beauty to any piece of art, be it a painting, a sculpture, any interior ranging from drawing room to bathroom and kitchen and also to buildings and landscapes.

Try to visualize a wooden, varnished sculpture and a sculpture carved out of a stone. What effects will they produce? In which settings can they be used? A layman with an eye for art can very well distinguish between the effects created by these two materials. Texture expresses the artist's meaning and accentuate the impact created.

Designers tap the potential of the textural qualities of building materials. In a bedroom, with traditional overtones, rough brick interior with the use of cast iron lampshade and course mats as rugs and carpets create such textural effects. But here one thing important to understand is that majority of us do not realise the capacity of texture to influence the impressions created and therefore a more detailed understanding of texture is needed. For instance shiny, reflective surfaces act in similar ways to a mirror, creating a more spacious effect. This means that painting walls and ceilings with gloss paint or walls fixed with mirrors will add to a room's apparent size whereas matt-finish paint will reduce its size.

Given below is a list of words which are often used to describe textures of various surfaces:
- blistered, bristly, bumpy, bearded
- coarse, corrugated, cracked, crinkly, crisp, crumpled, curly, corky

- delicate, dense, downy, dull, dusty
- feathery, filmy, fine, firm, flexible, foamy, frilly, fussy
- glassy, glossy, granular, groomed, gritty
- hairy, harsh
- lacy, leathered, lumpy
- marbled, messy, metallic
- perforated, pierced, plaited, pleated, polished, porous, powdery, prickly
- quilted, quiet
- ribbed, ridged, rippled, rough, rubbery
- sandy, satiny, scaly, shaggy, shiny, silky, smooched, smooth, solid, spongy, stiff
- thorny, tough, twisted
- uneven
- velvety,
- waxy, wavy, woolly, woody

Harmony in the use of Texture

We are living in an era, which defies tradition – an age of strong contrasts, when boldness, bravery and adventure are in vogue in dress and behaviour as well as in interiors. Today contrasts of textures are widely used. Inspite of the popularity of bold contrasts, some combinations appear to be more pleasing than others. Contrast is necessary in order to avoid sameness and monotony. However, a predominant texture idea should be apparent so that unity or harmony is achieved.

One of the first decisions to be made in furnishing a room or a house is the selection of the furniture wood or woods, for, other textures employed must be in harmony with the wood. Different kinds of wood seems to produce different kinds of feelings in an observer. Pine, oak, and teak suggest strength, and mahogany and rose wood suggest elegance. Mahogany requires delicate textures like fine silk, satin, velvet, roses, deep-pile rugs and light weight brassware to accompany it, whereas, with oak, coarses textures such as tapestry, large patterned linen, iron and parchment should be used. In the study of texture, it is helpful to analyse the significance of materials that have been combined in each of the decorative creations.

Modern use of texture is creative. Rooms are now composed in which areas of various textures on walls, floors, and furnishings are organised to rich, subtle effects. The texture of any single article is not considered separately, but as a contribution to the total effect of the room. When the full possibilities of this element are utilized, plastics, glass, metal, wood, cork, leather and fabrics will be composed into a symphony of texture. Combinations related to weight (thickness and thinners) and those related to firmness (crispness and softness) do not present peculiar problems, but combinations related to image, feeling or personality of texture should be analysed carefully. Delicate lace and fine embroidery harmonize with fine, sheer fabrics rather than coarse cotton lace and heavy crowded embroidery which create too great a contrast.

When making decisions regarding textural combinations, stand away from them. Distance diminishes the effect of the combinations present. When they are close, they may blend together and appear too similar in texture, or may not belong together because they are too different. Thus best results in creating harmony in textures can be achieved by observing them from at a distance rather than in close proximity.

Texture and Scale

Texture has definite physical dimensions – weight, size, bulk and shape. These physical dimensions are also visually perceived. All materials have some degree of texture. But the finer the scale of a textural pattern, the smoother it will appear to be. Even coarse textures, when seen from a distance, can appear to be relatively smooth. Only on closer viewing would the texture's coarseness become evident. The relative scale of a texture can affect the apparent shape and position of a plane in space. Textures with a directional line can accentuate a plane's length or width. Coarse textures can make a plane appear closer, reduce its scale, and increase its visual weight. In general, textures tend to visually fill the space in which they exist.

Texture and light

Light influences our perception of texture and, in turn, is affected by the texture it illuminates. Direct light falling across a surface with physical texture will enhance its visual texture. Diffused lighting de-emphasizes physical texture and can even obscure its three-dimensional structure.

Smooth, shiny surfaces reflects light brilliantly, appear sharply in focus, and attract our attention. Surfaces with a matte or medium-rough texture absorb and diffuse light unevenly and therefore appear less brighter than a similarly coloured but smoother surface. Very rough surfaces, when illuminated with direct lighting, cast distinct shadow patterns of light and dark. Contrast influences how strong or subtle texture will appear to be. A texture seen against a uniformly smooth background will appear more obvious than when placed in juxtaposition with a similar texture. When seen against a coarser background, the texture will appear to be finer and reduced in size. Finally, texture is a factor in the maintenance of the materials and surfaces of a space. Smooth surfaces show dirt and wear but are relatively easy to clean, while rough surfaces may conceal dirt but are difficult to maintain.

Texture and Colour

There are no hard rules about the relationships of colour and textures. Certain textural qualities are easier to emphasize with particular colours. Dark red, emerald green, purple, and gold suggest luxury and elegance of texture. The earthy colours of brown, mustard yellow, burnt orange and yellow-green seem to be more suitable with less refined textures. But one should not feel restricted by any definitive rules with reference to texture and colour. The overall effect is the most important consideration. For example, on a flat surface, some appear dull but on an interesting textured surface, the same colour can look more interesting. Smooth surfaces reflect light and do not absorb it. So the same colour may look brighter in an emulsion paint on wall than on velvet upholstery or on a silky brocade curtain material.

PATTERN

Pattern refers to any sort of extrinsic surface enrichment. It is a two-dimensional or three-dimensional ornament arranged in a motif form. Because patterns can be created by textures and forms, the entire arrangement of the room creates a pattern. Pattern is found in the shape of individual items of furniture and in their groupings, in wall paintings, parquet floor, architectural detail and light and shadow in the room. Pattern has movement and should be arranged so that it will flow with the rhythm of the room or the object it adorns (Fig. 3.7 – see colour plate 6).

Pattern can coordinate the entire decorating theme. The trend toward carrying patterns from one room to another—bedroom and bath, kitchen and eating area, living room and entry hall—is justifiably popular because it creates smooth effects and unifies the overall schemes of the home.

At times, people use cost as the criteria for selecting patterns. The cost of an article is no indication of the quality of the decorative pattern used on it. The finest designers are employed chiefly for expensive goods. However, we also find that their designs are adapted or duplicated or imitated in inexpensive materials. It is desirable to buy patterned articles and other furnishings and fabrics designed by famous designers. Manufacturers often underestimate the taste of the consumers and make articles decorated with cheap elaborate designs, which people buy because nothing else is available. This is commonly seen in silverware, wall paper, drapery, carpets and decorative items.

Pattern creates an illusion of depth and adds character and life to a room. A pattern is an overall design. A **'motif'** is an individual unit of pattern. Fabric design is often created when motifs are repeated in a manner to create an overall pattern. These may be considered formal – showing a regular or methodical repetition of the motif, or informal – having irregular placement of motifs.

Surface patterns add to liveliness and interest in a room. A dull and boring room may be brightenened up by the use of pattern in a desirable amount. Pattern should not compete with the major focal point in the room, and too much pattern will make a room seem busy. Too much of a pattern if used on floor, curtains and sofa too might give a restless and too intense an effect and lack restfulness. Opinions differ on how much pattern is desirable in a room, but it is customary to use pattern on atleast one-fourth of the total surface area. For example, if the walls are papered, then the ceilings and floor can be kept plain. Similarly, if the walls and carpet are plain then draperies and two-thirds of upholstery fabrics may be patterned.

Pattern can also be used to create illusion of spaciousness. If a small patterned carpet is used rather than a strong, bold design, then a feeling of spaciousness is created. Books, pictures, flowers and plants, all produce pattern in a room. It is here to note that, a large room can accommodate more pattern than a small one. When arranging pattern in an interior, an agreement needs to be arrived at. In different objects which may be used in a room, the types of pattern should comply, whereas size may be varied. Let us visualize a room to understand the impact created by pattern. If a bold pattern of an expensive rug is to be emphasized then inconspicuous stripes and textured patterns are desirable companies. Large patterned draperies should not have motifs of the same size in a room.

While choosing a patterned fabric, design of the fabric and colour are enormously important. Patterns range from huge flowers in rich blue and pink plus purples, mauves and apricots to paisley patterns in dark blue, deep magenta and burnt orange and colourful spots and stripes. If you are buying a fabric that is to be pleated/gathered as draperies/curtains, check how the pattern will look pulled together in their way. Some subtle designs come to life when used in a pleated or gathered form, while other patterns lose their impact.

UNITS OF DESIGN

There are in general, three types of motifs or units of design:

1. Naturalistic
2. Stylized
3. Geometric

Naturalistic motifs

They duplicate nature and look like pictures, usually of flowers, fruit, animals or sceneries. Such motifs or pictures are seldom suitable for the decoration of utilitarian articles. Excellent designs include floral, as they are grown in garden, strawberries that hang on tree, toys that belong to nursery, and are confined to frame effects. Or still other dishes and plates decorated with pastoral scenes and realistic fishes are dubious backgrounds for food. Some of the best natural floral patterns are those in which flowers are grouped as in the case of wall-papers. At times they are also grouped in definite stripes, blocks, or bouquets, or sometimes confined to frame effects. Naturalistic designs are appropriate in some period rooms, cottage rooms, children's rooms, particularly in draperies, upholstery and wall papers.

Stylized motifs

Stylized motifs are variations of natural forms. For example, when stylized floral or leaf motifs are used, they show imagination. Their motifs do not look like pictures of natural objects; usually the lines are simplified and conventionalized, sometimes they are distorted. The designer uses his creativity and imagination to produce a design depending upon the material available and the purpose of the article. Stylization alone does not insure high quality in design; however, in general, stylized designs are likely to be superior to the naturalistic ones. The most fashionable stylized motifs are ferns and other similar leaves. Mangoes, butterflies and fishes are the other objects which are often used in stylized forms, specially in textile designing. To have better effects, patterns are made of a combination of stylized and geometric motifs.

Geometric motifs

These are based on the pure forms of the circle, square, rectangle and triangle, although endless variations and combinations of them are used. Geometric motifs include plaids, checks, stripes and dots. They are the safest designs procurable for untrained consumers. Modern designers prefer geometric motif in the small amount of pattern that they use, as they are easier to work with. The Greeks also realized the value of geometric forms and developed it to a high degree. They have suggested patterns which have good proportions. More details of geometric patterns having good proportions are given in the next chapter on principles of design. A knowledge of the types of designs can aid a designer to choose and use them appropriately in areas to create variance in ideas and arrangements.

HOW TO SELECT PATTERN

Since most patterns available are poor, consumers should learn to discriminate good ones from the poor ones. An elementary course in design, books and periodicals on designs, and museum displays, or other art exhibitions can help a layman to analyse and judge patterns.

Most often a pattern accent is at its best against a neutral background. Usually, a bold pattern calls for a large massive area, but this guideline also needs to be viewed in terms of its exceptions. A large chintz can be charming on a small chair; a small print that appears in one colour from a distance may be attractive on a large sofa. Using the same pattern on many pieces is a good way to achieve harmony.

It is popular today to mix paints, but such mixing must be done with skill and a feeling for colour and design. Two patterns of the same colour may be effectively combined if they are of related styles and design scales. The same pattern in another related colour may also combine well.

Geometric patterns may be combined with florals, if they use the same colours. The dominant colour could then be used in solid areas of the room. A knowledge on the various means of beautifying the surface pattern can help a person to select patterns wisely.

Beauty in surface pattern is produced by a number of ways. They are listed as follows:

1. Excellent design in individual motifs or units: Out-of-the ordinary designs are most desired so that the work depicts an outstanding design. The patterns should be well-designed.

2. Fine arrangement of the units in a repeat pattern: It involves the grouping in a systematic manner which produces a rhythm. The units may be grouped in borders, stripes, checks, diamonds, ogives – in either regular or irregular plans. Arrangement of the motifs is such an essential component that the same unit may appear insignificant when used sparsely but distinctive when used in a compact scheme.

3. Definite Character: Every motif should have the ability to giving a certain identity or character to its presence. Thus a definite expressive quality is required, quaintness, speed, restlessness or whatever quality the designer expects to have. The character of a pattern is determined by the direction of the lines and by the sizes, the shapes and relation of spaces.

4. Honesty in technique: The design should suit the process used in its production. For example, fine detail should be avoided in a linoleum block print. Proper regard for the medium ensures honesty in technique.

5. Evidence of joy of the Creator or Designer: A design should appear to be a joyous expression of the creator, and not a laboured or forced piece of work.

6. Harmony of line between an article and its decoration: The design should fit the material it decorates and should express the same idea. The lines of the pattern should usually follow the lines of the article that it decorates, i.e. a circle fits better than a square on a round table.

HOW TO USE PRINTS AND PATTERNS

Beauty in pattern depends not only on having well-designed motifs, but also on arranging them well. Arrangement is so important that the same unit may appear insignificant when used sparsely, but distinctive, when used in a compact scheme. Therefore, it is good to start arranging the units into various groups as regular or irregular plans. In today's free-wheeling world of decorating, mixing prints and patterns is more the rule than exception. In many handsome traditional rooms, a single print is still used lavishly for upholstery, drapery, and even to cover the walls. But the newer way is to mix and scramble both patterns and colours with an uninhibited hand.

One can always start with very simple patterns perhaps in a single colour. If different patterns are used in the same colour scheme, it is safe to go on adding patterns almost indefinitely especially when the patterns themselves are very simple. Another easy way to scramble patterns is to use the same one in several different sizes. This works especially well with dots or checks and sometimes with floral prints, although these are harder to find. It is safe to keep the size the same and use a miniature floral pattern in several different colour combinations. Simplified paisley patterns, which are perennially popular, are very good mixers, especially with simple geometric patterns such as stripes and checks. Here the common denominator is colour rather than pattern; there should be some colour relationship between the two, although the colour scheme of both need to be identical. One of the two might be in black and white, the other in black and white with red.

Stylized flower prints and medallion patterns also mix well with stripes and checks. Again, there should be a colour relationship. Varied black and white patterns are especially compatible, and for some reason, more patterns can be mixed in this colour combine than in any other without a confusing effect. Two different prints with a related theme make an interesting combination if the colours also have something in common. Some related prints are designed that way. Even if they are not designed to relate to each-other, still look great together. They should be either in approximately the same scale or in a widely divergent ratio, with one print quite large, the other small. Two dissimilar prints in precisely matching colours are often extremely effective together, especially if the distribution of colour is approximately the same. This does not mean that the backgrounds have to match, but it makes the relationship more easily apparent.

PATTERN AND TEXTURE

Texture and pattern are closely related design elements. Pattern is the decorative design or ornamentation of a surface which is almost always based on the repetition of a design motif. The repetitive design of a pattern often gives the ornamented surface a textural quality as well. When the elements that create a pattern become so small that they lose their individual identity and blend into a tone, they become more a texture than a pattern (Figure 3.7 – see colour plate).

A pattern may be structural or applied. A structural pattern results from the intrinsic nature of a material and the way it is processed, fabricated, or assembled. An applied pattern is added to a surface after it is structurally complete. An example of the former can be a woven motif on a fabric surface and the example for the latter is a printed motif on a fabric surface.

SPACE

Most of us, at some time or the other, feel that we suffer from a lack of space in our homes, yet there are probably parts of the house or a flat that are rarely used. At certain times, some corners and alcoves that are dark or awkwardly placed are not utilized fully. On the other hand, there seems to be a clutter every where, that a house, whatever its size, does not create an impression of space. Therefore, the importance of space as an element of art has not been fully recognized, more so in the organisation of space as basic in modern architecture and in interior decoration.

Everyone of us is conscious of space. Any one who travelled from a city to a village is pleased to see a large area of open space around the house, whereas the same person feels choked in a crowded urban area. A sense of beauty of space makes us feel the need for large undecorated walls and floors which bound space without disturbing its effect. Some people use less furniture on floors and a few accessories/pictures on walls because the uncovered empty silent spaces help in appreciating space. On the contrary, complete coverage of space would lead to overcrowding or crampedness and the emphasis of beauty of objects used in these areas would be the last in the process.

The element, space, can be used successfully to increase the size of the room visually and to give the room a quiet feeling of rest and beauty. Large unadorned openings, black and white effects in rooms, encourage the eye to explore the distance beyond, specially when the same material or colour is carried throughout. Hence, a feeling of spaciousness is created by the way we decorate a room and the way we arrange the furniture in it. For example, light colours, low furniture, shiny surfaces, reflective images and well-planned storage, all contribute to a feeling of extra spaciousness.

Light colours appear to recede, and therefore, a room decorated and furnished in pale tones will appear much larger than the same room decorated in deep hues. Shiny reflective surfaces act in a similar way to mirror, creating more spacious effects. This means painting walls and ceilings with glass paints will add to a room's apparent size whereas a matt paint will reduce it.

A low ceiling will appear higher if you paint it white and high ceiling can be 'lowered' by the use of deeper tone of colour used on the wall. A pale carpet will make a room seem larger and more spacious. Lines that travel along the length of the room on floor or wall will appear to extend the space while those that travel across its width will shorten but widen the area. Any strong colour used as a border around the room – a skirting painted in deeper colour will enclose the space and will make it appear smaller. Other ways to develop the feeling of spaciousness are to use floor-to-ceiling glass curtains or painting the colour of the wall or placing illusionistic devices such as mirrors, paintings with deep perspective, and scenic wall-papers to suggest distance. Glass-topped tables will give an illusion of space because you can see through them and furniture that looks light, such as cane, add to the general impression of light and space. Mirrors are excellent visual space makers. A wall of mirror will make a room look twice its size.

Lighting is another device that can be used to provide a feeling of space. Lighting that emphasizes the ceiling can make the room look larger. Curtains with valance lighting or a luminous panel on one wall will make a room appear visually larger than what it is and hence, more spacious. Lighting placed under the sofa can make it appear to float, thus, less massive. Covering one of the walls with a large painting in cool colours like green, blue etc. or a wall paper with the picture of a natural scenic beauty or a landscape with lots of trees at different distances, can also take the eyes to travel beyond the picture wall, which can also provide an additional illusionary space in the room. To preserve a spacious atmosphere, the available space should be divided as little as possible, with a minimum amount of small-scale furniture with a few accessories. Furniture on wooden legs will contribute to a more spacious appearance than furniture with bases that extend to the floor. The clear plastic furniture available today creates a feeling of airiness.

Exposing as much floor as possible increases the room's spatial fluidity. If the floor is covered with carpeting, continuing the carpeting over the base of a sofa/window seat will increase the visual size of the room – some sofas, beds and coffee tables have an enclosed base of mirror; thus, the visual space is increased.

The implications of space are —

- A pleasant relationship needs to be maintained between space and pattern. Too much pattern is as bad as too little.
- Space may give a feeling of being exposed to a large area which is not provided with a boundary or fence.
- Space may give a trapped feeling if there is a high wall surrounding a small space.
- Gradual change of space is more pleasing than an abrupt change.

At the end, in addition to interior decoration, an architect and landscape designer should utilize the element of space. Therefore with good spatial arrangement a feeling of quietness and beauty can be created.

LIGHT

Light, as an element, is universally studied by people of every walk of life, since light is an art element as well as an utilitarian element. The emotional effect of light can easily be understood from a sunshine day which stimulates us to sparkle or from a dark day which makes us feel depressed or dull. People living in north hemisphere are unfortunate not to see sunlight at all during winter months and feel the gloom that darkness brings in. On the other hand, when summer comes, light becomes too brilliant and a welcome factor, because one feels physically active and in good form. Thus, in our homes, we should have available, but under control, all the light that we can use. A variety of moods and effects may be obtained by the clever use of light and shadow, bright and dark areas. Some people may prefer to have diffused soft lighting in some areas like dining, whereas perforated metal lanterns cast interesting shadows in areas like porches and verandahs.

Light is an integrated and important aspect of planning any interior. Among the various sources of light, day light is an important factor which contributes to the appearance of a room. For this reason, all plans of decoration becomes incomplete without the consideration of the amount of light that is made possible to enter a room and any effort of beautifying such a room becomes wasteful. It is important for an interior designer to consider the exposure, the number of windows, the amount of sunshine that enters the room, the trees/vines that shut out light and in what season of the year the room is used most, etc. Light can be provided either through day light or through artificial light. The modern artificial light, particularly electric light plays an important role in dramatizing the effect that a room reflects. It can bring rhythm and continuity to a room's furnishing by linking together various points of emphasis.

Modern artificial light is not only a remarkable functional utility, but also a marvellous flexible art medium. It should be used in a room in such a way that the effect produced is different in different areas. In a bed room or a drawing room, it should not be too loud, whereas the lighting in a kitchen or a study room should be more functional and localised. Light can also be used to highlight an important area or an important object, or subdued to camouflage a defect. It can decide the amount of emphasis that an area or an object requires or vice versa.

Light that emphasizes the ceiling can make the room to appear larger. Curtains with valence lighting or a luminous panel on one wall will make a room appear visually larger than what it is, and thus, more spacious. Lighting placed under the sofa can make it to appear to float; thus, less massive. The use of light and the great potentialities of artificial light is successfully utilized by the theatre artists and film industry. An interior decorator can learn from them and use these effects to his advantage in all his arrangements and creations.

Creating a design is an art. To make it scientific, there is a great need for understanding its fundamental components. A thorough knowledge of these components, namely the elements of art and design, would help a person to create beautiful things and enrich her surroundings.

Chapter 4

PRINCIPLES OF DESIGN

W hile trying to invent things, man discovered and assembled various materials by observing natural things or modified the existing things to suit his own requirements. By doing so, these things turn out to be artistically good, as they might have been chosen and arranged according to their natural forms. In due course of time, man started creating beautiful things not only by copying the nature, but also by following some guidelines. To create beautiful things consistently, he had realised that he has to select certain elements like colour, form, texture, pattern etc. and to arrange them in an orderly way to create designs that conform to certain standards. These guidelines for creating designs might have eventually taken the term, *principles of design*.

It can, therefore, be concluded that structuring of art forms is guided by established principles. They are expressive of the manner in which various design elements are arranged within a given a space – be it the artist's canvas or the confines of a home garden or a courtyard. It is also realised that certain clearly defined principles of design or arrangements are common to the space arts like painting, sculpture, architecture, handicrafts, and industrial, commercial and other related arts. Thus it can be said that the principles of art are not only actually the formulae for creating beauty, but also help in judging whether an object or an area of the house is artistically good or poor. Because a design is created by the selection and arrangement of form, colour, texture etc., the result is an expression of the person who selects and arranges them. In that sense, a home setting is a *design*, as it represents the ideas and personalities of the individuals who live in it. It is, therefore, important for an interior designer to understand some of the basic principles that might serve as a guide for creating *good* designs.

Objectives of design

There can be no rigid rules or formulae for selecting and arranging the components of a design because these would immediately preclude the spontaneous expressions of the individual. Designs that are stereotyped or imitative or lack individuality are likely to become monotonous and tiresome. Yet there are certain fundamentals that should be considered if the created designs are to achieve their objectives, namely—

- Beauty
- Expressiveness, and
- Functionalism

Creation of design involves the selection of individual design elements and the arrangement of them within a spatial enclosure to satisfy certain functional and aesthetic needs and wishes. This arrangement of elements in space includes the act of making patterns. No one single element in space stands alone. In a design pattern all the elements of design depend on one another for their visual impact, function and meaning.

It is of significance here to consider the design principles as the guidelines to the possible ways with which elements can be arranged into recognizable patterns. The principles can help to develop and to maintain a sense of visual order among the design elements of a space while accommodating their intended use and function. These principles, broadly speaking, may be thought of as measuring sticks which help to judge taste, and are discussed in detail in this chapter under the following heads—

- Proportion
- Balance
- Harmony
- Rhythm
- Emphasis.

PROPORTION

Every time two or more things are put together, proportions are established. The proportion thus created, refers to the relationship of one part to another or to the whole, or between one object and another. The relationship may be one of magnitude, quantity or degree which results in the correct or pleasing relation of parts of a single object. This principle of proportion is therefore called the *Law of space relationships*. It implies that the relationship between parts of the same thing or between different things of the same group should be satisfying. It deals with the relationships in size, shape, colour, light, texture and pattern. For example a chair has good proportions if the size and shape of the back, the size of the seat and the height of the legs are in pleasing ratio to each other, and to the size of the room in which it is going to be placed. This exhibits different sizes of chairs that can be selected according to the size of the room in which they are going to be placed.

Proportion is also the relation of parts of a design or composition to the other parts of a design or the composition as a whole. It indicates how an objective item is related to others with regard to its size, number and quality. It is the ratio of the comparative sizes and reveals that all space dimensions should be pleasingly related to each other.

The most important or dominant application of the principle of proportion in home furnishing and home planning is, however, the relationships of the area. The appearance of the exterior of a house is due primarily to its proportions, first the total mass which depends on the height in relation to the length, then the proportions of roof, walls and foundation and finally the relationship of doors, windows and other elements that must be organised into a unified whole. The shapes of rooms and of every article of home furnishing should be judged by their proportions. A room has good proportions if the width, the length and the height of the ceiling are in pleasing ratio

PLATE 7

A

B

Figure 4.4. Use of equal divisions of space in striped fabrics (monotonous: A) and National flag of India (interesting: B)

Figure 4.5. Variations and repetitions in designs for cushions

PLATE 8

Figure 4.8. Formal balance can be observed in the design of the bed whereas informal design is seen in the placement of a chair on one side of the bed and a pot with a plant on the other side

Figure 4.11. A sample Phulkari embroi-dery design for a cushion

Figure 4.12. Rhythm in interior designs

to each other and if the windows, doors and other features included are in pleasing relation to the room as a whole. Some people have an instinct for good proportion and whatever combinations they plan are sure to please the eye. It has been seen that even an untrained person has an inherent sense of good proportion. Others can learn it by understanding this principle.

In everyday life, we are constantly aware of proportion and are often applying this principle of design even though we are not always aware of doing so. The sofa is almost automatically placed against the long wall in the living room. A letter typed with due regard for margins or the paper is another example which is pleasing to the eye. These two examples represent space divisions that are pleasing and are not disturbing. When not done, the composition appears to be disturbing or lacking in some element. Thus when two or more things are put together proportions are automatically established whether good or bad.

Proportioning Systems

Any study of proportion must begin with the achievements of the ancients, because the standards they had set are followed even to this day. In the case of history, several mathematical or geometric methods have been developed to determine the ideal proportion of things. These proportioning systems go beyond functional and technical determinants in an attempt to establish a measure of beauty – an aesthetic rationale for the dimensional relationships among the parts and elements of a visual construction.

According to Euclid, the ancient Greek Mathematician, a ratio refers to the quantitative comparison of two similar things while proportion refers to the equality of ratios. Underlying any proportioning system, therefore, is a characteristic ratio, a permanent quality that is transmitted from one ratio to another.

Greek Proportion

Ancient Greek designers were the first to analyse space relationships and became masters of proportions and their art and architecture have for centuries been considered the epitome of perfection in space divisions. The Greeks through long study became very sensitive to fine relations in space. They formulated rules based on the proportions of the human body.

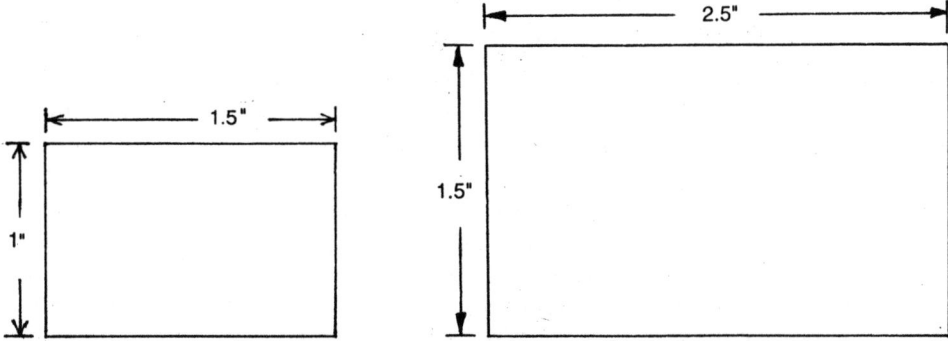

Figure 4.1. The golden oblong as created by the Greeks

The most familiar proportioning system is the *golden section* or the *golden oblong* established by the ancient Greeks (Figure 4.1). It defines the unique relationship between two unequal parts of a whole body in which the ratio between the smaller and the greater part is equal to the ratio between the greater part and the whole body. This proportion is found in any rectangular oblong which can be divided into two unequal areas, one of which is a square, so that the smaller one is to the larger as the larger is to the whole. This is the main reason for rectangles to be more interesting than the squares, and the Greeks discovered that the unequal sides of the most pleasing rectangles (i.e. for flat surfaces) have a ratio of 2:3, 3:5, 5:8, 8:13 and so on, while that for solid is a ratio of about 5:7:11. In the same way, any line can be divided at a point so that the shorter line will be to the longer as the longer is to the whole line. This division of a line is pleasing, because variety attracts the eye and monotony does not. Exact divisions of spaces into halves, thirds or quarters are less attractive than unequal divisions. This is why we find that the rectangles are more interesting than the squares. This basic law of proportion can be applied to everything in decorating or any kind of design and is the reason why the rectangle appears so frequently in the sizes of rooms, rugs, tables, pictures, mirrors etc.

Aspects of Proportion

As mentioned earlier, the principle of proportion is called the law of space relationships. According to Goldstein, there are three practical problems in proportion which confront us in everyday tasks. These are:

1. How to achieve arrangements which will hold the interest
2. How to make the best of given sizes and shapes
3. How to judge what sizes may successfully be grouped together.

There are definite means which can help to solve these problems. The suggested solutions for these three problems are:

1. In order to achieve arrangements that will hold interest, one must know how to create beautiful space relationships.
2. In order to make the best of given sizes and shapes, one must be able to produce a semblance of change in appearance, if it is desirable.
3. In order to judge what sizes may be grouped together successfully, it is necessary to grasp the underlying significance of scale.

Creating interest through space relationships

In any group of objects, the period of interest will depend upon the kind of things grouped together. When all of them are similar or the usual thing like a lamp or a door, they all get equal attraction at the same and then the interest subsides. But introducing an element out of the ordinary, such as an unusual lamp or door, an interest is immediately stimulated. If an arrangement is built on the plane of three equal divisions of one square each, the mind will record these squares without a pause, and the eye will not be arrested. Instead, if an oblong is mixed with two squares, or two oblongs with one square, one would have to look an instant longer that the total picture was recorded, and in that instant one would actually perceive and remember that group more clearly than the one composed of three squares (of equal size). This is an example and an answer to the question of how to achieve arrangements to hold the interest of the viewers. This is where the Greeks have achieved good results by creating the *golden oblong* which is the recognized standard for space relationships. This oblong is more beautiful than a square, because the equal sides make a square more obvious, and the unequal sides of an oblong make it to be more interesting.

Whenever we make combinations that are sure to please the eye, the best method is to adopt a standard, and then, by comparing the results of experiments with that standard, one will soon arrive at the point of having a *true feeling* for *fine space relationships*. The ancient Greeks, after centuries of striving for beauty, arrived at the point where nearly everything they made exhibited good spacing. The standard golden oblong they arrived at, measured approximately two units on the short side and three on the long (Figure 4.1). This oblong is more beautiful than a square because of their unequal sides in a ratio of 2:3. This Greek oblong has more beauty than a very long, narrow oblong, in which the breadth and length vary so greatly that they do not seem to be related.

The designer of today who goes to the Greeks for his inspiration is likely to gain beauty if he interprets their proportions in terms of his own problem, whereas, if he is intent only on copying their details, his creations are likely to be merely a collection of historic fragments. The most modest house can have the essential character of Greek art without having a single so-called *classical details*, if it is based on Greek ideals of *simplicity, fitness* and *fine proportions*. It is not always necessary that as a rectangle approaches a square it becomes less pleasing, and that the best results depend on being able to approximate Greek proportions. The use of Greek proportions (2:3 for flat surfaces, and 5:7:11 for solids) in the design of houses, and the division of spaces for arrangements in its interiors will add beauty to a simple room. Such delightful space relationships will continue to give pleasure to its owners.

Dividing the space into two interesting parts: Perhaps no art problem occurs so often (even where one does not realise that a question of art is involved) as the one in which a space has to be divided into two or more parts. Some of the typical examples for such situations are when:

- an address or a name is written on a post or a greetings card,
- dividing the wall space for hanging wall pictures,
- an embroidery design is positioned on a dress.
- a living room is divided for sitting/drawing and dining areas, or
- a rocking chair is to be placed in a group of drawing room chairs.

We might come across countless other situations, where similar problems occur and the same principle is required to be applied. If the particular division is to be into two parts, the most satisfying result is achieved where the dividing line or the object is placed at a point a little more than one-half and a little less than two-thirds the distance from one end or the other. However, this point should not be located mechanically, and these proportions are only approximate. Any position within the limits is potentially pleasing and there is no necessity for a stereotyped choice. As can be seen in figure 4.2, the point somewhere near 'A' would be the most interesting point within this space to place an important object or to divide the space on a horizontal surface.

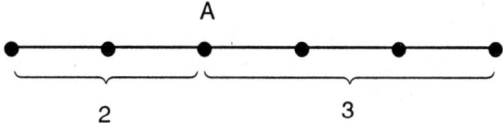

Figure 4.2. Point of division of space to place (A) an important object

Dividing a space into more than two interesting parts: There are a few possibilities by which a space can be divided into more than two parts, by means of lines or objects. In all, there are three such possibilities:

(i) All the spaces may differ.

(ii) All the spaces may be alike.

(iii) There may be a variation in some of the spaces and repetition in others.

(i) *All the spaces may differ:* When all the spaces allocated for each grouping differ (refer figure 4.3), it tends to provide the greatest variety. This type of spacing is excellent and is suited for relatively small areas or for a few spaces. But the effect produced may be confusing and inharmonious when used in large areas, because many of these divisions are seen and compared at one to one.

(ii) *All the spaces may be alike:* The space allocation may be equal for each grouping (Figure 4.4 – see colour plate 7). This results in monotony because of the repetition of lines, colours or space. But this is a common practice in most of the designs. However, one can produce or create interest through this method by introducing variety in the colour or texture in the repeated areas or objects, which is lacking in spacing. National flag of India is a perfect example of this type of space allocation where the use of three different colours provide

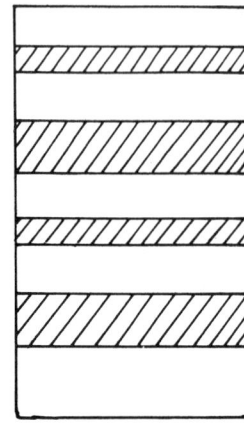

Figure 4.3. Dividing space into different interesting parts

variety, whereas it has equal weightage in terms of width of the bands. The monotonous equal spacing of colours is commonly seen in striped fabrics.

(iii) *There may be variations in some of the spaces and repetition in others:* A combination of the above two may be used where there may be a variation in some of the spaces and repetition in others. In figure 4.5, the design is repeated at intervals alternating with a space from which it differs in width. This makes the design to be more effective. In such combinations, it is also desirable to make one section or a design to dominate the other (Figure 4.5 – see colour plate 7).

Arrangement of objects and space relationships: When arranging objects on a shelf, one should make an attempt to secure interest in their heights and in the spaces between them. For instance, let us take the arrangements given in figure 4.6. The first arrangement is made by leaving equal spaces between the ends of the cabinet and the objects, and between the units themselves. Moreover, all the three objects are too nearly are of the same height. This arrangement (A) is monotonous because the proportions are poor. The objects are so placed that they divide the background into equal spaces and, at the same time, the heights are too much alike. To have better proportions in this arrangement, the central object is substituted by a larger object, and moving the objects from their original places has introduced variety both in the heights and the spaces (B). This is more interesting than the previous arrangement because the heights of the objects and the spaces between them show more varied proportions.

Arranging groups of objects within a larger group: It often happens that one has to arrange groups of objects within a larger group. Let us take the example of assembling several pictures which will harmonize with a particular wall space. It is generally true that if single units or objects in a group are to be viewed as units, they may be separated by spaces wider than the unit measure. But, if objects are to be seen as a group, the spaces between them should be smaller than the sizes of the objects.

The plan for arrangement of pictures on a wall as shown in figure 4.7 would help them to group well. There is less space between the objects than the width of the object. The two groups can also be seen easily together because there is less space between them than the area of each group. It is also interesting to note here that all spaces follow Greek proportions.

Producing a change of appearance

The second aspect of the principle of proportion is to create *illusionary spaces.* In order to make the best of given sizes and shapes, one must be able to produce a semblance of change in appearance. In most cases, lines are used to alter proportions. This can be illustrated by viewing the two oblongs shown in figure 4.6.

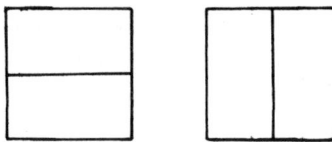

Figure 4.6. Change in the appearance of two oblongs of the same size

Use of lines to alter proportions: The illustration of the two oblongs of same size shown in figure 4.6 indicates the use of lines to alter their appearances. In one, a horizontal line has been drawn, and in the other, a vertical line. Where the eye is carried across the rectangle, it looks shorter and wider and where it is carried up and down, the effect is that of apparently increasing the height and decreasing the width. Horizontal lines thus seem to add width and vertical lines height.

A second effect may also be produced, which must also be taken into account. Vertical lines can be so arranged that they will carry the eye from one line to the next, while they still add height to an object, they will also add width. However, it is of interest here to remember that a vertical movement make an object to look taller and more slender and a horizontal movement has the opposite effect. A house with rectangular windows will appear to be taller as compared to a house with square shaped windows. This is because the vertical lines of the rectangular windows draw the eye upward and the repetition of square shaped windows with wide arches in the other building carry the eye horizontally.

When, for the sake of economy, a person plans a house that appears to be a square, it is still possible to overcome the disadvantage of this plan by the shape and arrangement of the openings. In a house where the windows and porch, are all happen to be square, it results in monotony from the emphasis laid upon this aspect of the house. However, a more interesting effect can be produced by the use of shutters on the windows, and variety can be introduced in the treatment of the door way. Although a single-storey house is less likely to appear as a square than a double-storeyed house, both types should be studied carefully so as to relate them to the landscape. By the application of some architectural devices such as unbroken roof line, variety in the designs and shapes of the windows, doors and arches, use of different materials or colours for the various areas of the house and its surroundings such as fence, garage etc., the appearance of a house can greatly be changed to the desired effect.

Effect of lines on the appearance of a room: It is now clear that the correct use of lines help to alter the proportions and countless puzzling problems can thus be solved. A room that is too low, can be made to appear high by painting the ceiling lighter than the walls and also by using vertical lines/stripes on the wall paper.

Rooms in which a part of the ceiling is slanting, are made to appear lower when the ceiling colour is brought down to the wall, the slanting surface and a lighter colour on the ceiling. Windows that are too short may have a long, narrow draperies and no valance, and the chair that is too low may have a vertically striped cover.

The placing of pictures and accessories may be used to emphasize height or width in a room. For example, a vertical hanging will produce an impression of height. A high room can be made to seem lower by carrying the colour of the ceiling down to the tops of the window, or having the ceiling darker than the walls, by the use of low book cases and furniture and by the suggestion of a horizontal movement in the design of the fireplace and the arrangement of all the furnishings in the room. A room that is unusually long can seem shortened by placing important groups of furniture at the central axis. The use of more than one rug also appears to decrease the size of the room. To increase the existing width of a room, a valance or a cornice board may be used across a group of windows and rugs may be so placed that their lines will carry the eye across the room.

SCALE

The third aspect of proportion is called *scale*. Scale is the capacity to select objects and arrange them so that they look well together. Under the general heading of proportion, scale is the proper turn when considering relative sizes without regard for shapes. The design principle of scale is related to proportion. Both proportion and scale deal with the relative sizes of things. If there is a difference, proportion pertains to the relationship between the parts of a composition, while scale refers specially to the size of something, relative to some known standard or recognized constant. Scale, in this sense means:

- that the sizes of all the elements making up the structure have a consistent, pleasing relationship to the structure and to each other, and,
- that the sizes of all the elements making up the structure have a consistent, pleasing relationship to the structure and to each other, and, that the size of the structure is in good proportion to the different objects combined with it.

A very small object will never look so small as when it is placed near a very large one. That is because the two sizes are not consistent. They accentuate each other by contrast, and are said to be *out of scale*. By following a consistent scale, it is possible to create *illusion* that cause astonishment when the actual sizes of objects are realized. An illusion to this effect can be stated from witnessing a well-staged puppet play. The puppeteer whose stage properties were in perfect scale with the puppets, could give his audience the ever-increasing impression of watching normal people and objects. Similarly, an architect, while preparing a model of his building project, makes them according to a particular scale, gives the viewer the impression of viewing the actual building because everything in that model are in perfect scale!

Scale in exteriors: When all the architectural details of a house are well scaled to each other and to the size of the house, it is successful, whereas it may appear to be a jumble of parts and not a single unit, when the scale is bad. Whenever a window and arch, a door or a porch is too large or too small, it will attract undue attention and destroy the effect of unity in the house. In all these cases, we can say that these parts of the house are *out of scale*. When all parts of the house are in perfect scale, it provides an appearance of *Unity*. Thus it is important to make sure that all the architectural details such as the chimney, windows, door way, arches, overhead of the roof and the sun-shades, are in scale with the size of the house. Thus the examples of good and poor exterior designs, are dependent upon how well or perfectly the parts of the house are *scaled*.

Scale in house furnishing: Any one who would select and arrange things to look well together, must develop a feeling for scale. For example, she should know that heavy furniture are not suitable for a room of average size because such furniture would seem to crowd the area. It is better to use a number of small pieces of furniture in such a room. If large pieces of furniture are to be used in such a small room, they should be as few as possible. At the most they should be upholstered in an inconspicuous colour and pattern. On the other hand, if the furniture seems too small for the room, it should be arranged in groups, so that the size of the group, and not the size of each piece, may become the unit for comparison. The maximum appearance of size may be given to a room through the use of furnishings comparatively small in scale.

Scale is judged not only by the size of the whole mass of an object, but also by the relationship of each part to every part, and to the whole mass (Refer Figure 4.9). The small scale which is applied in the plan of the exterior of the house and its parts, as mentioned earlier, is to be used in the home interior too. Two chairs of the same outside dimensions will appear different in scale if the arms and legs of one are very heavy and of the other, very light. It is necessary to consider every aspect in detail while furnishing the home interiors to achieve *perfect scaling* in their arrangements.

While furnishing the home interiors, one more factor needs to be considered. It is not correct to say that furniture, to be comfortable, must be huge. If any one believes in this factor, they may have to live in average-sized rooms crowded with bulky furniture pieces. If it is understood that comfort is more a matter of the design, construction, material and texture of the piece than that of its size, and that equal comfort can be obtained with smaller pieces, our small houses and apartments would show much better scale and would appear more attractive. Unusually large pieces of furniture call for more attention to themselves than to the room as a whole. After keeping the large furniture, it is necessary to gain floor space to place the additional pieces of furniture to make the room more enjoyable. There can be a sense of grace in the slender lines of the chairs when combined with a sofa which does not in any way interfere with the feeling of strength. When light pieces of furniture are used, they can be moved about easily to form any other grouping that is desired in the room, besides being large enough to be comfortable.

Fabrics, too, have scale. A room can exhibit a *class* when furniture and textures of a finer and lighter scale are used. Fabrics can be grouped under three classes—coarse materials suggesting large scale, fine textures suggesting small scale and an intermediate group which may be used with either of the first two groups as well as with objects of an intermediate scale. Fabrics also show scale in pattern. Large figures are suitable for large pieces of furniture to be used in large rooms, and small patterns are consistent with small pieces of furniture for use in small or average rooms.

Correct scale, or, in other words, consistency in size, is indispensable in garden designs, interior and exterior house design and in furniture design. In garden design it is necessary to choose plants and trees that are in scale with the grounds and the house. In exterior house design, the scale of the openings and features such as columns and coves in relation to the house are very important factors in appearance. In interior design, the length, breadth and height of a room are required to be in scale with each other. Similarly, the various parts of a furniture piece are required to be in scale with each other.

In home furnishings, the requirements of scale apply in four different ways. Each article must be in scale with the room containing it and also with the other articles in the room. The various structural parts of each article must be in scale with one-another and with the whole body. The decoration of each article must be in scale with it. A well-scaled table, for example, is one in which the proportion of the legs is right in size and shape for the top. Such a table should be placed in such a way that its size strikes a balance with the adjoining objects. Lamp shades should be neither too large nor too small for the base of the lamp they are shading. Patterns, too, have scale; a too-small motif may be out of scale on a large sofa.

Thus it can be observed that the principle of scale is used by everyone of us while placing objects consciously, or unconsciously in the day-to-day life. Everyone acquires a sense of scale in buying their own clothes. A tall thin person would like to avoid vertical patterns and narrow skirts while a short heavy person may not like to wear horizontal lines and avoid flat hairdos. The most common mistakes in scale are made in combining articles of inconsistent sizes such as large lamps on small tables, large bouquets in small vases, etc. Examples of violation of scale are so general that anyone can probably see several by looking around. Things that are not related in size should not be a part of the same group, for, the mind refuses to consider them together.

PROPORTIONAL RELATIONSHIPS

Most of us do not measure areas in our homes to divide the spaces according to mathematical ratios, but our eyes soon tell us whether or not the proportions are interesting and the scale is pleasing. Every individual tries to apply the sense of proportion while selecting and arranging the objects in a room; a rug on the floor, a sofa against a wall, a picture or group of pictures over the sofa, a table and lamp next to a chair, etc.

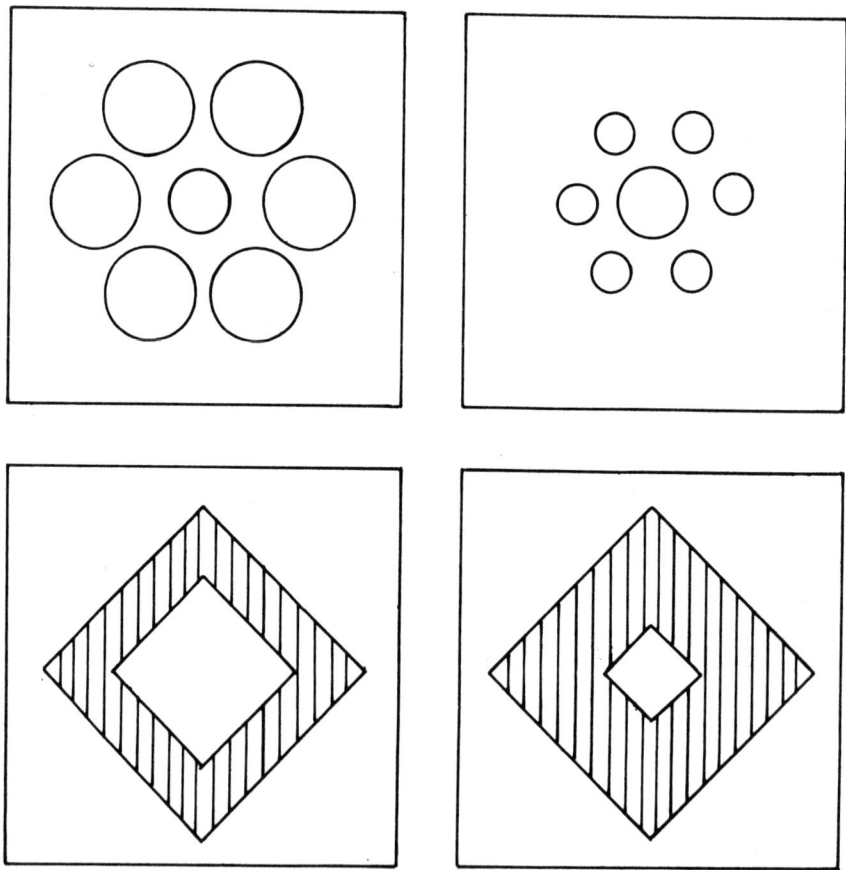

Figure 4.7. The apparent size of an object is influenced by the relative sizes of other objects in its environment

The size and shape of the room will certainly determine the amount of furniture and size of each piece. A very small room crowded with heavy, massive pieces is not likely to be either pleasing or functional. In modern rooms, we tend to use a few pieces of rather small-scale furniture to preserve an airy, spacious look. But again the design of the room must be carefully considered so that the furniture may not seem lost and insignificant.

Another kind of proportion is involved in the relationship of pieces of furniture placed in close proximity to each-other. It is not advisable for example, to place a massive contemporary coffee table in front of a delicate Victorian sofa. The proportion of one to the other would be all wrong and the eye would indicate this. There are no hard and fast rules for this kind of a relationship. Not only the size, but also the comparative weight and bulk of pieces placed near each other must be considered. In these cases it may be considered to be a matter of harmony rather than proportion. Colour, texture and line play significant part in establishing proportions. Strong, brilliant colours advance and will therefore, make a particular area more obvious. Textures that reflect light or the patterned areas tend to increase the importance of an area. Strong contrasts of colour and texture emphasize lines and forms. Vertical lines tend to standardize an object and make it to look taller. Horizontal lines make an object short and broad.

Proportions, therefore, are subject to the types and amounts of colour, line and texture in different areas. This inter-relationship of the elements may be used in many ways to emphasize desirable space divisions or to minimize those that are not so pleasing.

BALANCE

Interior space and their elements of enclosure, furnishings, lighting and accessories – often include a mix of shapes, sizes, colours and texture. How these elements are organized is a response to functional needs and aesthetic desires. At the same time, these elements should be arranged to achieve visual balance – a state of equilibrium among the visual force projected by the elements.

Definition of balance

Balance is probably the, most easily understood design principle. In simple terms, balance is 'rest' or 'repose'. In decorating, it means a state of equality, between two elements or two parts of a composition. *Balance can be defined as the restful effect that is obtained by grouping shapes and colours around a centre in such a way that there are equal attractions on each side of that centre.*

Balance is as fundamental in the visual arts as it is in life itself. Moreover, it is so simple that almost anyone can understand that a feeling of steadiness, repose and balance is the result of the equalization of attractions on either side of a central point. Balance must be attained in colour, texture, pattern and light as well as in weight. It is more pleasing to the eye when weights are so adjusted around a central focal point that they appear to be in a state of repose. A feeling of imbalance is almost always disturbing and unpleasant.

The 'weights' of the furniture and other objects in a room are determined by the size, shape, colour and texture all of which must be considered in adjusting the balance. Lines and shapes have visual 'weight' or visual magnetism. A heavy line attracts the eye more than a thin line of equal length. To create a psychological sense of balance, the thin line would have to be extended to a point at which its total mass, or visual weight equalled that of the heavier line.

The seesaw is, of course, a perfect example of physical balance, and the basic laws and principles affecting a seesaw may be applied to artistic balance. When two objects of equal weight are placed at opposite ends they must be equi-distant from the central point to balance each other. If one object is replaced by a heavier one, it must be moved closer to the center to balance the lighter one at the other end of the board.

Enhancing visual weights: A number of factors which may cause a form to appear heavier are texture, colour, size, decorative pattern, or placement. Because coarse-textured surface give a tactile feeling, they are visually heavier than smooth-textured surfaces. Colour is equally important in weight. In general, warmer colours, stronger intensities and darker values give a feeling of more weight. Size also adds to the visual sense of weight. Irregular shapes which engage the eye longer than regular ones seem more important and therefore also 'weigh' more in terms of balance. All things being equal, the object that is larger will always appear heavier. An object that has more decorative pattern will also appear heavier than the same object with very little or no pattern. Placement of these objects can thus affect the visual weight.

An object that is placed closer to the viewer will appear heavier than one placed to the back of the composition. This is an excellent guide to arranging objects on top of a table, chest/mantel. Ignoring this rule makes one side of a composition seem topsy-turvy.

Characteristics that will enhance the visual weight of an element and attract our attention are:

- Irregular or contrasting shapes
- Bright colours and contrasting textures
- Large dimensions and unusual proportions
- Elaborate details.

Our perception of a room and the composition of its elements is altered as we use it and move through its space. A room also undergoes changes over time as it is illuminated by the light of day and lamps at night, occupied by people and paraphernalia, and modified by time itself. The visual balance among the elements in a space should, therefore, be considered in three dimensions and be strong enough to withstand the changes brought about through time and use.

When applying this principle to design, one should remember that similar areas will seem lighter or heavier in different colours and textures. On the visual seesaw, therefore, if two objects are of the same size but one is bright yellow and the other grey, the brighter one will appear heavier. Thus while balancing them against a background one would place the object in yellow nearer the center of the wall while the less conspicuous object would be moved farther always. The brighter the yellow, the nearer it would have to come towards the center line, and the duller the grey, the farther it should go.

Types of balance

There are three types of balance:

1. Symmetrical, Bisymmetrical/Formal/Axial/Bilateral balance
2. Asymmetrical/Informal/Occult balance.
3. Radial.

Symmetrical/Formal balance:

Symmetrical balance results from the arrangement of identical elements, corresponding in shape, size, and relative position, about a common line/axis. Such a grouping might be a pair of love seats on either side of a fireplace or side tables/table lamps on the two sides of a double bed. If objects are alike and are equally forceful in appearance, they will attract the same amount of attention and therefore should be placed equidistant from the centre.

Formal balance is 'bisymmetrical' balance when the objects on each side of the center are identical and is 'obvious' balance when the objects are not alike, but are equal in their power of attraction. Formal balance is quiet, reposed, dignified and gives a sense of precision, which is readily apparent, especially when oriented on a vertical plane. This type of balance is usually used as it is a bit simpler to arrange and almost everyone can easily achieve this.

Symmetry is a simple yet powerful device to establish visual order. If carried far enough, it can impose a strict formality on an interior space. Total symmetry, however, is often undesirable as well as difficult to achieve, because of function or circumstance. Although formal balance has more static and stable qualities, it does not need to be dull and uninteresting. This type of balance may be achieved through arrangements in which the objects on either side of the central line are not identical but are of equal weight and importance.

It is often possible to arrange one or more parts of a space in a symmetrical manner and produce local symmetry. Symmetrical groupings within a space are easily recognized and have a quality of wholeness that can serve to simplify or organize the room's composition.

Symmetrical balance can be achieved as effectively in a contemporary room as in a traditional one. The type of balance present in the interiors of houses helps to determine the emotional effects created. Formal balances in a room naturally creates an air of formality. Therefore it is not the effect usually desired in a simple or small room or home or in any place that should have carefree, young and unusual air.

Asymmetrical/Informal Balance

Asymmetry is recognized as the lack of correspondence in size, shape, colour or relative position among the elements of a composition. While a symmetrical composition requires the use of pairs of identical elements, an asymmetrical composition incorporates dissimilar elements irregularly positioned to either side of a central axis i.e., eye movement to either side of a central axis is equal, even though the elements attracting the eye are dissimilar.

To achieve an occult or optical balance, an asymmetrical composition must take into account the visual weight or force of each of its elements and employ the principle of leverage in their arrangement. Elements which are visually forceful and attract our attention—unusual shapes, bright colours, dark values and variegated textures, must be counter balanced by less forceful elements which are larger or placed farther away from the center of the composition.

Radial Balance

The third type of balance, radial balance, results from the arrangement of elements about a central point. It produces a centralized composition which stresses the middle ground as a focal point. The elements can focus inward towards the center, face outward from the center, or simply be placed about a central element.

Informal balance is freer and less structured than formal, and thus requires more experimentation with effects to arrive at a desirable relationships. Asymmetrical balance is not as obvious as symmetry and is often more visually active and dynamic. It is capable of expressing movement, change, even exuberance. It is also more flexible than symmetry and can adopt more readily to varying conditions and function, space and circumstance. This type of balance is more subtle than formal balance and affords greater opportunity for variety in arrangements. Its success depends upon training the eye to recognize a restful composition.

There is more intimacy in informal arrangements than in formal, and a sort of chatty, conversational quality is likely to characterize a room where informal balance prevails. A room which combines both kinds of balance is more agreeable and interesting than one balanced all one way or the other (Figure 4.8 – see colour plate).

Balance in decorative designs:

In all types of decorative objects, both formal and informal balance in designing can be achieved. In a pictorial composition, figures on either side of the centre line can be made nearly alike that they attract the same amount of attention. When the light and dark colours are in practically the same relative positions and the figures are balanced so skillfully that, even though, both sides are not identical, one can get the impression of symmetry. Thus formal balance can easily be obtained in this way in any composition.

In the landscape compositions two trees of different sizes can be adjusted and placed sensibly to give a distinct impression of balance of the informal type. The composition can position the larger tree on one side, closer to the centre, and the smaller tree farther from the centre line. A steep house or a sunrise can be used as the focal or central point for this composition. Many of the early periods produced formal expressions in their art, and numerous examples of bisymmetric (formal) balance are found in their designs. Some of the designs created by the Persians have motifs adopted to formal designs. Because of their stateliness, they were often used in the rich fabrics that upholstered the chairs of the Renaissance. While the characteristic designs of the classical period are formal, the art of Japan is informal. Japanese artists understood the art of occult balance, and they acquired such expertise that their painting, prints and stencils are remarkable for their subtlety and spontaneity. They have mastered the art of perfect adjustment of unequal spots on either side of the centre line. The exquisite grace of the lines combined with the subtle balance of forms from these stimulate the imagination of any one who looks at it.

Balance in exterior designs

Just as the painter arranges his composition on canvas, so has the architect to balance doors and windows, porches and dormers around the central axis of a building. Whether he uses formal or informal balance, it depends largely upon the following conditions:

- The spirit of the age in which he lives
- The use to which the building is to be put
- The type of people for whom the building is planned
- His own personality

If we look back at the history, there were times when all buildings expressed mainly formal balance. Again, the bright fantasy of the times was echoed in the graceful occult designs in many of the venetian palaces. Still, a different expression was reflected in the seriousness of the Florentines in their stately, unadorned, bisymmetric palaces – the natural outcome of their lives and thoughts.

It can easily be illustrated how formal and informal balances appear in buildings. If a line is drawn through the centre of the house, it would be found that everything on one side is repeated on the other side. This house can be easily called as bisymmetrical or formally balanced. On the other hand, informal balance can be obtained in building the door way which can be placed near the centre of the house, and it can be balanced by adjusting the windows, the dormer and the light chimney at different distances from the centre.

Balance in interior design:

Balance is by far the most important principle employed in furniture arrangement. In placing the furniture of a room, the architectural openings must be taken into consideration. Large pieces of furniture such as sofa must usually be distributed around a room so that the walls and various areas of a room balance one-another. The halves of each single wall should usually be equally weighed with furniture, windows or doors. Opposite walls should be balanced against one-another.

The large pieces of furniture should be placed first, with regard to balancing centers of interest in the room. The smaller movable objects would then be arranged so that they will make convenient groups as well as balanced units. After the furniture has been arranged into group, the next step should be to obtain balance within each group. At the end, the room should give a feeling that the attractions are about equally distributed around the room. A well-balanced wall will have the same amount of attraction on both sides of its central line. A well-balanced room will have approximately the same amount of attraction on opposite walls. Although the two side walls may be somewhat heavier than the end walls, there should be the feeling that the attractions are equally distributed around the room.

All heavy pieces at one end and all light weight pieces at the other would certainly produce an unbalanced design. If function or some other reason dictates such an arrangement, colour and texture can be employed to re-establish a more pleasing equilibrium. For example a dining area at one end of a living room would probably have furniture that is lighter in weight than the heavier upholstered pieces at the other end. In such a case one might use a different wall cover, bright accents of colour on chair seats or some emphatic center of interest to add importance to the dining area. Any one interested in the meaning of designs will find that the kind of balance used in the arrangement of furniture and decorative objects helps to give an individual quality to a group. It also influences the character of the room. Bisymmetrical arrangements convey a feeling of formality, but it can be formality combined with simplicity and charm. However, if formal balance is carried to an extreme, it may result in effects that are cold and stereotyped.

To avoid such effects of formal balance, one can make an arrangement which illustrates the second type of formal balance – obvious balance. This is a symmetrical arrangement in which the balancing masses are not alike. Yet they have the same amount of force. When two different objects – like a vase and a lamp – attract the same amount of attention, these two objects can be placed at equal distances from the centre. This type of balance gives to a room an effect half way between the precision of bisymmetrial balance and the variety of the occult.

Unlike bisymmetrical balance, there is more intimacy in informal arrangements than in formal, and a sort of active, conversational quality is likely to characterise a room where informality prevails. If we compare the types of arrangements, we will notice that the effects are essentially different. There are freedom and variations in the uneven groupings, whereas in the other arrangement, it is quiet and more reserved.

While making efforts to arrange a balanced arrangement, one should continuously test both the halves to see that one half does not present much greater attraction than the other. While arranging the room, the four walls, with everything seen against them, must balance. If one did seem too heavy, it is necessary to add a brighter colour, a more striking shape, or simply more material to the weaker side, and to keep adjusting the attractions until the whole room looks restful.

It is not essential that all parts of the room agree in being either formal or informal in arrangement. For instance, one might use a formal arrangement on the fireplace, an informal grouping on the table and a combination of both on the desk. Thus, the room is a blend of both the types of balance. This lends more variety than if just one type of balance was used, yet a certain dignity comes from the repetition of some of the objects.

HARMONY

Harmony can be defined as consonance of the pleasing agreement of parts or combination of parts in a composition. It should form a satisfying whole. Harmony is the fundamental requirement in any piece of work in which appearance, as well as use, has to be considered. Goldstein defines harmony as the art principle which produces an impression of unity through the selection and arrangement of consistent objects and ideas. All the elements and principles of design blend together to create harmony. While balance achieves harmony through the careful arrangement of both similar and dissimilar elements, the principle of harmony involves the careful selection of elements that share a common trait or characteristic such as shape, colour, texture, or material. It is the repetition of common trait that produces visual harmony among the elements in an interior setting.

Harmony when carried too far in the use of elements with similar traits, can result in an unified but uninteresting composition. Variety, one the other hand, when carried to an extreme for the sake of interest, can result in visual chaos. It is the careful and artistic tension, order and disorder, unity and variety – that enlivens harmony and creates interest in an interior setting. The elements or components of an object must blend to present an unified whole, and each unit must contribute to the theme or mood of a design, as is felt in the case of musical notes.

Ofcourse it would be easy to achieve harmony by keeping all the variables the same, but using forms, colours, textures, patterns and lines that are similar would be too monotonous. Variety must be introduced to provide interest. Yet too much variety yields to confusion. A good design is neither monotonous nor confused. When all the objects in a group seem to have a strong *family resemblance*, the group depicts the principle of *harmonious selection*. When these form – friendly articles are so arranged that the leading lines follow the shape of the object on which they are placed, harmony has been secured in both selection and arrangement.

When all the elements of a design are harmonious we are aware of an integral whole that is more than merely a sum of the parts. This relationship of the various parts is responsible for the unity of the design, and one discordant note can destroy the totality of the effect. The variables should enhance the mood rather than contradict it. As one chooses furnishings, therefore, it is important to understand the philosophy and conditions that prompted the style and the reasons for the choice of materials, that can contribute to harmony in arrangements.

Today there is a very strong trend towards eclecticism and an almost indiscriminate use of elements. But while having a single element or two out of harmony with everything else in a room may give that room a distinctive and personal touch, having most of the major elements opposed to each other in size, mood, line and colour, produces confusion and an unpleasant sense of discord. Mixing periods and styles is not only acceptable, it makes for the most interesting kind of decorating. But the different periods and styles should all be in the same mood either formal or informal, happy and frivolous, or serious and dignified. Thus it depends on the designer as to how she expresses herself in any design that she creates and how it reflects her own perception, taste and imagination.

In both the fine and applied arts, the principle of harmony takes five aspects. These are the harmony of

    (i) line

    (ii) size

    (iii) texture

    (iv) idea

    (v) colour

Harmony of lines

It is possible to reduce the types of line in a composition to three main groups – lines which follow or repeat one another, lines which contrast one another, and transitional lines which soften or modify the others. Harmony can thus be achieved through repetition, contrast or transition of lines as shown in figure 4.9.

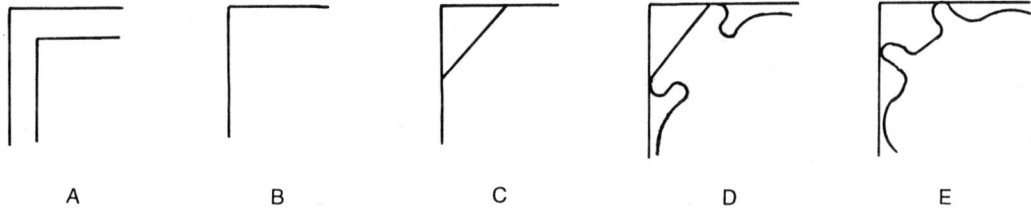

      A   B   C   D   E

Figure 4.9. The main types of lines: A – repetition; B – and C – contrast; D and E – transition.

The simplest kind of harmony occurs with the 'repetition' of a set of lines drawn within a corner (figure A). The figure B forms a 'contrast' when a horizontal and a vertical line meet at a right angle or a corner and are in opposition to each other. Although, any line that cuts across a corner from one opposition line to another is a transitional line, but a straight line, as shown in fig. C, is so sudden and sharp to result in a contrast. This type of line is called contradiction. The curved, easy and graceful lines as shown in figure D which leads from one line or shape to another, give harmony instead of contradiction. These lines draw away from the solidarity and severity resulting in a *transition* to harmonize opposing lines.

Harmony of lines and shapes

We all know, that a combination of lines result in shapes. Using the above three types of lines – repetition, contrast and transition – various shapes can be combined to result in a perfect harmony, as seen in figure 4.10.

While combining shapes with one another, it will be seen that shapes corresponding to one another are in perfect harmony. The most harmonious shape that can be put into a rectangle is another rectangle of the same shape, and a circle makes the closest harmony with another circle. Lines that oppose or contradict each other form contrasts in shapes and are therefore, the opposites of harmony (figure 4.10). Some examples of these contradicting shapes are triangles and diamond shapes within squares, oblongs and circles. Such combinations should be used only where extreme contrasts are desired. Transitional liners have a graceful, softening effect and have the power to bring together shapes which might in themselves be inharmonious.

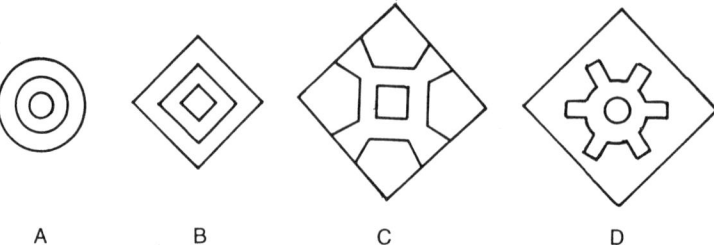

A   B   C   D

Figure 4.10. A & B lines repeat one-another to create shapes producing harmony through uniformity. C, lines contradict one another to create sharp contrasts. D, transitional or modifying, lines create shapes having unity with variety.

Contrasting lines and shapes have made a strong appeal to primitive artists. The sensitive adjustment between the lines and spaces in the composition of *Phulkari* embroidery of Punjab, is a typical example of Indian art where varied types of lines are employed to create designs on the materials. The regular repetition of the triangles alternatively with the bands of diagonal lines creates a surface design that is striking in this art. Besides, the extreme contrasts in lines that have been arranged with vigour and the use of transitional lines to connect the series of diamond shapes provide the necessary force which has been greatly modified (Figure 4.11 – see colour plate 8).

In designing any interior, a number of shapes and forms are used. The most important thing in such a designing is to attain an orderly arrangement i.e., a sense of order should be reflected. The objects especially the larger ones should follow the boundary lines of the enclosing shape whereas the smaller ones may be placed at slightly varied angles. If there are too many angles they may contradict each other and create confusion instead of an order or harmony.

Harmony of Consistent sizes

Every individual – whether a homemaker, an interior designer, or a fashion designer aspires for harmonious or consistent sizes. A school teacher will make the children of the same height to stand or sit in the same row. The same applies to interiors as well. For example, it is good to avoid placing of large lamps on small tables or a huge painting on a small wall. Placement of consistent sizes of furniture, pictures, and other decorative pieces are discussed in detail under the principle, *proportion* earlier in this chapter. Under proportion it was seen that when sizes which are too different are used together, they are inconsistent. *Scale* – one of the aspects of proportion – is allied to harmony in the sense of *harmonious* or *consistent* sizes. The understanding and application of the principle of proportion, as discussed earlier, will assure harmony of sizes too. It was made clear that a small room should be furnished with small sized furniture and large room perhaps, with heavy or bulky furniture. Even while placing decorative pieces or hanging wall pictures, similar rule – objects of similar or same or consistent sizes, would prevail not only to assure proportion, but also harmony of sizes. For more details, the section on the principle of proportion can be referred and applied for harmony in consistent sizes.

Harmony of textures

The sensation of texture, either through the eyes or touch, is gaining momentum these days in the decoration of the showrooms or the window designs. Texture has become an important consideration to the manufactures of objects made from all types of materials – from modern to traditional.

Thus it is seen that starting from a home-maker to interior designer or a sales person to manufacturer, traditional artists to modern day advertisers, are all interested in cultivating a feeling for harmony in texture. More often we keep hearing from people, *how does it look* or *how does it feel*. Therefore, texture may not only be explained as the way the material feels when the finger tips are run lightly along its surface, the sensation of texture is suggested also through the eyes. It is, therefore, no wonder that for people from all walks of life, interest in texture has become an important consideration. In our environment too, we can give as well as gain much pleasure through discriminating combinations of texture.

For obtaining success in many of the projects and schemes, it is important for the planner to recognize that very coarse textures have nothing in common with very fine ones. There is, however, a group of textures which occupy middle ground and may be used with either the coarser or the finer textures. One needs to be very careful in selecting the type of textures that go along well with each-other. It is a very obvious fact that the strong and sturdy oak will match with fabrics such as rugs, khadi, denim, or similar coarse textures or working up to the middle group such as printed lines, chintzes etc. On the other hand, the light, thin fine silks, satins, velvets (specially those with delicate designs) go out of harmony of texture with the sturdiness of oak. But these go along well with walnut, mahogany enameled furniture because of the fine satinlike grain of these woods and their smooth texture.

These textures have no relationship to the coarse group, but they have enough in common with the middle group to be introduced when it is desirable to take away from the thinness or the *overdressed* effect which may come with too much fine texture. A good example of poor texture combination sometimes seen in some places is cane furniture upholstered with in lustrous rayon. These textures are extremely unsuited to each other and are ridiculous to the person thinking in terms of consistent combinations. A group of well-related textures can be seen in the brick fire place and its accessories. Here the textures of the pottery and brass appear to have something in common with the texture of the brick, which seems to be related to brass, to iron and to sturdy pottery. Placement of accessories made of delicate glass or other similar fine textured objects will not be in harmony with the brick-textured fire place.

But with change in fashions leading to unexpected texture combinations, a person may have to compromise with contrasting textures. However, it is quite not so easy to contrast textures, because those that are crude and rough suggest an idea that does not blend easily with those that are highly refined as a result of skilled, delicate craftsmanship.

Harmony of ideas

An idea of how things are to be worked out and placed forms, a major consideration for any interior decorator. The earlier mentioned elements – size, shape, colour and texture need not only to have something in common, but also to have an idea behind their placement when presented together. A haphazard arrangement of similar things leads to confusion for both the home maker as well as the onlookers. The set-up should reflect the harmony and sincerity in the ideas of the decoration or furnishings of the house.

While providing harmony of idea in the exterior design of a house, the following can be taken as an example of a house. When their architectural details are used in harmony in a small building, they seem to be in accord and are all suited to the idea of a modest, unpretentious home. On the other hand, when the columns on the porch of a house do not harmonize with the house, there is lack of harmony in the idea of the classic orders, and the grace and the beauty of an ionic column may suggest a temple, an art museum or a palace. Such columns are not in keeping with small columns. An ordinary pillar support may be more appropriate for either the simplest or the finest dwelling.

Just as all the elements in the exterior of the house must agree, so must there be harmony and sincerity in the ideas suggested by the furnishings of the house. A cottage style should be reflected in the farm house settings while it should be a distinct and different setting in a mansion. Examples of harmony of ideas can very well be noticed in the State Emporium outlets. We will find rich Kancheepuram silk sarees and brass statues and diyas in a Tamil Nadu State emporium whereas wood carvings and woollen shawls would be displayed in the Assam State emporium. Whether we notice the State name boards or not, we will certainly realise that these art objects are typical to these States. Thus the State ideas are displayed in their harmonious settings.

In a formal setting, rich period furniture give a harmonious effect, whereas it is incongruous to introduce such furniture into a small informal house. In order to introduce harmony of ideas in the furnishings, some typical designs, textures and colour in draperies, upholstery, floor coverings etc. are to be assembled into a harmonious group. It should be observed that the idea of cost will not influence these characteristics or in their combinations, for inexpensive blended fabrics may provide the most formal and elaborate patterns. It is generally acceptable that both or all of the designs in the informal group may easily be combined with those in the intermediate group and all of those of the formal group could be used with any of the examples of the intermediate group.

Thus, in a nutshell it can be said that it depends on the designer or the artist as to how she expresses herself in any design that she creates and how it reflects her own perception, taste and imagination. Whatever mode is adopted in the decoration, the essential feature is to have something that goes along with each other or is in good harmony and looks pleasing and appealing to the eye.

Harmony of Colour

It is clear now that it is necessary to have an association among the objects in a group in order to have harmony. Similar to the other aspects like lines, shapes, textures, ideas and sizes, it is possible to carry association of colours to create harmony. When the colours having something in common, are combined together, they give us pleasure because they are harmonious. They give the impression that the colours combined, really belong together. A large number of colour harmonies/ schemes can be created, starting with a single colour in its various shades and tints to a combination of three to four colours. All these beautiful colour schemes give a single impression, because they are harmonious. More details on these colour harmonies are given in the chapter on *Colours: Its Use and Applications*, as colour itself is one of the most important elements.

UNITY AND VARIETY

Harmony and unity are often thought of as interchangeable terms. Harmony should not be restricted, however, to the generalized connotation of unity/consistency, referring more to the creative variances of design elements that are intended to heighten interest without becoming discordant. And, unity may refer simply to the spatial relationship of compositional parts.

Unity refers to the interdependence of constituent parts to form a whole; each component or element therefore contributes to the total being of something whether abstract or real. Organisation and singleness of purpose characterize unity, without which artistic endeavours appear fragmented, lacking clarity of expression. In its simplest form, unity is expressed by the repetition of identical element. It is important to note that the principle of balance and harmony, in promoting unity, do not exclude the pursuit of variety and interest. Rather, the means for achieving balance and harmony are intended to include in their patterns the presence of dissimilar elements and characteristics.

For example, asymmetrical balance produces equilibrium among elements that differ in size, shape, colour or texture. The harmony produced by elements that share a common characteristic permits the same elements to also have a variety of unique, individual traits. Thus asymmetrical schemes can organise a variety of shapes, colours and textures into their layouts. Similarly, colours that contrast each other are used in different proportions to create harmonious effect. The various contrasting colour schemes, when put in an orderly manner, not only provide variety, but also create harmonious effects.

Another method for organizing a number of dissimilar elements is simply to arrange them in close proximity to one-another. We tend to read such a grouping as an entity to the exclusion of other elements farther away. To further reinforce the visual unity of the composition, continuity of line or contour can be established among the element's shapes.

RHYTHM

The term rhythm suggests something that is graceful or sinuous to some people and to others it is something that is spontaneous, energetic or primitive. Thus we can say that rhythm is an organized movement in continuity. When the elements of a design are arranged to make the eye travel from one part to another, the design has movement. If the eye moves smoothly and easily, the motion is rhythmic. Rhythm is measured motion, whether in written or spoken language, music or decorating. In music, it is achieved by regular recurring groups of accented tones. In decorating, it is achieved by the regular recurrence of accent feature or elements. In art, rhythm means an easy, connected path along which the eye moves in any arrangement of line, form or colours. Rhythm, then, is related movement. Rhythm is vital to good designing because it helps to hold the varied elements of a composition together whether that composition is a single object, such as a lamp, or the contents of a whole room.

In a perfectly plain surface, there is no movement, there is simply a resting place, and the eye remains quiet at any point where it happens to fall. The moment that pattern is placed upon that plain surface/space, or an object is placed against it, the eye will begin to travel along the lines suggested by the object or the pattern, and at that moment, movement is created. This movement may be organized and easy, and thus rhythmic; or it may be very restless, distracting and lacking in rhythm.

This principle of rhythm is extremely important in producing unity, because it makes the eye sweep over the whole design before it rests at any particular focal point. The principle of rhythm is an exciting one to work with because the effects are interesting and dramatic. A few simple tricks of decorating will provide an easy, graceful motion. If the eye jumps from one spot to another instead of moving smoothly, the result may be most disturbing.

Rhythm may be exemplified by the ripples generated when a stone is cast into a pool of water. The concentric circles enlarge in a cadenced or measured form of repetition resulting in rhythmic movements.

Rhythmic movement is obtained in two ways. The two types of motions that can be attained are:
- Regular measured rhythm
- Variable rhythm.

Regular measured rhythm is the simplest and oldest way of producing harmony and order. It is the basic element in music, dance and poetry, and it is important in architecture and interior decoration. It is a system of regular movement accenting such that is found in a row of duplicate columns or in striped fabrics.

Variable rhythm is found in the irregular intervals of dissimilar parts. This rhythm may carry the eye along smoothly flowing lines or it may force the attention abruptly here and there in order to convey the desired emotional effect. This type of rhythm is employed to attract the eye throughout a painting until it is seen as a whole. Such rhythm unites all the articles in a group of furniture and also connects each group with adjoining groups. Variable rhythm dominates in natural landscaping or in diagonal design types.

More intricate patterns of rhythm can be produced by taking into account the tendency for elements to be visually related by their proximity to one another or their sharing of a common tract.

The spacing of the recurring elements, and thus the space of the visual rhythm, can be varied to create sets and subsets and to emphasize certain points in the pattern. The resulting rhythm may be graceful and flowing, or crisp and sharp. The contour of the rhythmic shape or pattern of the individual elements can further reinforce the nature of sequence.

While the recurring elements, must, for continuity, share a common trait, they can also vary in shape, detail, colour or texture. These differences whether subtle or distinct, create visual interest and can introduce other levels of complexity. An alternating rhythm can be superimposed over a more regular one, or the variations can be progressively graded in size or colour value to give direction to the sequence.

There are several methods of obtaining rhythm in a design. The methods by which rhythm can be produced are through:

    – Easily connected or continuous line movement

    – Repetition of shapes, colour, line etc.

    – Progression or gradation of sizes, colour, lines, light etc.

    – Radiation

    – Alteration

Rhythm through Continuous Line movement

The continuous lines compel the eye to follow the directions they take. This powerful quality may be employed in various ways to control the movement of the eye. Of course the design of a room is usually composed of many different lines, but a predominance of one type will cause the eye to move from one object to another in that direction.

Continuous line has a flowing quality and compositions that show rhythm. Rhythm through continued lines are likely to be made up very largely of curves. While all other forms of rhythmic movement are seen in the natural patterns, the rhythm to be found in the continued movement of line can be seen in sea-shells (Figure 4.12 – see colour plate 8).

This enlargement of the spiral of a shell brings out the beauty in the sequence of its line movement and in the rhythmic gradation of its spaces. There is a swinging movement throughout the entire picture, and no matter where the eye enters, it is carried along by the suggestion of an easy flowing arrangement of lines so that the view travels over the whole object without the least sensation of hindrance.

In interiors, it is most often used in moldings, borders, and framing elements such as windows and rugs. The continuity of the line may occasionally be broken, but the gaps are small and the eye moves on to the next section in a rhythmic manner. The top of picture frames may be approximately the same height as major openings in the room. The bottom of the frame may be near the height of the lampshade. End tables reach some major division of a chair/sofa, either the same height as the seat or the same height as the arm rest. Of course to have this line perfectly measured, would make the room look geometric and less interesting, but a sharp variation in the continuous line will call attention to the difference.

Some typical examples of this type of rhythm are seen in laces and saree borders. In all lace articles, the free swing of the undulating line is continuous and repeated to lead the eye in an undisturbed movement and produces rhythm. In other words, if the lines of the pattern go in such discordant directions, then the design lacks rhythm and the eye becomes fatigued in the attempt to follow them.

Rhythm through Repetition

Repetition as a principle of art, is closely related to rhythm, for its use results in rhythm. Repetition is generally necessary in producing beauty; it is the simplest way to achieve order. The theme is repeated in almost any form of art for the purpose of making a composition. In home decoration, colours, lines and shapes used in a room should be repeated for the sale of unity. Repetition is also the outstanding principle employed in the creation of surface patterns. This principle is fundamental in nature as well as in man-made beauty; it is the basis of design in flowers, leaves, shells and other natural objects, and so is the case in all man-made objects such as furniture, equipments etc.

Repetition is the easiest way to develop rhythm. Repeat a colour, a line, a design motif or an object more than once in a composition and a sense of design and purpose begins to develop. When anything is repeated, the eye is carried along from one motif to the next, and the interest mounts. When a shape is regularly repeated at proper intervals, a movement is created which carries the eye from one unit to the next in such a way that one is not conscious of separate units, but of a rhythmic advancement making it easy for the eye to pass along the entire length of the space.

The design principle of rhythm is based on the repetition of elements in space and time. This repetition not only creates visual unity but also induces a rhythmic continuity of movement so that a viewer's eyes and mind can follow a path, within a composition, or around a space.

The greatest enjoyment of rhythmic sequence can be found in all natural objects. As seen earlier, the sea shell shows the beauty that can be gained when repetition is carried along in such rhythmic measures. The enjoyment of this shell is spontaneous, and when observed closely it exhibits the subtle variation in the spacing of lines which causes us to welcome the repetition of the shapes. In addition, the sequences in the contour of the shell gives its form a progressive movement that delights the eye. In the lace design, there is a rhythm which seems almost to have melody. The regularity of repetition of the design forming the edge of the lace sets off the more varied repetition of the other units.

In securing rhythm through repetition, one must be careful to avoid monotony in spacing, for good proportion is necessary accomplishment to repetition if beauty is to result. Moreover, when intervals are too far apart, the movement will lack rhythm.

The simplest form of repetition consists of the regular spacing of identical elements along a linear path. While this pattern can be quite monotonous, it can also be made useful by establishing a background rhythm for foreground element and in defining a textured line, border or trim. Repetition of shape will also cause the eye to move in the same direction. For example, a series of pictures mounted in frames of similar shape will cause the eye to travel from one point to another. Colour provides an excellent means of producing rhythm. Repetition can be used without incurring monotony. For example, if the same colour is repeated often, variety can be created by using different shades, textures or patterns.

The principle of rhythmic repetition has a number of practical applications. Combined with good spacing, it makes pleasing effects as can be seen in the following:

- stitching rows of buttons and braids on a dress
- placing tucks in a skirt
- placing groups of designs on a wall
- repeating dots/squares/prints in an embroidered design
- arranging similar chairs in a verandah
- placing similar shaped flowers in a bowl
- repeated designs on a saree border
- repeating same pattern in a rangoli design

Repeating a shape a number of times not only provide rhythm, but also gives an effect of repose. Sometimes if it becomes difficult to use a design as a single unit, it can be made successful when it is repeated at close intervals, as it results in rhythmic motions.

Rhythm through Progression or gradation

Another way of obtaining rhythm is through progression or gradation. A progression through a series of intermediate steps will carry the eye from one end of the scale to the other. This principle may be applied through gradual changes in line, size, shape, light, pattern, texture and colour.

Progression is more lively and dynamic than repetition; it is perhaps more easily used with accessories than with large pieces of furniture. It is also more difficult to manage the large pieces of furniture, as it carries the eye more daringly than rhythm by repetition. While a regular progression of sizes may be satisfying enough for scallops on lace and embroidery, one enjoys a more varied progression when large pieces are involved. Used unsuccessfully, it may cause the design to look like a stepladder since progression of sizes create a rapid movement of the eye. However, gradation brings in interesting eye movements, as can be seen in the case of borders of temple sarees from Kancheepuram, Tamilnadu.

An example of the misuse of gradation is seen in the arrangement of art objects or pictures against a wall in a series of steps that carries the eye up towards the ceiling, and hence away from the part of the room around which one would like to have the interest centred. While a series of steps is undesirable because it leads the eye to the wrong place in the room, a group of objects in which there is no variation in height may be monotonous. An example to this effect can be traced to a graded seating arrangement in an auditorium or a cinema theatre as compared to a huge hall or a class room where there is no gradation in their seating arrangements. In order to avoid both extremes, one should use a series of varied heights. When there is too much of gradation in the use of objects, there can be a break by introducing a smaller object which can be placed in the middle as a central object at a level lower than the other objects.

While making a flower arrangement, one can use flowers of different sizes or flowers in different stages of blooming, i.e. from bud to fully open flowers and make a graded arrangement seen. Gradations of colour are used in some fabrics and canvas painting. The eye will travel from the more dominant tone to the more subdued.

Radiation

It is the type of movement that grows out of a central point of axis. It is a method of obtaining organised movement like the radiating rays of sun which is a typical example of this kind. In India, this is frequently used by the womenfolk to decorate their homes through the Rangoli—a decoration for the floors, usually at the centre of the entrance or courtyard. Radiation may also be observed in the diverging lines which form the pattern of snow crystals and some leaves like those of a palm tree. Radiation is used very commonly in designs for store displays and by the person who makes designs for embroidery, since it is the plan for many geometric patterns. At times, the rapid action is restrained, around the outside of the radiating lines by means of a heavy band. Although diverging lines do not tend to carry the eye smoothly from one part of a design to another, they are sometimes useful in creating a particular effect. Radiation is frequently employed as a basic design in lighting fixtures, structural elements, and many decorative objects. A round dining table in the center of the room with seating chairs radiating out from it creating rhythm through radiation, is an example in creating rhythm in furniture arrangement.

Alternation

Any element can be alternated—black and white, warm and cool, tall and short, large and small or light and dark. In nature, we see rhythm in the alternation of day and night or the dark and light stripes of the zebra. Without destroying the whole, a surprise difference can thus give variety. For example, when black and white stripes are alternated, a surprise of two black stripes provides interest without destroying the unity of the design.

It is also possible that one can find arrangements in which all three kinds of rhythmic movements are used. This is true in the cases of large arrangements. All artistic designs in the objects such as laces, handicrafts, embroidery etc. show this combination. For example, variety in laces can be secured through the use of the simplest elements – the dot and the line – and combine them in these three ways. In a lace band, the rise and fall of the line throughout creates rhythm by means of the continuous line, all over dots can create rhythm through repetition, and the rhythm through progression of sizes is seen in the wave-like lines. In this case, the spaces between the dots should have an interesting proportion when compared with the size of the dots.

Rhythm in exterior design

It is generally seen that the principle of rhythmic line movement comes into frequent use in the design of the house. It can easily be recognized if one starts from a point where there is no apparent movement. In the outline of a square house, the horizontal and vertical lines are of equal force, and so they balance each other. While comparing this with the house roof and sun-shades in gable shape, it is seen that there is a great deal of movement in the lines of this type, and if it is to be pleasing to the eye, the movement should be rhythmic. When many gables are used, the architect should be careful to have the angles of these gables, their placing, and the variation of their sizes similar enough to have an easy eye movement from one part to the other.

In the house design, there is generally a lack of rhythm in the arrangement of doors and windows on the sides of the house. While working out a house plan, one usually begins with the arrangement of the rooms and then places the doors and windows to secure light and air, and an easy traffic movement. While this factor is planned according to the interior design, its exterior design is ignored with respect to providing rhythm in their arrangement. However, when both the interior and exterior designs are planned and considered simultaneously, the provision of rhythm in their arrangement can easily be made.

Rhythm in interior design

It is essential to determine as to how much movement will be enjoyed in the design of spaces in the long run and which are the strategic places for the rhythmic patterns and arrangements. Extreme care needs to be taken, since rhythm can create an uneasy and wasteful movement which is undesirable. Instead of unrestful feeling, it is desirable to have complete absence of movement. It is a common observation, that bold, rhythmic lines may look perfect on a small area but when it is radiated on larger areas, the result is chaos and disturbance. Thus, one should try to use such rhythm on smaller areas such as on curtains, table covering, pictures etc. rather than on wall-papers and rugs.

White selecting the furniture, a person prefers to have something that suggests stability rather than movement. The general preference is choosing either straight lines or restrained curves. The ideal design for furniture is one where there is enough of the straight line to give dignity and stability, and enough of the rhythmic curve to relieve the severity of the design. It is not difficult to achieve rhythmic movement in furniture arrangement. The rhythmic movement should be gained not only through grouping the large pieces of furniture but each separate object in the room should also be examined for its line movement before it is finally placed.

The rhythmic arrangement of furnishings in the rooms can convey an impression of livableness. On the other hand, there is a scattered, unsociable effect in a room where the furnishings are placed without regard to line movement. One of the fundamental principles in the arrangement of furniture is that it be grouped according to its use.

An interior created with regard to line movement suggests sociability and hospitality to anyone who steps into the house. One of the fundamental principle in the arrangement of furniture is *everything in its place and a place for everything*. A knowledge of shape harmony will lead one to place the main lines of each group so that they will conform to the lines of the room, but one needs to know how to control the movement of the eye to be successful in creating a pleasing interior. The line of the furnishing should be such that they lead an onlooker to direct the eye towards the centre of interest or emphasis, where the eye rests for a while. This may be a window, a picture or a desk. A cleverly planned room can have a centre of interest on every wall where the eye moves in a rhythmic manner from one wall to another creating a unifying effect in the whole set-up (Figure 4.12 – see colour plate 8).

If it is desired that the floors form a quite base for the room, the same good judgement should be used in the selection of the design for the floor coverings as for the walls. As with the wall coverings the most useful carpet designs are those which show merely a vibration of pattern and colour. When the amount of desirable movement for walls are carpets has been decided, one is ready to think of the design of the furnishings and it should be suitable to its purpose.

EMPHASIS

Emphasis may be called the excitement factor-because that's the quality it adds to a room. Emphasis is the art principle by which the eye is carried first to the most important thing in any arrangement, and from that point to every other detail in the order of its importance. Thus emphasis assumes the coexistence of dominant and subordinate elements in the composition of an interior setting. A good design needs leadership or some particular note that attracts the eye first. Emphasis is the principle that direct us to have a centre of interest in any arrangement, and a dominating idea, form or colour in any scheme. A design without any dominant elements would be bland and monotonous, inspite of its merits of having good balance and proportion. Its elements in perfect harmony, and are rhythmic. If there are too many assertive elements, the design would be cluttered and chaotic, detracting from what may truly be important. If each part of a design is given importance in the over all scheme, the arrangement lacks emphasis, and because of this, it fails to attract and to give a sense of accomplishment and enjoyment. Thus it is essential to keep one idea as the main centre of interest.

Centre of Interest

If possible every room should have a centre of interest, which is the most important point in the room and should be worthy of the attention given to it. It is always easy to plan any arrangement if one picks a focal point and start from there. The starting point, or interest should be the most dramatic element in the room. It is not always desirable to emphasize the same feature in summer and winter or even by day and night. A large room may well have secondary points of interest too.

The centre of point should be such, which immediately attracts the attention of anyone entering the area – walking into such a room is bound to give a better impression than walking into a room that lacks any features impressive enough to attract notice. When there is no focal point, the eye slides around from one part of the room to another and boredom rapidly sets in. If nothing attracts the visitor's interest, the person becomes restless and feels an urge to move on.

When emphasis – the art principle is applied, the eye is carried first to the most important thing in any arrangement, and then from that point to every other detail in the order of its importance. Wherever any object is selected or arranged with reference to its appearance, this principle of emphasis is used, and the success of the result depends upon the knowledge of the following: ·

- What to emphasize
- How to emphasize
- How much to emphasize
- Where to place emphasis.

WHAT TO EMPHASIZE

Whenever we have the question of choosing or arranging of things, the most common answer is *keep it simple*. It is therefore, clear that the best quality of any object is its simplicity, next to appropriateness for its purpose. Irrespective of the objective in the choice of the objects, the general standards for judging them are:

(i) suitability to purpose

(ii) simplicity, and

(iii) beauty.

When we analyse the historic art for its standards of beauty – the two best periods of art,

Greek and Japanese – simplicity is the most important single factor. To achieve this kind of simplicity, it is for us to understand emphasis and subordination, meaning to subordinate less important details in an arrangement so that they become supplementary or supporting accents rather than competing centres of interest. When we fail to choose an outstanding feature in the room or costume or any other arrangement, and subordinate the others, and put equal emphasis upon all, the resultant effect is chaos. It is, therefore, necessary to decide and select a centre of interest as the first step in emphasis is *what to emphasize.*

To achieve an impression of clarity in any arrangement, whether it is in a room, a house design, a picture, or a costume, the designer would find it helpful to form a rather definite plan. While planning, one should classify the material and arrange them according to what is considered to be the most important and the least. In each field of decoration, the most important features may vary, but the one that should have the least emphasis is likely to be the same – it is the background against which objects are to be seen. From this observation, the most important concepts in art can be drawn. As a general rule, backgrounds should be less conspicuous than the objects to be seen against them.

What one chooses as the focal point of an arrangement depends on the type and purpose of the room. If the room boasts some outstanding architectural feature it should be utilised in the placement of furniture. Occasionally some structural feature of the room almost automatically become a centre of interest. A fireplace, picture window, a soaring ceiling, a wall or a window – any of these architectural assets can become the furniture-arranging focus. Windows are natural focal points in a room. In each field of decoration the most important aspect is the background against which objects are to be seen and should be less conspicuous than the objects to be seen against them.

When there is no built-in center of attention, or anything remotely like them, it is a simple matter to create one. A wall of books, a group of paintings, strongly patterned wall paper, a hand painted screen are all easily managed and ideal take off points for a living room plan. A particular furniture grouping or even one piece of furniture can be a focal point. In a bedroom, for example, the bed is often the center of interest with all other furnishings as subordinates. Some decorators distribute the emphasis so that the observer will give attention to the accessories, the furniture, the floor and the walls, in this order. Sometimes, however, this order is deliberately reversed in order to withdraw attention from things that should be minimized. For example, nondescript furniture should be eclipsed by an attractive wall paper. A large area of pattern might provide a center of interest. Although the floor is not the place to focus attention, some floor covering, especially oriental rugs, are so conspicuous that they cannot be ignored. If such a rug must be used, the only solution is to subordinate everything else in the room. Almost every room includes some feature or furnishing that makes a natural focal point. The table or buffet in a dining room, the bed in a bedroom, or the desk in a library is usually the most noticeable feature, and the logical candidate for being emphasized.

But something 'different' in the way of decorating is one of today's status symbols and the result is an 'anything goes' attitude towards emphasis. Huge paintings, supergraphics of all sorts and sizeable pieces of sculpture are high on the status list at the moment, and can be used as focal points themselves to add to the excitement of an emphasized area. Conversation pits and avant garde furniture made of innovative materials are sure to be scene-stealers. Unusual fabrics, either vividly colored or strikingly patterned, can be used for emphasis, and lighting any specific part of a room you please.

HOW TO EMPHASIZE

There are several ways to attract attention to the important part of the room. These include repetition, unusual size, and contrast in texture, hue, value, intensity, line, space, form or pattern. The arrangement of object or the use of space and light may help develop a focal point. The unexpected is also a way to draw attention to an area of interest. Sometimes several methods are used to develop the center of interest. In each case, a discernible contrast must be established between the dominant element or feature and the subordinate aspects of the space. Such contrast would attract our attention by interrupting the normal pattern of the composition.

There are several means by which one may create emphasis or attract attention, and the most important of these, as suggested by Goldstein, are the following.

- By placing or grouping of objects
- By the use of contrast of colours
- By using decoration
- By having sufficient plain background space around objects
- By contrasting or unusual lines, shapes or sizes.

It is not, however, absolutely necessary that we apply any one of the above to create emphasis. Sometimes, even all five of these methods may be combined in one single design, as in the cases of a house, a room or a very large store window. In any case, it is enough to use any one, two or three of these means to obtain the necessary force for emphasis.

Emphasis through grouping or placing of objects: One can produce an interesting arrangement of objects by combining and grouping them well. If it can also provide a good amount of relevant detail without creating any confusion, it will result in an organized design that can create unity among its various objects. For example. in a drawing room setting when the sizes of all furniture are maintained at the same level – the sofa, the settee, side table, central table etc. and a rocking chair close to the centre of the setting is a slightly bigger size and placed at a central point, the entire setting will appear to be a single unit. However, the emphasis is placed on the rocking chair since it is a part of this setting, and is slightly different from the rest of the furniture. Thus grouping or placing this chair along with the other drawing room furniture, has obtained a proper place for the chain in this well organised arrangement. The rocking chair on its own would not have got so much of attention as it is getting in this group of furniture arrangement.

Similar arrangements could be followed in placing or grouping of decorative pieces or in town planning. In town planning, the principal group could be community hall or an entertainment centre, and the subordinate group could be the houses, the landscape/trees with their little group of people. The big/tall hall along with the houses, trees, landscape and the people, all make up one unit. The details of the subordinate group along with the main building merge together because each one of them is relevant to its context and put into their proper places. In such a combination, emphasis is placed on the main building hall, which is different from the other building. Similarly a well planned centre of interest can easily be obtained even among the various decorative accessories in a drawing room.

Emphasis through contrasts of colour: One of the most striking means of calling attention to any object is to place it against a background with which it contrasts, since the eye is quickly attracted by strong contrasts of light and dark, or by contrasting colour. The use of light to focus attention is generally seen on stage performances. The main centre of interest or the main characters is focussed with strong lights and the light effect is subdued on the rest of the performing artists by focussing it with gradually deepening shadows at the background. It is interesting to note here the interesting effects of the use of light and dark with value contrasts.

The manipulation of light and shadow to produce centre of interest is widely used in open air theatre programmes. *The son et lumiere (sound and light) programme* at Red Fort, Delhi is a typical example of this kind. The stage is set at various places and is not fixed at one place. The attention of the spectators is directed towards the spot where light is focussed. It is to be enjoyed for the drama of its gradations light from one point to another through the manipulation of light and shadow to produce beautiful and subtle pattern of the story.

A valuable source of inspiration can be gained by the careful study of the effects of light and shadow for creating centre of interest. If striking contrasts of light and dark are to be used in any decorative scheme of a considerable size, they should be tied together. This can be done by combining with them a large amount of intermediate values somewhere between these two extremes. The arrangement that shows equal amounts of light and dark would be as confusing as two equal centres of interest in a picture. The final effect of a good composition should be that of a dark scheme accented with lights, or of a light scheme made interesting through its dark notes.

Similar to black and white notes created by the use of lights and shadows, the use of the other contrasting colours, can help to create centres of interest. In an arrangement full of green foliages, red roses would stand out since red and green are contrasting colours and therefore, complement each other. The other complementary or contrasting colours are blue/orange and purple/yellow. When used appropriately, these colour combinations help in obtaining centres of interest.

Emphasis through the use of decoration: There is no doubt that decoration is an useful means to create emphasis. Nothing could be more desirable and acceptable than to have all decorations for indicating a house where there is a party or a wedding or any other celebration. One can easily make a guess in a colony about a house in which there is a celebration. The house would have been decorated as a means of gaining emphasis.

Earlier in the second chapter on structural and decorative design, it has been said that there is a need for a fine form or structure in any object before decoration is even considered. When a satisfactory form is obtained, the worker may decorate it in such a way that the beauty of the structure will be enhanced. It is necessary to set a good standard in the structure as well to make use of decoration as a means of gaining emphasis. Besides showing how the eye is attracted to a pattern, it goes further and proves how complete satisfaction may arise from a good standard and an economical use of the right kind of emphasis.

Sometimes decorations are used for no other purpose than to please the eye. If the object is really beautiful, it has every reason to exist, and one frequently brings an object of this sort into decorative arrangements in order to lend a certain note of emphasis, either through its colour or pattern. Jewellery is a good example for this type of emphasis in dress. A gold chain and a pendant may be, for example, rich in its jewels and intricate in pattern. The choice of proper dress with which to wear this jewel then becomes the art problem. A heavily patterned dress may shield the beauty of the chain and pendant. Similarly a good and well-patterned vase may be a decorative object by itself, but unsuitable to be used as a holder for flowers.

While judging a good surface pattern, it is necessary to consider the way it is to be used. The surface pattern that is good for a good background, should have two main characteristics. First, the design covers the surface rather closely, and second, there is very little contrast between the lights and darks. We often face the problem when we go to the market for buying household linen, carpets, rugs or even dress materials. When the choices are made with a complete understanding of how much emphasis need to be secured through pattern, one is not likely to make unwise selections and realise later that the material that looked attractive in the shop, does not fit well when used in a room or as dress. The two things of greatest importance, then, in selecting surface pattern for backgrounds, are value contrast and the amount of plain space around each figure. It is now clear, that, first, if the objects are packed together closely, they attract less attention than when widely separated; and secondly, if there is a strong contrast between light and dark, an object is much more conspicuous than when the values are very much similar. If these two points are kept in mind, we will not make wrong choices in selecting a surface pattern to serve as a background. If the design is flat, it covers the surface well, and if close in values, it will be quite in effect.

Emphasis through plain space around objects: When a person learns the value of using plain spaces, he is able to bring in a change in the choice and arrangement of objects. It is often seen that certain uncrowded schemes begin to produce a peace of mind while a feeling of confusion and unrest is caused by crowded arrangements. Psychological experiments have shown that an individual has the capacity to enjoy only a limited number of things at one time, and that when this amount is exceeded, he actually sees less rather than more. While making efforts to produce emphasis, two important things are to be kept in mind. First, an object gains importance when it is separated from the things around it, and is given enough plain space for a background. Second, when objects are placed close together, they are seen as a group and not as individual units. The use of plain spaces is one of the most important considerations in emphasis, because plain backgrounds bring out the quality of every object seen against them. For example, a person standing alone against a plain background is easily noticed while the same person in a crowded area is difficult to be seen in isolation.

Emphasis through the use of contrasts or unusual lines and colour: It is natural that anything that is unusual, or different from the rest of its group, attracts our attention. This type of emphasis is applied in underlining words in a letter or capitalizing them in print. The various items of a newspaper show the power of emphasis through unexpected sizes, colours and shapes – some are in capitals, some in big or small letters, some in italics or some in boxes. Thus the headlines of news items exhibit various levels of emphasis, depending upon their significance. The force is gained through a change in size or shape or by an unusual line, or colour.

Similar effects can be seen in a shop window, paintings, or in a library where books with their cover pages varied from light to dark, from bright to dull, and of different designs. Among the various books on a shelf, the first book to attract the attention may have a big check pattern in dark blue and white colour, the second of bright orange, and the third an unusual figure against a plain background. The remaining books may have had nearly equal forces of attraction. The books which attracted attention might have a very striking and unusual contrast of light and dark, a dominant or conspicuous colour, and an usual pattern set off by plain space. Thus one can realise that anything unusual in line, shape, colour or size becomes emphatic. However, the designer has to take care and realise the power of the unusual constantly, as the attention of his audience may wander or actually may be diverted to pay attention on some purely irrelevant detail, and the actual significance may be lost completely.

Sometimes, all the above mentioned methods may be combined in one single design, as in a design for a building, a room or a very large store window. In order to know what attracts the eye the most, a person can glance quickly at a shop window or a table full of new books in their paper jackets. First, one should note the order in which these things catch the eye and then the special feature the objects possess to make them outstanding.

HOW MUCH TO EMPHASIZE

To get a general idea on the amount of emphasis that is desirable, one has to deal with different fields in a different way, because the suitable amount of emphasis varies with every problem.

The success of a beautiful decor will result only if the focal points are distinct and not by a chaos due to a number of focal points. To know how much should be emphasized, a scale may be imagined where the greatest amount of force that can be used with good taste for each of these types will come at different points in the gradation. The backgrounds, are at a point near zero in the scale and over this, the various focal points are graded. For a good impact in the work, one should stop short of the full amount of possible emphasis and leave something in reserve. However, it is important that the simplicity and reserve of decoration should not suggest barrenness in the design. Equal amounts of two or three focal points divide the interest and makes a house appear disorganized. If one focal point prevails and the other is used merely for emphasis, unity results. It is a common experience, that rooms which are overemphasised create a feeling of uneasiness. Such dramatic schemes may be used in such areas as team room or game room where one spends only a short time. The amount of emphasis vary with the room and with the people who live in it. The Greek proportion of emphasis of 2:3 is the most pleasing one. It means that a very small amount of striking pattern should be used as compared to the simpler pattern, along with some rhythmic repetition, and a subtle variety in the light and dark values of the objects. Keeping these points in mind, one can create objects of arrangements having the characteristics of a masterpiece. As mentioned earlier, *"simplicity" is the key word* or the answer for the question of *how much is good to emphasize*.

WHERE TO PLACE EMPHASIS

Once the object for centre of interest is decided, then arises the question "where to place emphasis'– The point where the centre of interest is to be placed, depends largely upon this factor i.e. from the place where it is going to be viewed. If an object on a horizontal plane is to be seen from all sides, it should be placed in the centre of the space, with borders of equal width on all sides. This is desirable, since the object looks the same from all sides/directions. The placing of a cover and a bowl of flowers on a table is an application of this kind of a plan (Fig. 4.13 A, B and C).

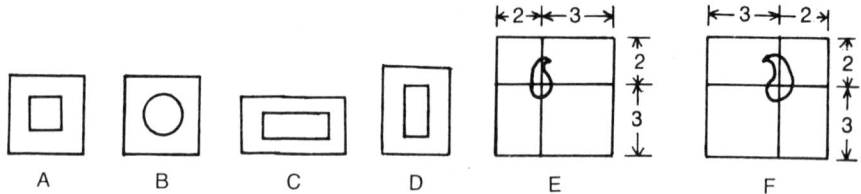

Fig. 4.13. Placement of objects of centre of interest

The centre of interest looks well if placed slightly above the exact centre when it is to be seen in a vertical position, as in the cases of a mounted picture, an invitation card or a design on the cover page of a book or on a printed page (Figure 4.13 D). In this case, the lowest margin is wider than any of the other sides in order to overcome the optical illusion which makes objects in this position appear to drop in space. The size of the margins would vary according to the emphasis of the object being used. If it is very striking, it may need a large mount. A sheet of written or printed matter would follow the same relative proportions, but the margins would be smaller. When planning compositions, it should be remembered that the greater the emphasis of the object, the larger the plain space around it should be.

While planning the position for centres of interest where the lines of the object carry the eye off toward the right and toward the left, it is observed that the eye enjoys more resting place on the side towards which it is being led. The positions for these centres of interest were determined in the following manner; the top and one side of the mount is divided into five equal parts; because the relation of two parts to three results in beautiful spacing (remember Greek proportions), lines drawn through points two or three made interesting divisions in the rectangle; the points where these lines cross each other give four points on which centres of interest could be placed. It is thus observed that the centre of interest should be above or below the mechanical centre, and to the left or right, depending upon the direction in which the lines of composition carry the eye. It is not necessary actually to measure these spaces, but an approximate division will be as pleasing as an accurate one (Figure 4.13 E and F).

When more than centres of interest is needed in a composition, we can use the following plan. Let us take the example of a method for placing three centres of interest in a composition, to be seen either vertically on a flat surface, or on horizontal place, where depth and volume are involved. It is often necessary to group three objects, and if one understands this simple plan, it will help to suggest arrangements of informal balance, such as arranging flowers, making embroidery designs, placing objects on a stage or a store window or a table.

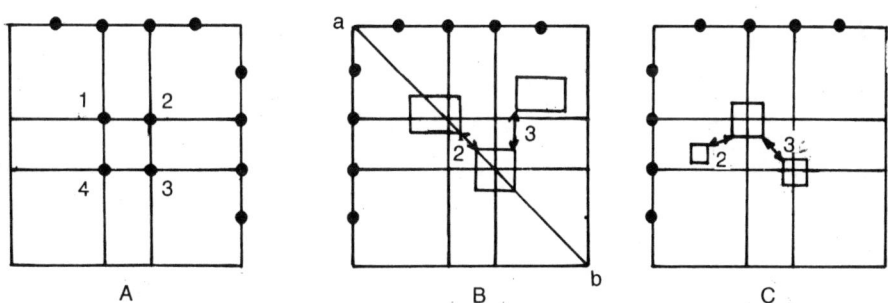

Figure 4.14. Points of centre of interest

To start with, the plane surface is divided into five equal parts both ways and keeping the ratio of 3:2, there are four spots of intersections, upon any one of which a main centre of interest might well be placed to avoid the mechanical centre of the composition (Figure 4.14 A). Two objects may be placed at points corresponding to A and 1 or 2 and 4. These relative positions are chosen so that both objects will not be on the same straight line. In figure B, 1 and 3 were the first two spots. In placing the third spot, one should avoid the following three classes of lines:

(i) any horizontal line,

(ii) vertical line running through either of the first two spots, because the arrangement would be monotonous if there were too many parallel lines,

(iii) the diagonal line that runs through the centre of the first two spots, because there would be a lack of rhythm because the eye would immediately be carried out of the picture.

The third spot, therefore, can be placed any where in the space except on the lines mentioned earlier (Figure 4.13C). The next step is to decide the distance to place this third spot. As per the principle of proportion, it should not be of the same distance from either spot, as the space between 1 and 3, and that a position that is more than one-half this distance, but less than two-thirds will be interesting. If the distance between spots 1 and 3 is of 2 parts, then the third spot will be in an interesting position of it is placed about 3 parts away (Figure 4.13`````````````````````````````` B). If it seems desirable to have it closer to the other, the space between 1 and 3 can be considered as the three-part of the ratio, and the third spot would then be placed only two parts away, as shown in figure C. In a three-spot arrangement, one of the spots will be the chief centre of interest, and the other two will be subordinated.

Thus it is seen that all art problems call for good taste and can be solved by the application of fundamental art principles to the selection of objects, their use in a particular place, and their arrangements. These principles, as mentioned earlier, are the measuring sticks, against which good taste is judged.

INTERIOR DECOR

Colour: Theories and Qualities
Colour: Uses and Applications
Lights and Lighting
Walls, Wall Finishes and Ceilings

Everyone of us would like to create a "home-like" effect while designing the home interiors. It all depends upon the basic idea of having a pleasant atmosphere in the home. The most important factor in accomplishing this is by designing interiors to suit our needs and tastes. Our homes, clothes, pictures, furniture and furnishings, all mutely exhibit our living standards and personality. They explain what is liked and what is not liked by the people. They make the family members feel comfortable and at home. They also make the home more functional, whether it is aesthetically good or poor. Thus the way the home interiors are done, affect the family living to a great extent.

It is equally important to keep in mind continuously the idea of home and its interior as a setting for a happy family life. Home interiors and exteriors should not only meet the functional requirements of indoors, but also should provide comfort. While planning, it is, therefore, necessary to pay special attention to utilize every available object so that the idea of the house as a home is not lost. In this respect, colour, lighting and the house structures like walls and ceilings, play a significant role.

Colour is the most significant and fascinating tool to work with in decorating interiors. If used carefully, it can do wonders, as it removes the monotony of life and enriches the beauty of the objects. An average person uses colour for various purposes and comes across many situations to use colour as well as to judge the effects of colour upon one another. An understanding of colour theories, and the knowledge of the effects and application of colour would enable a practical worker to achieve success in its use. Unit II of this book, therefore, starts with a detailed account on colours, distributed into two chapters.

Light is an important element in determining the beauty and comfort of a home. The basic function of a lighting arrangement is to illuminate the form and space of an interior environment. It allows its users to undertake and perform various activities in a comfortable and efficient manner. Thus good lighting design is a factor to determine not only its aesthetic aspect, but also its functional efficiency. Therefore, in this unit, a chapter on various factors of lighting is included.

Similar to colour and lighting, walls and ceilings in a room, contribute in designing the interiors. They provide a major aesthetic background on which the additional factors for decoration can be placed. Besides providing safety and insulation against climatic changes and dampness, they prove to be the status symbols on account of the elaborate decorations on the walls and ceilings of a house. They may either enhance the beauty of the objects or diminish the impact of beautiful objects that are placed against them. Thus they are the indicators of beauty through the ways and means by which they are built and furnished. Keeping this in view, a complete chapter on walls, wall finishes and ceilings is incorporated in this unit.

Chapter 5

COLOUR : THEORIES AND QUALITIES

The appeal of colour is universal. One of our greatest enjoyments in life is our ability to use it aesthetically. It is almost impossible to imagine our world without colour. Colour removes the monotony of life and enhances the beauty of objects.

Colour is one of the most fascinating tools to work with in decorating. It can do wonders if used with imagination. Colour is often the decorators' favourite element, probably because it is important in establishing the mood and the personality of a home. Colour brings an atmosphere into our homes through our conscious choice of a few fashionable colour accents.

Each one of us, be it a small child, an adult or an elderly, love to see colour. We develop a sense of appreciating colour right from childhood. For instance, a red toy is a matter of attraction for an infant. Colour, thus, plays a great role from the beginning of our life. Later we may become very choosy about colour when we select fabrics of our clothings and household furnishing. We are very careful in selecting colours for the walls of our living room, bedroom, or drawing room. We are all very anxious how the colour of draperies, tapestries and carpets match to the colour of the surrounding walls as colour leaves an impression of our expressions on others.

SOURCE OF COLOUR

Colour is an inherent visual property of all forms. We are surrounded by colour in our environmental settings. The colours we attribute to objects, however, find their source in the light that illuminates and reveals form and space. Without light, colour does not exist.

When white light falls on an opaque object, selective absorption occurs. The surface of the object absorbs certain wavelengths of light and reflects others. Our eyes apprehend the colour of the reflected light as the colour of the object. Which wavelengths or bands of light are absorbed and which are reflected as object colour, is determined by the pigmentation of a surface. A red surface appears red because it absorbs most of the blue and green light falling on it and reflects the red part of the spectrum; similarly, a black surface absorbs the entire spectrum; a white surface reflects all of it. This distribution of colour in light rays is reflected as a rainbow on a rainy day when the light ray passes through a water droplet, which acts as a prism.

83

A surface has the natural pigmentation of its material. This colouration can be altered with the application of paints, stains or dyes which contain colour pigments. While coloured light is additive in nature, colour pigments are subtractive. Each pigment absorbs certain proportions of white light. When pigments are mixed, their absorptions combine to subtract various colours of the spectrum. The colours that remain determine the hue, value and intensity of the mixed pigment.

APPROACH TO STUDY OF COLOUR

To understand colour and to know why it is better to choose some colours rather than others for a particular use, it is necessary to learn something of the nature and language of colour.

The study of colour may be approached from any one of five angles; that of physiologist, the chemist, the physicist, the psychologist and the person who works with pigments. The physiologist is concerned with the way in which the eye receives the sensations of colours. The chemist studies the chemical properties of the natural and the artificial colouring materials used for the manufacture of dyes and paints. To the physicist, the significance of colour is merely its wavelengths and intensities. The psychologist shows how a person is affected by the colours he sees and how colours are affected by one another. Those who mix paints and dyes, find that mixtures of coloured pigments behave differently from mixtures of coloured light; they differ also from the way the colours of material mix in the eye. Each of these fields has given us a group of simple, easily understood facts to enable us to work with colour with success.

Most of the studies in colour occur in three fields – colour in light, colour in vision and colour in pigment. The colours which complement each other in the three media of light, sight and pigment are of importance in the study of colours.

Interior Designer's approach to colour

An interior designer is concerned with all these and also the way of using these studies for a particular subject or client. For instance, from the physicist she would learn how to use properties of colour to make a proper colour scheme for a room; from the chemist, she would use appropriate pigments and paints for the job; from the physiologist, she would learn how to use certain colours for certain purposes; from the psychologist she would learn how the colours like yellow or orange cheer up and green and blue calm; from the painters and artists, the creation of colour effects, moods and atmosphere. Unlike the other specialists, a designers' quest is in all directions. She has to go back into the past and delve into the colour schemes the ancient Egyptians or Greeks used, or keep her eyes open in the present to observe what colour is popular or liked, or peep into the future to foresee and create futuristic colour schemes.

DIMENSIONS OF COLOUR

There are three basic properties or qualities of colour which may be called 'dimensions of colour' These are:

- Hue
- Value
- Intensity/chroma

Hue is the attribute by which we recognize and describe a colour, such as red or yellow. Hue also refers to the name of a colour. Value is the degree of lightness or darkness of a colour in relation to white and black. Intensity or chroma refers to the degree of purity or saturation of a colour when compared to a grey of the same value. In other words, it qualifies the brightness or dullness of a colour. All of these attributes of colour are necessarily inter-related and it is difficult to adjust one attribute of a colour without simultaneously altering the other two.

HUE

Hue is used to describe a kind of a colour, and is practically synonymous with the term *'colour'* itself such as red, yellow, blue, green, etc. The difference between red and blue is the difference in their hue. If a person wishes to *change the hue* of a colour, he will mix it with some of a neighbouring or adjacent hue. For example, some red added to blue will change its hue to purple. A change of hue may be accomplished by dyeing, or by putting a semi-transparent fabric over the colour. Some very interesting effects may be obtained by this process.

In the pigment colour chart, the hues fall into two large groups, one on either side of the vertical line (refer Fig. 5.4). Hues on the right side near the blues are the cool hues, and those one the left side around red and orange, are the warm. Red and orange are the warmest of all the colours, and they seem to advance the most and thus, be the most conspicuous. Blue and blue-purple are the coldest hues and they seem to recede and become inconspicuous. Green is between heat and cold, but it gets warmer as it grows yellowish, and becomes cooler as it grows bluish.

Hues and seasons

Certain hues seem to be particularly appropriate to the different seasons of the year. Window decorations and advertisements may be made to suggest the seasons, if colours are chosen according to the following plan:

Spring   : Starting with blue, through blue-green to green

Summer : Green, yellow-green and yellow, approaching a yellowish-orange towards the end of summer.

Autumn : Orange, red and red-purple

Winter   : Purple, blue-purple and blue.

Effect of hues

Warm hues are more cheerful and stimulating than cool hues which are calm and restful. This quality of a colour, i.e. warm and cool, can successfully be used to counter such effects in a room. A room which is facing south west or west, receives high intense sunlight during the day, and this makes the room to remain warm, especially during summers. A cool colour scheme with blues, blue greens etc., can help to counter this effect. Similarly a north or north east facing rooms becomes very cold during winters, since it does not receive any sunlight at all during the day time. A warm colour scheme of orange, red and yellow, is appropriate, since it would provide an atmosphere of warmth in those rooms.

When used on an enclosing place of space, cool hues appear to recede and increase apparent distance. They can, therefore, be used to enhance the spaciousness of a room, and increase its apparent width, length or ceiling height. White, being the lightest and coolest is always preferred for ceilings of all rooms. Any warm colour used for the ceiling might give a feeling of advancing towards the persons who are present in the room! Warm hues thus appear to advance and suggest nearness. These traits can be used to diminish the scale of a space or, in an illusory way shorten a room's dimensions.

VALUE

The value of a colour refer to its 'darkness' or 'lightness'. There are many degrees of light and dark, ranging all the way from white or black. By adding white, a colour is lightened, and by adding black, a colour is darkened. White has the highest value and black the lowest. If the colour chart is compared with the value scale, it will be seen that the hues change gradually in value with the lightest at the top and the darkest at the bottom as depicted in Table 1 below.

Table 5.1: Value Scale and Step in a colour wheel.

Value Scale	Value step in colour wheel
White	White
High light	Yellow
Light	Yellow-Orange/Yellow-Green
Low light	Orange/Green
Middle	Orange-Red/Green-Blue
High dark	Red/Blue
Dark	Red-Violet/Blue-Violet
Low dark	Violet
Black	Black

So, by adding white to a colour, we obtain a lighter colour i.e. a *'tint'*. By adding black to a colour, we obtain a darker colour, a *'shade'*. When grey is added to any colour, its tones are obtained. Thus, technically speaking, pink is a tint of red and maroon is a shade of red. The term *'tone'* refers to a range of tints and shades of a colour. Tones are obtained by greying the colour. For instance, in the above example, greyish red is the tone of red. See the value chart given in Table 5.1. In interior design, mostly tints are used. Pure colours and shades are used sparingly (Figure 5.1 – see colour plate 9).

Implications of Value

1. Light values seem to increase the size of objects. Small room may be made to appear larger if furnished in lighter colours. Also, light values create the impression of distance. A low ceiling can be made to have a raised effect by painting it white.

2. White and other very light values reflect colour and seem to intensify colour of objects seen against them, and vice versa.

3. Black and dark values seem to decrease the size of an object. Therefore, a small room should not be furnished in dark colours. Instead, dark values seem to be making a large room to appear small. A large room where light colours are used in their furnishing, may seem empty. Instead, dark coloured furnishings can help this room to appear adequately furnished. Besides, darker values of a colour are particularly appropriate for floors and rugs because they give to the room an impression of stability. In store display, dark values should be used below, rather than above light values, for if they are seen above the light colours the display will appear unstable. As it appears in Table 5.1, dark value shades are to be used at lower levels and lighter values tints at higher levels. It is no wonder that all room ceilings are done with white paints which is the lightest colour!

4. In home furnishings, close values are agreeable if many colours are to be used. Close values produce quiet effects, strong contrasts have the opposite result. Although, many colours add interest to a room, but if used too much, the effect of unity may be lost. Therefore not more than three or four values are to be used at one time in a room.

5. To bring attention to an object which is aesthetically good, it should be placed against a background of very different value to emphasize the object. But, when an object is not beautiful, and least attention is desired, it should be placed against a background very near its own value, to have a subdued effect.

INTENSITY OR CHROMA

Intensity or chroma is the saturation or purity of a colour, i.e. it represents its brightness or dullness. In other words it shows the presence or absence of grey or dullness. In a nutshell it can be said it is the property describing the distance of the colour from grey or neutrality line as in the case of the lemon which is brighter than the banana though both of them are of the same colour. A colour in its purest form has the greatest brilliance or intensity. Therefore, intensity is that quality of colour that makes it possible for a certain hue, such as red, to whisper or to shout, or to speak in a gentlemanly tone.

The colours at full intensity are very striking and form brilliant and interesting effects. The colours in lower intensities are more subtle and are enjoyed in large areas, whereas colours in higher intensities are to be used in smaller areas. Similarly, larger the area to be covered, the less intense the colour should be and the smaller the area, the brighter the colour may be. Colours in background are usually painted somewhat dull, grayed or neutralized because it is easier to live with them in large amounts. Saturated colours or colour with pure intensity are often reserved for small areas like hall-ways or cloakrooms where people do not spend a great amount of time. Intense or bright colours are also used for accessories and for flowers in an arrangement since they draw more attention than other colours.

Changes in the intensity of a colour may be brought about through mixing the complementary colour which lies opposite on the colour chart. When complementary colours are mixed, they neutralize each other and when mixed in certain proportions destroy each other and produce gray or neutrality.

A colour may be emphasized in the following way

1. By placing it next to its complement e.g., – red against green or orange against blue.
2. By combining the colour with another neutral colour like black or white which will emphasize colour more than does grey, or black.
3. By repeating near it, a large amount of the same hue in a lower intensity. For example, a little bright green surrounded with dull green would become more emphatic.
4. By repeating in some other part of a composition a small note of the same hue in a brighter intensity.

Texture and Intensity

Texture plays a very important part in colour use. Rough surfaces dull the intensity of a colour because their surface reflects in tiny accents and throw little shadows. On the other hand, plain or shiny surfaces show a clearer and brighter colour, and such a texture aids in increasing the intensity of a colour.

PIGMENT THEORY

A surface has the natural pigmentation of its material. This colouration can be altered with the application of paints, stains or dyes which contain colour pigments. While coloured light is additive in nature, colour pigments are subtractive. Each pigment absorbs certain proportions of white light. When pigments are mixed, their absorptions combine to subtract various colours of the spectrum. The colours that remain, determine the dimensions of the mixed pigment.

Object colorants, such as paints and dyes, are the means to modify the colour of illuminating light, which we interpret to be the colour of the object. In mixing the pigments of paints and dyes, each of the attributes of colour can be altered.

The hue of a colour can be changed by mixing it with other hues. When neighbouring analogous hues on the colour wheel are mixed, harmonious and closely related hues are created. In contrast to this, mixing complementary hues, hues directly opposite of each other on the colour wheel produces neutral hues.

The value of a colour can be raised by adding white and lowered by adding black. Lightening a hue's normal value by adding white creates a tint of that hue; darkening the hue's normal value with black creates a shade of the hue. A normally high-value colour, such as yellow, is capable of more shades than tints, while a low-value colour, such as red, is able to have more tints than shades.

The intensity of a colour can be strengthened by adding more of the dominant hue. It can be lowered by mixing grey with the colour (tones) or by adding to the colour its complementary hues.

THE PHYSICIST'S THEORY OF COLOUR

The science of physics deals with colour as a property of light. Within the visible spectrum of light, colour is determined by wavelength; starting at the longest wavelength with red, we proceed through the spectrum of orange, yellow, green, blue and violet to arrive at the shortest visible wavelengths. When these coloured lights are present in a light source in approximately equal quantities, they combine to produce white light—light that is apparently colourless.

The physicist regards the seeing of colour as a psychological sensation which provides a more or less crude and inaccurate indication of the relative intensities of the wave-lengths to which the eye responds. Because the colour sensations which these wavelengths arouse in the eye are only approximately correct, colours in physics are denoted by their intensities and by their wavelengths rather than by colour names of all the colours. Red has therefore the longest wave length and violet the shortest. The physicist determines the quality of light radiation by spreading the wavelengths present in light into a spectrum, and measures the intensities of the different wavelengths. By this method a clear and physically complete analysis of the quality of light radiation is obtained.

It is found that, in the mixing of coloured lights, all the hues, as well as white, can be secured from three primary colours, but they are not the same three colours the painter must use as primaries in pigments to secure all the hues. In coloured lights, red, green and blue-purple are the three primaries instead of the red, blue, and yellow of the painter. The three secondary colours, as they are seen in light, are yellow, red-purple, and blue. These secondaries are produced as follows: yellow light is secured by mixing red and green lights; red-purple results from a mixture of red and blue-purple; and blue light is made by mixing green and blue-purple. The terms "magenta" and "violet" are often used in this field to indicate red-purple and blue-purple.

The Figure 5.2 shows the colours that are complementary to each other in light. These complementary pairs are red and blue; green and red-purple; and blue-purple and yellow. Each of these pairs of coloured lights will neutralize each other when combined and will produce white light. This may be demonstrated by throwing lights of complementary colours upon a white screen.

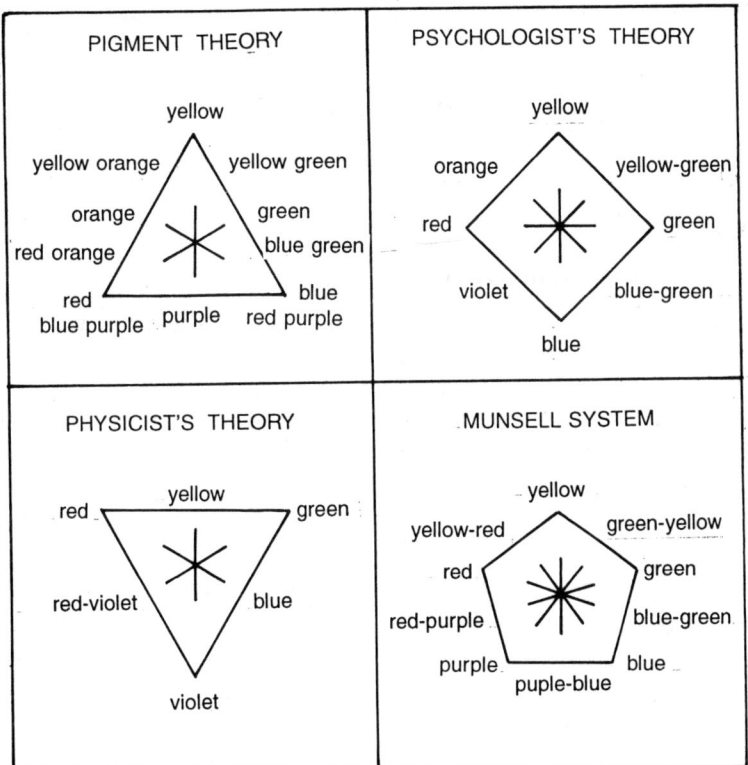

Figure 5.2. Colour Theories

When one understands how coloured lights affect colour in materials, colours may be emphasized, changed, or made actually to vanish when the proper lights are thrown upon them. Window decorators and lighting experts in the modern theaters accomplish wonders in the transformation of backgrounds and costumes through the application of coloured lights.

PSYCHOLOGY OF COLOUR

The psychological value of a colour is a very important aspect of the impact of a work of art. These values are more or less bound by traditions in religious and folk art. They are almost free in present day art expression. Here, colours are instilled with meaning by the expression of the total, not by the colour of individual parts.

Psychological Aspects of Colour

The sensation, which we experience while seeing form and colour, is a psycho-psysiological creative process. The eye receives energy information which is sent to the cerebral cortex as nerve impulses. Here the translation of information from the material world appears suddenly. This translation comprises our world of visual experience, which is overpowering, three-dimensional and colourful. This psycho-psysiological creation of the world outside of ourselves takes place in ourselves.

To a professional colour consultant, the psychological conscious and sub-conscious reaction to colour are tools with which she creates colour balance for objects and spaces from the colours available. It is evident that use of colours can create certain changes in our environment or in an object. This is due to psychological perceptions of colour plus an association by experience or observation.

ASPECTS OF COLOUR IN VISION

There are a few aspects of colours in vision:

(i) Colours affect the mood of an individual

Colours have a strong effect on emotions. Cool colours have quietening influence. On the other hand, warm colours have a cheerful, comforting effect which may increase stimulation and excitement. Colour, because of its emotional effect, is largely responsible for atmosphere in the home. It is capable of smoothing or irritating, depressing or cheering, welcoming or repelling, charming or boring. For example, yellow has the effect of cheerful gaiety, optimism in the home. It gives the effect of sunlight. Similarly blue is associated with water and sky, giving a cool effect.

(ii) Colour has dimension

Colour can make an object appear larger or smaller. A light coloured object appears larger than a dark, coloured object. A white cube of the same dimension appears larger than a black cube.

Cool colours carry an illusion of distance and thus objects with cool colours seem to recede and on the other hand, warm colours seem to advance. It is possible to make a room look shorter than its actual length by using warm colours at the end of the room. The effect of colour is very important in interior decoration, since a square room can be made to appear oblong with receding or cool colours on three walls and an advancing or warm colour on one wall of the room.

(iii) Colour has weight

Colour can make an object appear lighter or heavier. A white or pastel coloured object will appear lighter than a similar object in a dark colour. In home interiors, dark colours should be preferred for carpeting or for floor mats while lighter colours should be used at higher levels. White being the lightest of all, is always used for painting the ceiling to give an impression of being away from the head!

(iv) Colour has movement

Colour can make an area or object appear nearer or farther away. Light blues and violets tend to recede and bright or dark reds, yellows, orange appear to advance. This aspect is of value in decorating through illusion of making a small room appear larger or a large room seem smaller by simply changing the colour.

(v) Colour has temperature

Temperature is a the quality of colour which gives the feeling of warmth or coolness. This quality is inherent in colours. For instance, red/orange/yellow give the sensation of warmth because they are associated with the sun and fire. But even in cold countries where sun appears almost white, red/orange/yellow convey the same feeling. With the juxtaposition of other colour/s, we can change the temperature of a colour, eg: if we place yellow with green, it looks cooler; but with red, it looks warmer.

Warm colours lead to activity and cool colours lead to repose. So they can be termed either as active or as passive colours also.

Relativism of temperature

Every colour cannot be categorised into warm or cool. Purple, supposed to be a cool colour, is a mixture of blue and red. But when it has a predominance of red, it would be warm, and the predominance of blue in it will make it cool. All this points to the fact that deciding factor to categorise the temperature of a colour is the predominance of either red or blue. Any colour that contains or bends towards red, is warm and any colour that contains, or bends towards blue, is cool. This also establishes the fact that red is the only warm colour and blue is the only cool colour, their center parts being black and white respectively in the range of natural colours. From this we derive that a shade is relatively warmer than a tint.

(vi) Effect of colours on each-other

The colours lying opposite to each other in the colour chart are called complementary or contrasting colours. Complementary colours are yellow versus purple, red versus green, and orange versus blue.

Sometimes, the effect of one colour upon another is so strong that in the eye of the observer, non-existent hues will appear. After gazing at a colour for about half a minute, a new colour called 'after image' may be produced, The after image is a complement of the original hue and can change the appearance of the colour even while one is looking at it. This is because of the fact that when the sensitive nerves of the eye which permit us to see colour become tired of looking too long at a hue, the nerves register the colour less and less vividly. The original colour becomes duller and an impression remains in the eye of its complementary colour. Similarly, if one looks fixedly at a spot of bright orange colour for about half a minute and then looks at white surface, a bluish-green spot will appear on the white surface.

COLOUR ASSOCIATIONS AND COLOUR SYMBOLISM

Emotional and symbolic significance have been given to colours through the ages. Many of these colours are linked with culture and period and can only be understood in the light of historical change. We associate red with anger, green with envy, yellow with jealousy, etc. All this shows the keen observation of human emotions and how they are affected by colour.

Some of these concepts, which were represented by certain colours, are based on the elements such as air, earth, fire, water, wind, and the seasons. Concepts come from the fight to survive, from religion, from heraldry and later, from the views of the anthropologists.

SOCIOLOGICAL FACTORS

Colours were at first natural, as evidenced in early cave paintings. As men progressed towards civilization, colours became symbolic. Symbolism was applied to colour when man's hopes and fears were represented by certain hues:

SYMBOLISM OF COLOURS

White	light-purity, joy and glory
Red	fire and blood-signified charity
Blue	heaven-signifies truth
Green	nature-signifies hope of eternal life
Purple	signifies sorrow and suffering
Black	signifies death, mourning

These historical definitions of colour remain in effect to this day.

Another historical meaning of colour is found in heraldry. In medieval times certain colours became associated with human qualities and became the basis of coasts of arms and identification of families and different stations of men.

Yellow or gold	honour and royalty
Silver or white	faith and purity
Red	bravery and courage
Blue	piety and security
Black	grief and sorrow
Green	youth and hope
Purple	high rank and royalty
Orange	strength and endurance
Red-Purple	sacrifice

Emotional Effect of Colour*

People like colour and associate it with different events in life. Receiving compliments or having a good time while wearing a particular colour may create a favourable response to it. There are reactions and moods to be associated with colour. Our emotional reactions to certain colours are partly due to the symbolic meanings that have become associated with them. In the early Christian churches, colours were used to convey definite ideas to the people who were unable to read. White was employed for innocence, black for evil or death, grey for penitence, red for love or martyrdom, blue for sincerity or hope, and the other colours for equally definite ideas. The colour symbolism instituted by the church has been carried on by the theater up to the present time.

Colour, because of its emotional effect upon us, is largely responsible for the atmosphere of a home. It is capable of soothing or irritating, cheering or depressing, charming or boring, welcoming or repelling. A colour changes its emotional value if its hue, value, or intensity is changed. An important thing for a decorator to know is how to use colour for its emotional effect. Different colours excite different emotional responses, and again some persons are more sensitive and more stimulated than others.

Yellow, which is the colour of the sun and artificial light, has an effect of cheerfulness, gaiety, buoyancy, optimism, exultation, sympathy, and even prosperity. It almost sings and shouts. For centuries it was considered a sacred colour in China. In home decoration, yellow is indispensable,

---

*Rutt. H. Home Furnishing.

because more than any other colour it gives the effect of light. It supplies sunshine, even on a grey day. The modified yellows, such as buff, cream, ivory, beige, ecru, pale lime yellow, and pale banana yellow, are useful wall colours because they have the happy faculty of pulling together and harmonizing colours used in draperies, carpets, and upholstery. Yellow is a friend to the person with a limited income because it has the power of making inexpensive cottons, linens, and woollens to look beautiful. Accessories of yellow are usually needed in north rooms. Gold, which is a type of yellow, is also useful; for example, a gold screen would add cheer to a dull room of the more elegant type. Yellow is also being used effectively for exterior house trimming, especially for shutters.

Orange is the most vivid hue that exists. It possesses the qualities of both red and yellow, and in its pure state it is so warm that it should be used only in small quantities. It expresses energy, spirit, hope, courage, and cordiality. Neutralized forms of orange such as peach, rust, cedar, and copper which are often used in home decoration radiate hospitality and cheer. They should be featured in autumn decorations.

Brown, is the most useful of all colours. Brown walls are effectively used with natural wood furniture and light beige rugs, or with one striking colour such as turquoise or cherry red. Chocolate, burnt cinnamon, or other red-browns are more usable than yellow-browns. Brown is traditionally associated with ideas of humility, tranquility, and gentleness.

Red is the colour of fire and blood. It is expressive of primitive passion, war, vigour, power, movement, aggression, boldness, and love. Red is one of the most beloved of colors. An explanation for this may be that red is the colour of fire, and, since for untold years the fire at the mouth of the cave of primitive man was his protection and comfort, his descendants may have inherited some of his feeling of pleasure in its colour. In decoration, red gives the impression of splendor, warmth, hospitality, and exhilaration. It is cheerful, but not restful, and so must be used discreetly. Cool reds like magenta harmonize with blues and purples. Warm reds like tomatoes or firecrackers harmonize with yellow and orange. Reds are usually grayed, but cherry red and Chinese red are used without modification. Certain rich reds are used freely in Italian and Spanish rooms. Dark, dull, raspberry red has proved to be a successful colour for carpets. Pink, one of woman's favourite colors, should usually be grayed if used in large quantities. If pink and blue are used together the pink should be slightly orchid and the blue should have a violet cast.

Purple is made of red and blue, which possess quite opposite characteristics and when mixed cancel each other's effect, so that purple is somewhat gentle and vague. It suggests mystery, dignity reflection, mourning, philosophical musing, and twilight. Originally, the pigment came from certain shellfish and was so rare that only royalty used the colour—hence the term royal purple. Some artists avoid purple and the common diluted purple known as lavender.

Blue is the colour of clear skies and deep water, and so is associated with coolness. It expresses distance, spaciousness, loftiness, dignity, calmness, serenity, reserve, formality, restraint, lack of sympathy, and coldness. In decoration it acts as a check or an antidote for too much warmth. Blues are not so friendly with one another as other colours are and therefore have to be selected with additional care under both daylight and artificial light. Blue is a very important decorative colour as it is usable in large areas. Since blue is not an aggressive colour it does not have to be neutralized as much as some of the others. Modified or Persian blue walls are now used with light modern furnishings, especially in bedrooms and dining rooms. Pale slate blue or pale grayed turquoise blues are refreshing wall colours which seem to add spaciousness to small rooms. Medium blue and white combine well in a two-tone scheme.

Table 5.2 Emotional Effects of Colours

Colour	Symbol	Nature	Normal feeling	Contrary feeling
Yellow	Sun and artificial light	Warm	Cheerfulness, gaiety, buoyancy, optimism, exultation, sympathy, prosperity	Unhappiness
Orange (Red + Yellow)		Warm	Intellect, energy, spirit, hope, courage, cordiality, hospitality, cheer, stimulating	Fickle
Brown	Autumn	–	Practicality, tranquility, gentleness, restful, cosy, manliness and masculine, handsomeness, hardwork	Triviality, unfinished, undefined, decay, tiredness, boredom, dejection
Red	Fire and blood	Warm	Stimulation, war, vigour, power, movement, force, advancing, vital, aggression, boldness, love splendour, warmth, hospitality, cheerful.	Turmoil
Purple (Red + Blue)	–	–	Mobility, gentle, vague, mystery, twilight, dignity, royalty, grand-ness, shy, ceremonial, richness, spiritualism, pride, wisdom.	Decadence
Blue	Clear skies and deep water	Cool	Peace, distance, spaciousness, loftiness, dignity, calmness, holiness, serenity, reserve, formality, restraint, happiness, hope, truth, honor, repose, soothing.	Depression, lack of sympathy
Green	Foliage	Cool	Wholesomeness, rest, refresh-ment, life, spring, hope	Poisonous, envy, jealousy
Black	Night	–	Solidity, mystery, wisdom, sophistication	Vacuum, secrecy, illicit, foul, evil, horrible
White	Moon	Cool	Clarity, serene, luxury, delicacy, feminine, peace, honesty, reliability, trust	Ambiguity
Grey	Monsoon/ cloudy day	Warm/ Cool	Variegation, gentle, serene, age, maturity, experience	limitation, dullness

Green is the colour of grass, leaves, and vegetables and naturally suggests rest, cool shade, refreshment and all pleasant things. Green is considered beneficial to the eyes, nerves, and dispo-sition. Some colorists say that green has negative qualities as well as positive ones and that it suggests envy, jealousy, and ill health. It is composed of yellow and blue and appears warm if yellow predominates or cool if blue is preponderant. Greens must be used with caution, for green ceilings and walls may reflect an unbecoming colour on the persons in the room. Greens that are employed on the exteriors of houses or on garden furniture of fences should be warm in quality so that they will harmonize with the colour of the foliage.

The **neutral colours** are **black, white, and grey.** The term neutral, however, usually refers not only to these but also to all the **tans, beiges, sand colours, natural wood colour,** and **brown,** which have no very definite colour of their own. Such colours are most valuable in home furnishing, for large background areas are often neutral in colour. The true neutrals are cool in effect.

Black suggests mystery, wisdom, or sophistication and in decoration can be employed to create dramatic or other extreme effects. To a decorative scheme of dark colours it adds spirit and interest, but in a light colour scheme it gives too much contrast and makes other colours appear faded. Small accents of black are often effective. Black carpets are sometimes satisfactory with Modern or Oriental furnishings, but they require constant care as they show all marks and dust. Black furniture, particularly Oriental pieces, is sometimes used in rather luxurious settings.

White is a recurring favourite between periods of colourful decoration because of its serenity and coolness. White and off-white are generally approved for both exteriors and interiors of houses. Some unusual interior colour schemes expressing luxury, delicacy, and femininity are based on white floors and carpets. An inexperienced person might well use white walls throughout and entire house. White is valuable for the opportunity it gives to display other colours. Pure white is best with cool colours whereas cream and off-white are usually more harmonious in warm schemes.

Grey, which is produced by mixing black and white, has no particular character of its own, although in light tints it seems gentle and serene, and in dark shades dignified and restrained. Greys may be warm or cool. A pale, warm grey containing either yellow or violet makes a pleasing wall colour. A dark grey wall can serve as an advantageous background for etchings or drawings. Grey stain on woodwork and furniture is more unusual than brown and is agreeable where a cool effect is desired. Grey is a valuable colour in Modern decoration as it permits emphasis to be placed on the form of objects by minimizing their colour. Grey should be accompanied by some brilliant colour to counteract its neutrality. A dominance of grey in a home may indicate a lack of imagination on the part of the owner. Pale grey is a pleasing colour for exterior walls of houses; white houses look well with roofs and shutters of slate grey.

COLOUR SYSTEMS

We can now understand why a colour is called blue, green, red or violet. But we may not understand when a colour is called by the name 'astral', 'canopy' or 'space'. This creates a great confusion in selection unless we go through the shade card. Even when the basic or primary colours are named like blue, green or red, we just do not know how much dark or light it could be, or whether red is purpled or mixed with orange or green is the pure green or mixed with yellow or blue. To avoid all such conclusions some suitable method is required for systematic designation of colour. Various colour systems have been devised to classify and use colours which are eminently suitable for the purpose they are devised. Two of these which are of significance in the context of home interiors are:

1. The Prang Colour System
2. The Munsell Colour notation.

THE PRANG COLOUR SYSTEM

The system developed by David Brewster is probably the best known colour system and is often referred to as the Prang System (Figure 5.3 – see colour plate 10). The simplest way to understand colour relationships in this colour system is to study a colour wheel based on three **primary colours** – yellow, blue and red. These three hues are called primary in the Prang System because they cannot be obtained by mixing other pigments, and also because the other colours are obtained by mixing these three colours in varying proportions. Theoretically, starting with five cans or tubes of paint – the three primaries plus black and white – one could build an entire range of colours. However, this is not practical, because to achieve the exact gradations in colour, it requires precise mixing of different pigments.

The **secondary colours** are orange, green and purple. They are obtained by mixing two primary colours in equal quantities. For example, mixing of yellow and red together in equal quantities make orange colour. (Refer Table 5.3).

Table 5.3: Primary and Secondary Colours

Primary colours		Secondary colour
Yellow	+ Red	Orange
Red	+ Blue	Violet
Blue	+ Yellow	Green

Primary and secondary colours together form 'normal colours'. When a primary and a neighboring secondary colour are mixed, an 'intermediate or a tertiary hue' results. In appearance, intermediate is halfway between two colours. There are six intermediate hues namely – yellow-orange, green-yellow, blue-green, blue-purple, red-purple and orange-red as illustrated in table 5.4.

Table 5.4: Intermediate Hues

Primary colour +	Secondary colour	=	Intermediate hue (Tertiary colour)
Yellow	+ Orange	=	Yellow - Orange
Orange	+ Red	=	Orange - Red
Red	+ Purple	=	Red - Purple
Purple	+ Blue	=	Blue - Purple
Blue	+ Green	=	Blue - Green
Green	+ Yellow	=	Green - Yellow

Some of the popular colours like pink, brown, lavender, beige etc. are not in the list of above colours but each of these can be described accurately by using the name of the hue with which it matches in the spectrum. A mixture of two of two tertiary colours is **quarternary colour**. For example tertiary yellow combined with blue gives quarternary green.

Colour notation in the Prang System

In the Prang notation a colour is expressed as follows, hue, value, intensity. Hue is indicated by the

PLATE 9

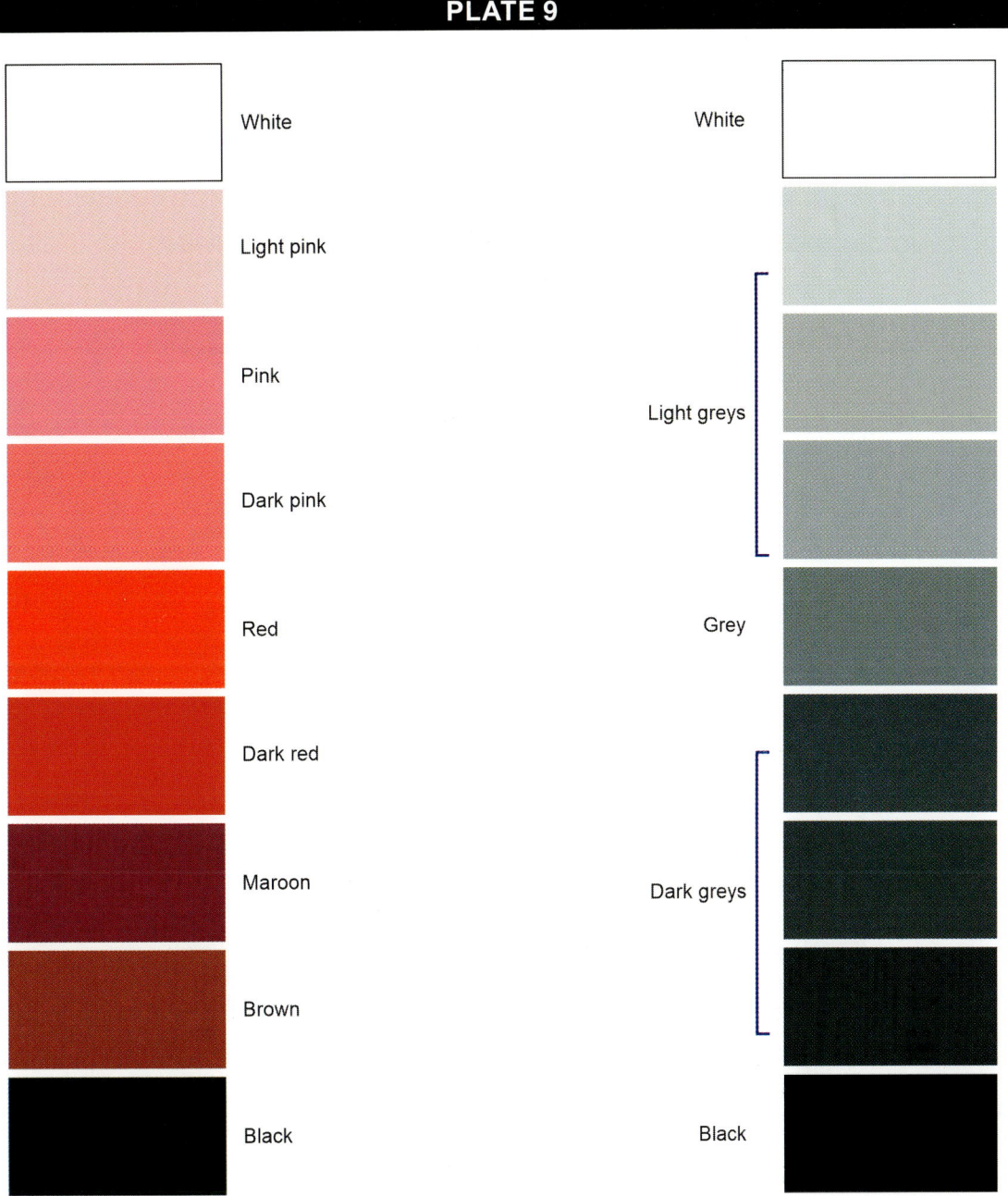

Figure 5.1. Value chart

PLATE 10

Figure 5.3. The Pigment
Theory Colour Circle

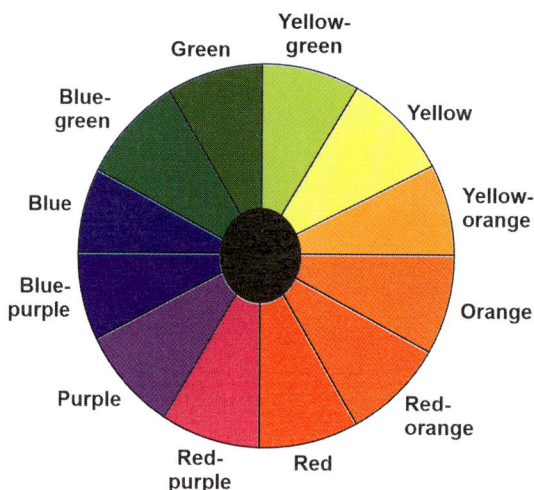

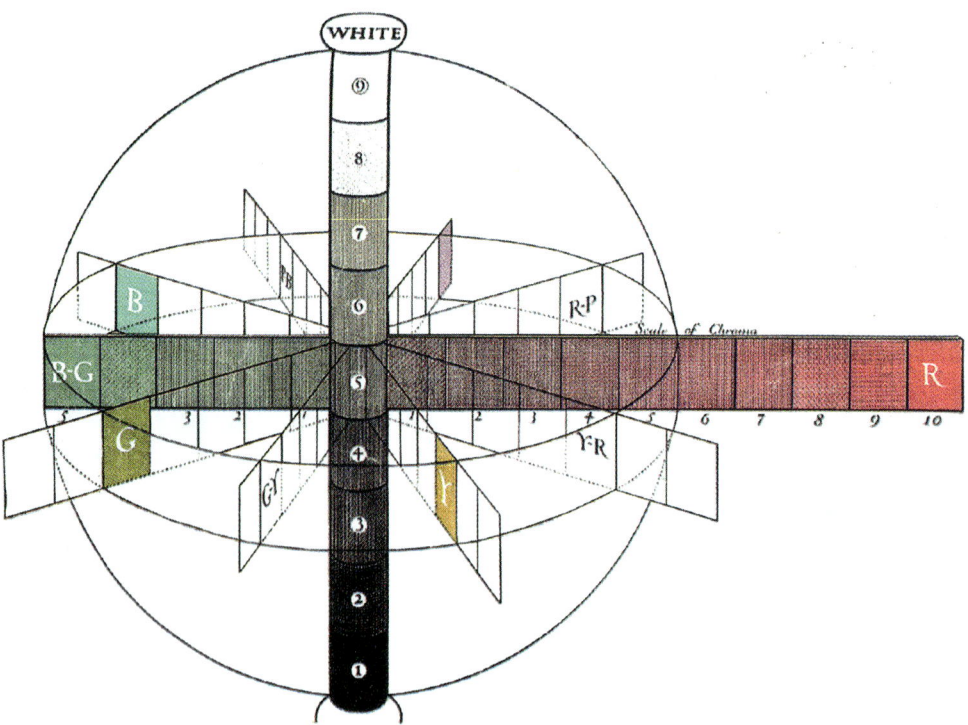

Figure 5.4. Munsel Colour System

name or initials of the colour, as Red or R. Value is denoted by the name or initials of the step to which it corresponds on the value scale, such as Low Light (LL), or Dark (D). Intensity is expressed by a fraction that shows its degree of neutralization as 1/4 N, or by a fraction showing its degree of intensity, as 3/4 I. Thus, a red of fullest intensity, in the value in which it is seen on the colour chart, would be written RHD Full intensity. Red that is high dark in value and one fourth neutralized would be written RHD 1/4 N or R HD 3/4 I.

THE MUNSELL COLOUR SYSTEM

A.H. Munsell worked out a colour system that eliminates much of the guesswork in colour study. In the Munsell plane, the dimensions of colour are shown upon a sphere (Fig. 5.6). The hues appear around the circumference of the sphere. Values in neutral grey are shown upon a vertical pole – the axis of the sphere. The 'North Pole' is white and the 'South Pole' black. As the hues become lighter in value, they are placed higher on the sphere, and as they grow darker they appear lower, toward the 'South Pole'. Chroma, or intensity is represented by paths or arms running from no-colour, or Neutral Grey, out to the circumference and beyond it. (Figure 5.4 – see colour plate 10).

Munsell found that if hues were in proper balance around the sphere and this sphere were rotated upon its neutral axis at a high rate of speed, the hues would blend together to form a neutral gray. He decided to use five principal hues in order to make use of the decimal system. The chosen hues were named Red(R), Yellow(Y), Green(G), Blue(B) and Purple(P). The hues intermediate between these were named Yellow-Red(YR) Green-Yellow(GY), Blue-Green(BG), Purple-Blue(PB) & Red-Purple(RP).

Instead of the twelve-hue circle we now see ten major hues divided into five principal hues and five intermediate hues. Munsell did not include the colours violet and orange because they represented the names of flower and fruit respectively.

Numerals are used to designate the hues lying between the principal and intermediate hues. Principal blue would be 5B; intermediate hue 6 blue-green would be 5 BG. The hues lying between 5 BG and 5 B are designated as follows: 6 BG, 7 B6, 8 B6, 9 B6, 10 B6 (the mid point between 5 B6 and 5B), 1B, 2B, 3B, 4B, 5B. the 6 B6 has little more blue in it than 5 BG; 9 B6 is four steps from 5 B6 and six steps from 5B; 10 B6 is just half way between B6 and B. The other steps – 1B, 2B, 3B, 4B – all show more blue and less blue-green until they read 5B.

Values, in the Munsell plan, have numbers. Ten steps are charted between black and white. Absolute black (which the eye cannot see) is O, and is written NO/. Absolute white is N10/. Halfway between black and white is Middle value, or N5/.

The full strength of the weakest hue-blue-green determines the circumference of the circle, and all other hues extend beyond the circumference in the degree of their relative strength. The scale of red chroma is written R/1, R/2, R/3 etc. Chroma is an almost grey but is recognizable as a warm grey and each succeeding step is nearer to the strongest visible red.

In the Munsell notation, colour symbols are expressed as follows, hue value/chroma. Thus the five principle hues as they appear in the fullest intensity now obtainable, with permanent pigments would read as follow: R4/14, R8/12, G5/8, B4/8, P3/12. The five intermediate hues would read: YR 5/12, GY 7/10, BG 5/6, PB 3/12 and RP 4/12.

A study of colour theories and qualities is essential for planning colour schemes for a room. It requires a great deal of knowledge and thought for achieving success in their use, because we find infinite variety of colours both in natural and artificial objects.

Chapter 6

COLOUR : USES AND APPLICATIONS

We can develop colour sense through observing and analyzing coloured objects. Many objects with the greatest colour harmony can be found in nature. The amount of ab sorbed and remitted sun energy (which we experience as colour) necessary for good functioning is in agreement with the harmony of the life rhythm of living nature. We observe the colour of air and earth as a whole, for example in a landscape. We observe the colours of plants and flowers as a whole, in their harmonious surroundings. We can see where each colour belongs, between which other colours, upon the schematic cross sections of the colour perception space. We can note the kinds of complementary contrasts or only saturation contrasts, mutual surface relationships, and the influence of distance upon colour changing, by observing and studying colours in detail.

HARMONY IN COLOUR

If colours are like the notes of a musical scale, then colour schemes are like musical chords, structuring colour groups according to certain visual relationships among their attributes of hue, value and intensity. Harmonious colour combinations are those which produce a pleasing effect when placed together. The impression is created that all the colours belong together and yet sufficient variety has to be there to avoid monotony. Some important guidelines that are useful in creating colour harmonies are:

- There is a natural harmony among the warm colours and a similar one in cool colours. Therefore, warm colours should be combined with warm colours and cool colours with cool ones. Discord is likely to result if warm and cool colours are combined together.

- The greyed warm hues which are somewhat advancing have a tendency to unify the colours placed against them and therefore make a better background than the cool hues which have a tendency to recede when separate colours are placed against them.

- The most beautiful colour schemes are those which give a single impression; an impression of warmth with, perhaps, a note of coolness for variation or of coolness with an accent of warmth.

98

Colour co-ordination is magical. With colour, one can create any atmosphere or mood by studying and understanding the principles of colour harmony. These principles have been studied intently by philosophers, scientists and artists. From this, were developed several formalised colour schemes including the related and contrasting colour harmonies.

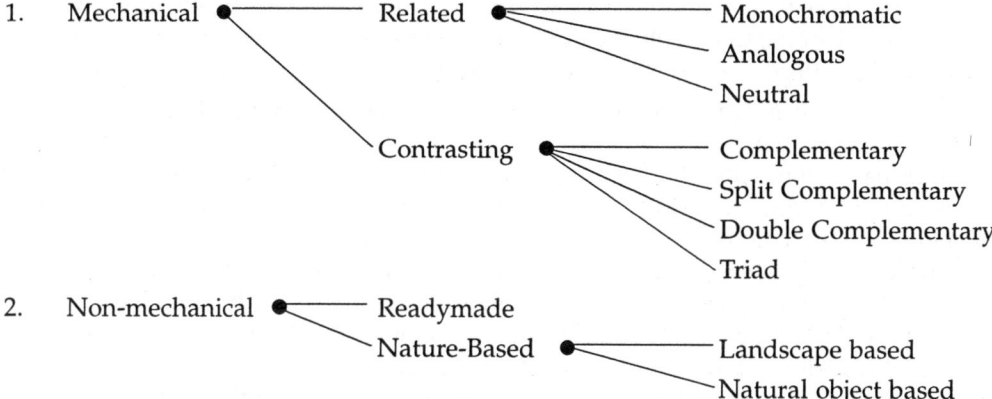

Colour harmonies can be created in two ways, namely, mechanical and non-mechanical. The colour harmonies under the first category are further divided as related and contrasting colour harmonies. The related colour harmonies include mono-chromatic, analogous and neutral, and this type of colour harmonies give a lot of freedom to its creator. A large number of objects with a variety of tints and shades are available in the market these days. Therefore one can successfully combine them to create 'ready-made' or non-mechanical colour harmonies. In the second category, they are nature-based and ready-made. Landscape and other natural object based colour harmonies are included in this category.

A natural way to approach the planning of this colour scheme is to choose some existing design or pattern and key the various colour areas to it. One might select a vase, a picture, a rug, or a fabric with a combination of colours that is appealing. Post cards and cards from museums, as well as photographs of a favourite scene, may suggest colours to use in a room setting. It is not necessary to match colours exactly, but the colours in other areas must be related to those in the design. Different values and intensities should be used.

NATURE-BASED COLOUR SCHEME

Like everything else, anything based on nature is successful for ever. It has a sort of rightness about it. Nature provides bountiful ideas and these can be utilized successfully to make a variety of colour schemes and harmonies. We can divide nature-based colour schemes into two categories, as landscape-based, and natural object based.

Landscape-Based colour Scheme

If we observe the colour of sea, sky, clouds, mountains, land covered with plants and trees and earth or barren land, we find that they do not have sharp colours and are large in area. The sky occupying the largest area is light blue; water in the sea/river/canal also has the reflection of the same light blue; mountains are generally covered with either snow or light brown; trees and plants covered with green leaves. Earth is of dull brown or ochre colour and trees are green.

But, with all this, one will find a fantastic variety of fruits, flowers, birds, insects (like butterflies and beetles), reptiles, animals, fish etc. in bright colour combinations, though these are small in size. To find colours that appeal to you, arrange flowers or vegetables until a combination of their colours satisfies you.

Again, observing very carefully, one will notice that nature never duplicates a thing in its entirety. It creates variety. The dull matt brown of soil is repeated with a different, a bit-smoother structure in the dark of the trees. Soft smooth yellow petals of a sunflower are repeated in the matt yellow wings of a butterfly which is repeated in the bright shining yellow of the sun.

Since our microcosm (small world) is a reflection of the macrocosm (big world), the lessons learnt from nature can be, and have been incorporated in interiors profitably. Japanese call it 'shibui' which has become popular even in the Western Countries. According to this, we can use larger areas in light or high-key, small areas or areas to be accentuated in bright or lower-key, and repeat a colour with variations for example – a dull brown rug can be repeated in a smooth, shining leather chair and that can be repeated in highly polished bronze pieces. **Shibui** echoes nature's use of bright accents in small quantities, against large dull backgrounds. Brilliant tulips are seen in small amounts, and nature usually paints large areas of the sky and earth in dull colours. This can be a decorative guideline – large areas of dull colours, dark colours on the floor, and light colours used on the ceiling and accents reserved for use in small amounts in other areas of the room.

Nature Object-based Colour Scheme

Nature has created various creatures with variegated colour schemes. Observe a butterfly, or a fish, or a flower. This will give the cue to a colour scheme. A butterfly might give a colour scheme of lemon/golden yellow, brown, black and white. Or a flower might lead to a colour scheme of lilac and white accentuated with violet and green. Or a panda might decide a colour scheme of black and white and with accent provided by brown or a rose shrub will give two distinct colours – red and green. While we find green covering larger areas are leaves and stems, and red is seen in smaller account are the rose flowers. The yellow pollens and some dry brown twigs provide accents and backgrounds, respectively.

MECHANICAL STANDARD COLOUR SCHEMES

A person who is conscious of colour may look toward reference materials, such as magazines, books, pictures, fabrics for good colour harmonies and adapt them to the need but unless one has a great deal of practice in combining colours, it is helpful to follow the 'standard harmonies'. These schemes are based on the Prang colour wheel, and are the most easy to use. The mechanical standard colour schemes/harmonies may be of two types; as –

- Harmonies of related colours,
- Harmonies of contrasting colours.

While knowing about the different colour harmonies, it is important to remember that every scheme is good in its own way. One colour scheme appear satisfying in one instance and may be disappointing in another. To be successful, each combination must fit the place and purpose for which it is intended.

Harmonies of related colours

Related colour schemes are based on common hues and tend to create a restful, quiet effect. Colour combinations giving the most pleasing effect are likely to be those schemes having harmony or unity. They give the impression that all the colours really belong together, and yet, at the same time, there must be sufficient variety to avoid monotonous combinations. Colour schemes sharing common characteristics with one dominating colour are used in the related colour schemes. Colours are closely situated to each other on the colour wheel and have one colour that is common or *related*. They are based on either a single hue or a series of analogous hues, and promote harmony and unity. Variety can be introduced by varying value and intensity, including small amounts of other hues as accents, or by introducing shape, form and texture into their combinations.

While selecting colours in a related colour scheme, it is to be remembered that interest is gained mainly by strong contrasts of brightness and greyness. The dominant hue should be supported by a fairly large area of a related but greyed colour, and a relatively small area of pure, bright related colour. It is also better to avoid using two primary hues in the same related colour scheme. For example, a scheme ranging from red to blue will probably produce a harsh glaring effect, unless one of the primary hues vary greatly in both value and intensity. All these considerations will call for an organized plan to decorate the entire house. Related colour schemes are based on common lines.

According to this criterion, the following are the related colour schemes:

- Monochromatic
- Analogous
- Neutral

Monochromatic Colour Scheme/One hue colour harmony

Monochromatic means 'of one colour'. Under this scheme only one hue and its different values and intensities are used. The popular scheme of beige brown, and orange is a truly, monochromatic scheme, because the tans and browns are simply tints and shades of grayed orange. A brilliant form of orange may be used as an accent or even for larger areas, but if any other hue is introduced here, the harmony is no longer monochromatic. Another example is that red can be used with its various intensities giving vermilion, carmine, crimson etc. and various values in high-key giving Alps red, peach, flesh tint, Indian red, brown, burnt sienna etc. The range may vary from almost white to almost black and from very bright colours to dull colours. Black and white are neutrals that can be used to add spice and interest to the scheme.

Monochromatic colour scheme can easily and successfully be used in small rooms, but not ideal for large rooms. This colour scheme is the simplest and easiest to use for any beginner in interior decoration. Any beginner can just pick up prints, checks, stripes and plain materials in different tints and shades of a colour and use them for furnishing a room.

While using a mono-chromatic colour scheme i.e., using a single colour, one can create a feeling of warmth or cold in a room. For example, when the tints and shades of blue is used, it makes a room exclusively a cool one, while the use of tints and shades of yellow or orange makes the room purely a warm one (Figure 6.1 – see colour plate 11).

Another factor to be considered in the use of this colour scheme is the way the different tints and shades are used. Normally the lightest of the colour (highest value) should cover the largest area and at the highest level, the mid value of the colour at the middle and the darkest at the lower level and so on. For example, when the tints and shades of blue are used in a bed room, it should have the following pattern:

Ceiling – White

Walls – Sky blue (light blue)

Draperies – Royal blue

Bed cover, Table cover etc. – Stripes/checks/Prints in royal blue, turquoise blue and grayed blue against a white background.

Floor covering – Navy blue

Accent – A wall picture/flower arrangement/lamp shade in rust/orange/Mustard orange colour.

Mono-chromatic scheme is the simplest of all other schemes to use, because they are based on only one hue. However, too much of one hue can become tiresome and monotonous even though it may be a favourite. The better alternative is to use it by itself in one room and combine it with another colour scheme in the rest of the home. To avoid single-colour monotony, mono-chromatic schemes are often sparked with neutrals – black, white, grey or beige – in many varied textures. For obtaining good effect in the simplest combination for a monochromatic scheme is the use of one tint, one shade and one pure hue as for example, pure orange, peach and brown.

While selecting a monochromatic colour scheme, the main consideration lies in the selection of the shades and tints of the colour in the room in which they are to be used. For example, yellow colour is an excellent choice for a dining room because yellow is a cheerful, light reflecting colour. Warm hues should be selected for rooms with cool exposures and cool tones for rooms that receive lots of sunlight. A careful thought, and a complete analysis of a room in terms of its use, exposure, size etc. is necessary for deciding the single colour to be used in any room.

This scheme, when more of the tints of a colour is used, helps to give a spacious feeling to an interior and provides unity for a composition and a quiet background for objects and people within it. The scheme also offers a designer reasonable chances of success. The only drawback is that being a single colour, it easily becomes boring and therefore is seldom used in its true form. The key to success is to provide enough contrast through pattern, texture, furniture shape, etc.

Analogous Colour Scheme/Adjacent colour scheme

Analogous colour scheme is one that combine colours that are adjacent to one another in the colour wheel. In this colour scheme, colours situated next to each other on the colour wheel of a Prang colour system, are used. It generally uses one colour as its major force with the two neighbouring colours as secondary forces, eg., yellow with yellow-green and green or blue with blue-green and blue-purple. Once again, with different values and intensities, any adjacent colours may be used in combination with interesting effects. Because of their position on the colour wheel, they are often referred to as related colour schemes, as they all contain a common line in the colour wheel (Figure 6.2 – see colour plate 11).

Analogous colour scheme is much more interesting than a monocromatic colour scheme. It is also a simple and easy colour scheme to choose and adopt even for an inexperienced interior decorator or an untrained housewife. Unlike the previous colour scheme, adjacent colour scheme can be used in any room irrespective of its size. It is suitable for both small and large rooms. All that is required is to choose the colours lying adjacent to each other in a colour wheel. While choosing the adjoining colours, one can choose either all warm or all cool colours, if it is desired so. For example, a combination of yellow, yellow orange, orange and red orange makes up a warm adjacent colour scheme. A combination of blue purple, blue, blue green and green provides a cool adjacent colour scheme. At the same time, a combination of both warm and cool colour scheme is also possible in an adjacent colour scheme. When the adjacent colours of blue green, green, yellow green and yellow or blue purple, purple, red purple and red are combined, they provide a colour scheme which has both warm and cool colours. Thus this colour scheme provides a wide variety of choices to an interior decorator. Whatever may be the choice of colour, the point to be kept in mind is that the lighter colours and their tints to be used at the higher level, mid values at mid level and then the darker value/shade at the lower level and so on, as was mentioned earlier (Figure 6.2 Anologous colour scheme).

Analogous colours are harmonious because they contain a portion of the same colour through-out. Analogous harmonies are quiet and restful and show more variety than one hue harmonies but still more suitable for the entire room unless change is provided by texture. It is also better to list the four adjoining colours in one quarter of the colour wheel and then select the first and last colours and one of the two mid colours should be combined to make a choice of three final colours. For example, when the four colours in a quarter of the colour wheel is blue, blue-green, green and yellow green are chosen, blue and yellow green are chosen first, and then either blue green or green is selected to make an anologous colour scheme with three final colours.

Neutral Colour Scheme

The neutrals are black, white and grey. But in interiors, wheat, beige, off-white and the high values of each hue are also considered as neutrals. As they do not represent any single colour, they all can safely be used in neutral colour schemes.

A neutral colour such as beige or brown with an accent colour for breaking the monotony can be used in this colour scheme. It has its plus points in the fact that it presents the interior with neutral background which enhances the objects and people in the interior, and for that reason is very helpful for the people wishing to display their collection of objects of art or merchandise. Colour schemes built entirely around neutrals are very restful, but texture and patterns must be used to prevent the design from becoming dull, boring or monotonous. Therefore, this colour scheme should be used in limited areas or for limited periods in any room (Figure 6.3 – see colour plate 12).

The neutral scheme became popular because of its adaptability to both traditional and contemporary style. A neutral scheme allows for easy colour accents changed by the simple process of altering colour efffects on small objects.

Black-and-white schemes are highly dramatic because of their strong contrast. For example, a scheme could be developed with more than two-thirds of the area with white and less than one-third with black. Some grey could be introduced in a pattern. Whatever may be the combination, such a colour scheme should always be used with caution and in limited areas.

Harmony of contrasting colours

Contrasting colour schemes are based on opposing hues and tend to be stimulating and balanced because they include both warm and cool hues.

Differences in colours are the bases of the contrasting colour schemes. Two colours, which are placed extremely opposite on the colour wheel, one might be cool, and the other warm, are chosen for the colour scheme. The colours give great variations from highly saturated to a normal colour effect. When the two contrasting colours are placed side by side, one appears to advance while the other seems to recede, producing a vibrant effect. Example of this are red and green, blue and orange, yellow and purple. This vibration can be counteracted by toning up or toning down one colour or separating them with a related or a neutral colour. When such colours are used in their fullest intensities, they emphasize each other. A brilliant red held next to a brilliant green will appear more intense than when viewed separately. Such combinations are common when a forceful use of colour is desired. Such high intense colour combinations might also produce 'glares' if the surface is smooth and shiny.

Although brilliant complements are frequently used in decoration, more interesting combinations of colour result when the hues are varied a bit. A soft maize or pale yellow with a deep plum purple, for example would probably have more appeal than a brilliant yellow and a brilliant purple used together. Instead of brilliant orange and blue, the grayed oranges or copper tones might be used with a blue that is also decreased in intensity. Using tints of each colour, would produce a combination of peach or apricot and light blue – still a complementary harmony but one that is easier to live with than the colors used in their most brilliant forms. Pink and dark green is also a softer combination than bright red and green. Neutrals may be added to enhance the contrasting colour scheme, and to strike a balance between the two contrasting colours used in a scheme. The contrasting colour schemes can further be divided into four classes, such as:

- Complementary
- Double Complementary
- Split Complementary
- Triad

Complementary Colour Scheme

It is based on two colours found opposite to each other in the colour wheel. These colours need not necessarily be used in their pure form and they can be used in many values or intensities to have variety. A complementary colour scheme can be developed from a monochromatic colour scheme accentuated with a complementary colour or an even balance of two colours.

On a Prang or pigment twelve colour circle, the complementary colours are:
- Yellow and Purple
- Red and green
- Blue and orange
- Yellow green and red purple
- Blue green and red orange
- Blue purple and yellow orange.

Thus, in all, they make six complementary colour schemes.

A complementary colour scheme is a stimulating, vivid and bright colour scheme. It provides a combination of both a warm and a cool colour, thereby results in a balanced combination. Between the two colours, one can be subdued and the other, dominating. Normally a cool colour is used in larger quantity or in larger areas as compared to the warmer one, to have a balanced effect. This colour scheme is highly suitable to such areas like living, nursery, teenagers and children's and drawing rooms. This colour scheme may not be suitable in a dining room to avoid the attention detracting from the food, more so when the room is a small one. When used in large rooms, they can be suitably mixed with a few neutral colours to avoid that extra stimulating effect of the contrasting colours.

Combination of opposite colours are more difficult to use than those of neighbouring colours as in the case of an adjacent colour scheme. When they are well-done, they are richer than related harmonies and more satisfying to the eyes in rooms, window displays, or any purpose where large amounts of colour are to be used. The addition of contrasting colours to a colour scheme is like adding *spice* to a dish. It is no wonder, advertisers of all kinds of products, prefer such a contrasting colour scheme in all their creations! (Figure 6.4 – see colour plate 12).

Complementary colour schemes may be the most pleasing, or they may be the least satisfactory, depending upon how they are used. The reddish hues like red, red orange, orange red purple etc, in particular, need careful handling, because they are much stronger than their complements. Either they or their complement, should be dull, or very light or dark, or else only a small note or the opposite colour should be used. Of all the contrasting colours, red and green are perhaps the most difficult to combine beautifully. Their tints such as the pink and light green lack character when placed together. The complements as seen in the Munsell colour chart-red with blue-green or green with red-purple are likely to be more pleasing.

Double Complementary Colour Scheme

This colour scheme is somewhat similar to a complementary colour scheme – the only difference being the number of colours used – two in a complementary and four in a double complementary. If a narrow 'X' is superimposed on the colour wheel, then there will be two sets of complementary colours with which to work. Two adjacent colours and their complements when used together form double complementary harmonies like the combination of – purple and red purple with yellow and yellow-green. In using a double complementary harmony, there should be an outstanding hue which should be in largest amount and be the dullest of all colour. For example, as in the case of the above example, purple can be considered to the line having these qualities. Use a brilliant form of one, like yellow, in a small area as an accent. The remaining two colours i.e. red purple and yellow green can be used in the remaining area.

This colour scheme is exciting, lively and sophisticated. Besides, it also provides more variety than a simple complementary colour scheme. Since this colour scheme provides a combination of both warm and cool colours, they can be successfully used in any room size, whether big or small. They provide a good background in a nursery or a child's room. They are also ideal for use in common rooms like living or a lounge. When sufficient care is applied in the amounts of various colour used, they make a stimulating and lively atmosphere. Some of the examples of a double complementary colour schemes are:

- Yellow and yellow orange with purple and blue purple
- Yellow orange and orange with blue and blue purple
- Orange and red orange with blue and blue green
- Red and red purple with green and yellow green.

Split Complementary harmony

In this colour scheme, one colour is combined with the two colours on each side of its complement. One might select a hue and combine it with the colours that are on either side of its complement as though placing a narrow-angled "Y" on the colour wheel. As the term implies, one "splits" or divides the complement of a hue into its component parts, and while using these parts the complement is omitted. A true split complementary scheme is a harmony of similar colours with a note of a contrasting colour. The amounts of the different values and intensities should be adjusted as in any other contrasting colour harmony (Figure 6.5 – See colour plate 13).

This provides three colours to work with. An odd number of colours combined together always yield a good effect as a colour scheme in a room. By varying their intensities and values some interesting combinations can be worked out. For example, if orange is chosen, the split complements used are blue-green and blue-violet and these colours in varying values on different textures would provide variety in their use. Similarly, the following combinations can also be worked out to prepare split complementary colour schemes.

- Yellow with blue purple and red purple.
- Yellow green with purple and red.
- Green with red purple and red orange.
- Blue green with red and orange.

It is a harmony of similar colours with a note of contrasting colours and make a brilliant combination. It is also a lively and stimulating colour scheme. However, the warm hues in combination with the cool hues need careful handling because of their strong and advancing effects. When executed properly, this colour harmony would produce a pleasing effect.

Triads

An equilateral triangle placed on the Prang colour wheel will point to three equi-distant colours that form the triad. Turning the triangle will point out different combinations. In the Prang Chart, there is a Primary triad – when the three primary colours fall at the tips of the triangle (red, blue and yellow) and a binary or secondary triad (green, orange and purple). There are two intermediate triads – yellow-orange, blue-green and red-purple while the other is yellow-green, blue-purple and red-orange.

On the Munsell colour chart, the triads are as nearly as many equilateral triangles as can be made with the ten hues in the circle. The hues forming the points of the triad triangle should always be atleast three hues apart from each other on the colour chart. Thus every triangle will have three spaces on two of its sides, and four spaces on the third side. Thus the following colour combinations can be seen:

- Yellow, red-purple and blue green
- Blue, red-purple and yellow green
- Yellow, blue and red-purple.

The combination of these colours can be balanced by choosing the colour with the lowest intensity, for the largest area, the next larger area with the intensity slightly higher than the previous and the smallest area with the brightest colour among the three. In the case of the triad, yellow-orange, blue-green and red-purple, the order of use as mentioned in the above areas, will be yellow-orange, blue-green and red-purple for largest, larger and small areas, respectively. These colours balance each other because they all have same impact on the colour nerves in the eye. Their combinations give pleasure, because they are so varied in the degree of brightness or intensity (Figure 6.6 – see colour plate 13).

Triads are the richest of all colour harmonies if well used, but they are the ones which need most careful treatment. In intense tones, such combinations might be suitable for a playroom or a nursery room because these are closing colours. But when these colours are subdued, the combination becomes the one that is often found in elegant rooms.

Dominating colour scheme

Keeping the colours found in the Prang and Munsell colour wheels, a number of colour harmonies are created and they have been discussed earlier in this chapter. While making combinations of colours and their use in the rooms, certain guidelines are to be followed and these are also discussed in detail under every colour scheme. In general, the colours used are distributed in the various sections of a room according to their intensity. The duller the colour, larger is the area in use, the brightest the colour, the smallest is its use, and so on. However, at times in a colour combination of two or more colour, one colour can be used in larger quantity, and the rest in more or less in equal amounts. For example in the use of a primary triad colour scheme yellow may be used in larger quantity, and the remaining two colours, red and blue in a smaller quantities. This use of colours creates a dominating colour scheme.

To create such unusual combinations in terms of quantity and intensity, an experimental attitude is necessary to achieve success in the use of colours in the home. The colour arrangement of the interior of an entire home should be planned as a unit. An unified multiple colour scheme for the home can be created and executed for each room separately, starting from the living or drawing room. If the planned unified colour scheme centres around a single colour, then it becomes easier for the interior decorator to create dominating colour scheme.

Colour Dissonances

In modern markets, a customer finds a large variety of colours their tints and shades. The advancements in textile technology has created so much variety that an interior decorator finds it a challenge to put them all in her creations and combinations. The colour harmonies or schemes we had discussed so far have the limitation of the use of colours found in Prang and Munsell colour wheels only. Since more choices in colours are available adequate attention should be made to dissonances as much as to harmonies. Such colour schemes may be difficult to create initially, but they can be made stimulating with experience. An interior decorator has to develop an experimental attitude to be successful in creating new and varied colour combinations as colour dissonances.

A courageous use of colours in creating colour dissonances, would help an interior decorator to develop colour sensitivity. A colour plan for the entire house in which traditional colour schemes as well as colour dissonances are used, can provide more excitement and variety. In addition, it permits more considerations in the use of new colours that are available in the present day markets. Such colour combinations are likely to produce pleasing and astonishing effects when aesthetically arranged and distributed. The decorative ability of a decorator is also highlighted, and she can exhibit her extraordinary skills in manipulating the use of colours.

When a person is unable to visualise and make her own colour combinations, she can adapt the colours that are seen either in pictures or fabrics or an advertisement in a magazine and then fit them into the colour scheme for her room. The enterprising decorator, thus keeps her eyes open for good colour suggestions, studies them closely and makes minor changes to suit her needs by identifying and discovering the colour aspects that make the original source material more beautiful. Instead of sticking to a dull stereo-typed colour schemes, such courageous use of colour can guide a person to develop colour sensitivity and be successful in all her creations!

Procedure for making a colour scheme for a room

The general procedure for preparing a colour scheme is as follows:

1. Consider the colour schemes of the adjoining rooms.
2. Decide on the background colour.
3. Procure samples of fabrics, wall papers, and paints.

4. Select a colour scheme from a fabric, wall paper, or picture.
5. Plan the colour proportions. (large, medium, or small areas)
6. Plan the values of the chosen hues (light and dark)
7. Plan the variations in intensity of the chosen hues (bright or greyed)
8. Draw floor plans locating the furniture as areas to be coloured.
9. Draw wall plans locating areas for draperies and furniture.
10. On the plans write colour names or paint the colours.
11. Assemble the completed colour scheme, making water-colour samples of colours not otherwise supplied.
12. Calculate the quantities of the materials required.
13. Collect the necessary material.
14. Execute the plan.

COLOUR DESIGN AND TREATMENT

The three basic divisions of colour in designing anything are subjective colour, objective colour and natural colour.

1. Subjective colour

Homes are usually subjective. When one person has the task of making colour decisions or presiding over diversified family opinion, the result shows all the preferences, taboos, training and education of a small group headed by one person. Even when an experienced interior designer is consulted she should study her subject very well and know more about the family before proposing any colour scheme.

2. Objective colour

In objective colour the use determines what the colour should be. Objective colour may also be referred to as functional colour. Colours increase the visibility as their chrome increases, but all colours increase or decrease in visibility depending upon their surroundings. It is generally agreed that the following colour combinations are visible in this order.

Table 6.1 Base and Visible Colours

Base Colour	Most visible colours
White	Green, Red, Blue and black
Yellow	Black and Red
Red	White and green
Black	Yellow and While
Blue	White
Green	Red

These should be used whenever safety or attention supersedes artistic use. For example, a red fire extinguisher is highly visible against a white base, but it should never be placed on a green wall unless white paint is painted around it.

3. Natural colour

Coloured stones, feathers and furs, dyes from berries and bark, and coloured earth were instrumental in man's earliest attempts at the use of natural colour. Architects, philosophers and poets have continuously used the beauty and balance of nature. Nature is the most inspiring aspect, and, therefore, it is not surprising that the colours in their natural forms have their lasting impact.

FACTORS AFFECTING THE USE OF COLOUR SCHEME FOR ROOMS*

Since colour is a source of universal pleasure, every one of us should have beautiful colours in our homes and learn to enjoy them. It is necessary for all of us to have some knowledge of colour so that they are used to one's delight and to fortify them against dullness elsewhere.

While planning colour schemes for the rooms, it is necessary to consider them for the entire home as one unit to obtain an unifying effect. A colour plan for the entire home in which each room is independent of the others, has less unity than a related plan. It is wiser to start planning from the living/drawing room first, then dining room and finally the bed rooms. Once these areas are planned, then the plan for the passages, corridors can be carried out. The colour schemes for these areas can be different from the rooms to produce variety.

When an interior decorator visualizes a colour plan for a room, she should consider a number of factors like, the room, its size, shape and exposure, the mood, style, current fashion, personal preferences, furnishings on hand, use of the room etc. The success of a good colour plan depends upon how much attention is being paid to these aspects. A complete account of these factors and their impact on the colour plans are discussed in the following paragraphs:

The Room

The foremost factor in the selection of a colour scheme for any room, is the room itself – its size, shape and exposure. Analysis of a room is therefore, the first factor in making a colour scheme for it. While considering the size, small rooms gain spaciousness and appear larger with light, cool and receding colours in the background and on the furniture too. On the other hand, the warm advancing colours reduce the size of the room and make a room to seem smaller and cramped. In large/big rooms, warm/advancing colours are more appropriate than a cool/receding colour scheme. A receding colour scheme will not only make the large room to appear larger, but may also give a feeling of being empty and under-furnished. Therefore, an analysis of the room in terms of its size, the amount of furniture and furnishings it requires would help to give exactly how a room need to look like. A well-planned colour scheme can help to alter visually the size of the room, if it is desired so. Similarly, a long narrow room can be made to appear wider and of better proportion if the end walls are decorated with warm advancing colours and side walls with cool receding colours. Atleast if one of the end walls are done with warm colour, this wall will appear to be closer and can help to reduce the length of the room Figure 6.7 – See colour plate 14).

While the size of the room can be altered by using colours cleverly, the room's shape can also be changed in similar ways – a rectangular room can be made to appear like a square one and a square to be a rectangular one. This can be obtained by alternating receding and advancing colours on the four walls of the room, thereby changing the shape of the room. Thus the room can be made to appear in better proportions.

---

* Rutt: Home Furnishing, Wiley Eastern Pvt. Ltd. New Delhi.

Another aspect of the room that affect the choice of colour is the amount of sunlight that it receives. Room with too much light need dark subdued colours. The rooms facing west, south and south-west receive the maximum sunlight, and therefore, become very warm, especially in summers. In contrast to these rooms, rooms in the eastern and northern direction receive comparatively lesser sunlight. These rooms are cool, specially in winters. To counter this effect, cool colours in former rooms and warmer colours in latter rooms should be opted. Thus a wise choice of colour schemes can help to change the warm or cool effects to provide a feeling of comfort in the rooms. However, before making a final decision, it is important to know whether the room will be used for quiet or for more active pursuits.

Mood

Mood of a room is an important consideration and can be expressed through and in colours. One has to decide which mood needs to be developed in the room – whether it is to be active or subdued, masculine or feminine, formal or informal, sophisticated or practical. A delicate feminine room can be done with dusty pink or peaches while a masculine room in wine red, brown or navy blue. A living room which is an activity – centred room, would require a colour scheme with electric blue, lemon and white with a small quantity of bright red. A elegant room, similarly, can be done with the neutral shades like grey, brown, beige, etc. along with the toned or greyed blues or pinks. Thus the colour scheme also exhibits the mood of the room.

Style

The style adopted in a room determines the use of colour. Certain styles, such as traditional or conventional, are best expressed by the authentic colours of the natural dyes that were used with the original furniture. These colours are rich and of medium value. Much of French decor requires soft, delicate colouring. The luxurious themes of eighteenth-century English styles call for rich, elegant tones of claret red, emerald green, gold and plum colour. The modern style has made use of bold, brilliant colours, often in combination and often as accents with white or neutral backgrounds.

While preparing colour schemes, it has to be found whether the person prefers contemporary or traditional style, or French decor, etc. Sometimes, the style with which a house is constructed and its furniture style also affect the choice of colours. For example, a cottage or a farm house constructed with typical traditional furniture items might require a colour scheme that matches with this kind of setting. Similarly, the room in an apartment in an urban area can have informal colour combinations that are available in the market. Thus the colour scheme has to be done in combination with the style of furniture, house construction and the items used in a room.

Fashion

Fashion is an important aspect of civilized living. It influences or in a way, limits all the things we use or enjoy. It is, therefore, not surprising that colour schemes are also affected by fashions. One has to keep up with colour fashion by visiting shops or reading magazines. In fact, it can be said that a person will find in the market, only those colours which are in fashion, though she might prefer other colours for her house. Thus, fashion in a way limits the choice of colour only to the available ones in the market. Thus, availability of colour according to fashions, is another factor determining the colour scheme of interiors. Draperies, fabrics and curtains are available in a wide range of colours in a modern market, but large furnitures are available in limited colours and this may restrict the colour schemes one desires.

Keeping up with fashion means almost constant redecoration, since colour fads, wall-covering designs and ideas for using materials change rapidly. The 1960's and early 1970's for example, saw an upsurge of interests in purples. Then browns, in many shades from beige to peach became the rage – only to be threatened immediately by an onslaught from sharp lettuce green and deep strong pinks! And while on the one hand there was increased interest in plain textures and natural colours, on the other, there was a Victorian revival. But, some changes do bring in a variety in the choice of colour for the rooms and break the monotony.

It is not easy to change the colour scheme of a room often to suit the changes in fashion. But it is possible to predict certain cycles or trends in fashion by keeping in touch with the fashion magazines. They would not only serve as a guide, but would also help us in preparing the colour schemes in anticipation of future trends.

Personality

Personality of people also influences their colour choice. Women who are too feminine may choose soft cool colours like pinks, peaches. An informal room may call for colours that are bright, strong and stimulating. A room that is meant for relaxation and repose should have colours that are quiet and restful. Children's rooms may have colours that are either bright or delicate, such as red, royal blue or yellow. The colours chosen should be those that the members of the family enjoy viewing. Even if one is not particularly colour conscious, there may be colours that one intensely dislikes. Colour preferences are not always clearly defined; learning what they are, may take some probing. It is generally believed that children prefer light colours, youth vivid colours, and others prefer soft colours. However, strong preferences of any one individual can be limited to that person's room only.

At times a person's personality is shaped by her profession, place of living, like urban, rural etc. Such factors determine a person's personality and her choice of colours. An expert can easily judge the profession or residential place of a person, by seeing the choice of colours in their rooms!

Possessions

For any one, who is setting up a house for the first time, the exercise of furnishing and colour plans begin from scratch. It is not difficult to choose whatever they would like to use in their homes. But, in the case of families, who are settled, and are in the middle of their family life cycle, it is seen that their possessions limit the choice of colours. Carpets, furniture etc. purchased earlier have to be used and cannot be thrown away in these days when these items have been bought at high prices. Some of the families may not like to throw away certain items because of their sentimental values like family possessions, gifts and others bought on special occasions. Some of the items may also have been in good condition and therefore, last for a long time. There is no point in discarding these to purchase fresh ones and incur more expenditures. Therefore, these items are to be taken into account while selecting fresh ones and suitable combinations are to be created while doing so. These families have to add or subtract whatever articles are necessary according to what they already possess.

Use of room:

Among all the factors that influence the colour choice, the most important one, perhaps, is the use of the room. The use of room influences the colour scheme to a great extent, because colour can help to emphasize the purpose of the room in addition to contributing to its efficiency and comfort.

The use of a room decides in a way its colour scheme. Colour can help to emphasize the purpose of a room in addition to actually contributing to its efficiency. The time of day when a room is occupied most should be a factor in its colour scheme selection, and naturally a room should look its best at that time.

The *entrance hall* often has much the same type of coloring as the living room, although in larger houses it is frequently more impersonal and dignified in colour than the living room. In an apartment or small house where an entrance hall is only a passage-way, it should be decidedly decorative and colorful. Since a hall has very little furniture, interest should be provided by colourful treatment of floors or walls. In a two-storey house, the hall is the transition point between the first and second floors and so may well contain colour ideas of both floors. An interior hall should be bright and novel in colour and pattern, as it is usually dark and also unfurnished.

A *living room* should express cheer and hospitality along with restfulness and relaxation. Therefore its colour scheme needs to be cheerful but not overstimulating, and characterful but not obtrusive; fairly light, warm colours are usually the most desirable for living rooms except in a summer home or in a tropical home, where a simple cool scheme such as white walls and a blue ceiling and floors is refreshing. For a temperate climate, white, brown, and coral with walls of pale grayed lime make a cheerful colour arrangement, which could be varied for summer use by substituting natural coloured matting for the brown rug, using striped green and white slip covers to conceal the coral, and substituting thin white curtains for the draperies. In a living room it is desirable to keep the colour interest and contrast on the general level of the occupants and furnishings, and not on the floor, walls, or ceiling.

An *outdoor living room* opening directly off the indoor living room should employ some of the interior colours, or others expressing the same mood as the interior. At the same time the colour of the exterior of the house and the green of growing plants must be considered when outdoor furnishings are chosen. An additional factor is the idea that outdoor colours should be few, simple, direct, positive, and cool. White is sometimes satisfactory for outdoor furniture, but it should usually be accompanied by one colour, such as chartreuse, leaf green, or the hue of the shutters or roof.

The *dining room* in a formal home is naturally in the same character as the house so its colour scheme is conservative and dignified. Most dining rooms are informal. However, they provide opportunities for pleasant surprises in colour and decoration. The unexpected causes a lift in spirits and incites a light merry mood that is conducive to a happy time and a good appetite. A novel colourful decorative scheme does not become tiresome in the dining room because the occupants do not remain there for long.

Colours such as lettuce green, shrimp pink, butter yellow, lemon yellow, watermelon rose, or tomato red have refreshing and delicious implications. The addition of white is desirable, particularly if white is used on the table. Trite old colour schemes like blue backgrounds for mahogany furniture should be avoided. Unusual experimental schemes, such as natural light wood or plastic chairs with a jade table, and pale pink backgrounds, are much more stimulating. Some successful dining rooms feature the garden idea with sky-blue ceilings and white walls, gay colours being supplied by house plants, tropical fish, and birds.

A *kitchen* colour scheme should be cheerful, light, and bright. Cool colours are thought to counteract the heat of cooking. The colour of the sink, range, and refrigerator, which should all be alike must be the basis for the colour scheme. When these three articles are white or pale grey, they are most easily fitted into a room design. Colored utilities are likely to become tiresome; therefore it is well to depend on less permanent articles to supply colour. White or light walls are usually best

PLATE 11

Figure 6.1. Monochromatic colour (blue) scheme in a room

Figure 6.2. Analogous colour scheme

PLATE 12

Figure 6.3. Neutral colour scheme in a room

Figure 6.4. Complementary colour scheme (blue and orange) in a room

PLATE 13

Figure 6.5. Split complementary scheme

Figure 6.6. Triad colour scheme

PLATE 14

Figure 6.7. Use of colours to alter the room proportions

for visibility. Natural wood is also desirable. One vital colour, such as a pure primary or secondary colour, may be used in interiors of cupboards, for furniture, in curtains, and possibly on the floor, ceiling, one wall, or on the wood trim. Other colours should be supplied by dishes, pots, and plants. The colours used in the kitchen should be stimulating enough to make the cook feel creative.

The cottage or farm house type kitchen takes gay decoration well. Painted ornamentation in peasant fashion may be copied in bright colours such as the Swedish yellow and bright blue. Colourful painted cartoons of family or local significance and favorite recipes painted on cupboard doors are personal and appropriate. There is no excuse for an ugly kitchen, for almost everything in it can be painted.

A *bedroom* colour scheme is usually more personal than any other; its dominating colour might well be the favorite colour of the occupant. In deciding upon this colour the exposure and the amount of light in the room should not be overlooked. For example, a woman who is fond of soft chartreuse might use it in a north bedroom on walls and ceiling, in combination with white wood-work, full white net curtains, and cherry-red bedspread.

A master bedroom used by both husband and wife should contain colours expressive of both. A man's room or a boy's room should be masculine, with rich characterful colours, possibly combined with natural wood. A woman's room or a girl's room should be bright and feminine and harmonious with her own coloring. A pretty, colourful room helps to develop a girl's personality. A guest room should be decorated in rather impersonal colours since it needs to please occupants of either sex of any age. Children's rooms are often finished with playful colourful decorations placed at the children's eye level. When bedrooms are treated as sitting rooms the colours should be darker and less personal than are customary for bedrooms in general.

Closet colour schemes should be cheery. The colours may be the same as the colour of the bedroom ceiling or walls or in pleasing contrast to them. Natural wood walls are not only attractive but also very convenient for attaching hooks and shelves.

Bathroom walls and fixtures should usually be white, for hygienic as well as aesthetic reasons. In general, coloured fixtures have proved to be tiresome. However, one bright clear positive colour is needed in a bathroom; it should be used on the floor or on the upper walls and ceiling and in towels, mats, curtains, and shower curtains. Bathrooms are often decorated in colours that suggest water, like green, blue, violet, or grey. One common mistake found in bathrooms is the use of wall tiles that almost but not quite match the floor tiles.

Game-room colour schemes should be bold and vigorous. The whole effect should be amusing, with colourful backgrounds supplying interest, since furniture is scarce. The walls might be painted in a warm, bright, solid colour such as coral or lemon yellow, or two walls might be painted in stripes, checks, or plaids. Mural paintings and decorations suggesting a circus, a ship, or a garden might be effective.

In planning the colour for the whole house, it is better to begin by sketching each room. Separate sheets can be prepared for each room and rough sketches of its shape, window placement, and furniture arrangement can help to provide a base for filling them with suitable colours using pencils or crayons. Once this is done for every room, they can be spread on the floor in sequence. A close analysis will give an idea about the strengths and mistakes in the scheme. The sketch can then be re-done to remove the mistakes. These sketches can be carried at the time of shopping for paints, wall paper, fabrics etc. Such an organised plan can help in achieving the expected results in colour plans for the rooms as well as for the whole house.

APPLICATION OF ART PRINCIPLES IN THE USE OF COLOURS FOR A ROOM*

There are certain guiding art principles in the use of colour. All the principles of design apply to colour use and also contribute to beautiful colour effects.

Balance

It is the first essential principle for good colour arrangements:

(i) Large areas of colours should be quiet in effect, while small amounts may show strong contrasts; the larger the amounts used, the quieter the colour should be and smaller the amount the more striking the contrast may become.

(ii) Value is also important in colour balance. If there is a difference in value, there must be corresponding change in amounts used in order to give the effect of repose. Thus, a large amount of light value will balance a small amount of dark value, or small amounts of dark balance large areas of light.

(iii) Complementary colours or the colours which are opposite to each other on the colour chart form a natural balance because they complete or complement each other (Figure 6.4).

(iv) Colour or values can be balanced by repeating some of the same colours or values in various parts of an arrangement and this repetition, sometimes called crossing, has a tendency to give a feeling of rest.

Proportion

The proportion of one colour to another must be taken into account. Colour combinations are more beautiful when the amounts are varied than when they are equal. In any arrangement if the colours used are equal in their power of attraction, then the Greek proportion of 2:3 or 5:7 will be good. However, if the colours are very different in their forcefulness, they should be arranged according to the "law of areas" and the brighter colours are used in smaller amounts.

Harmony

Whenever there is unity or harmony in the colours used, such combinations give greater pleasure. They give the impression that all the colours really belong together, and yet at the same time there must be sufficient variety to avoid monotonous arrangements.

Earlier in this chapter a list of related colour harmonies was discussed. It can be seen there that there is a family likeness – a natural harmony – among the warm colours, and a similar kinship or unity among the cool colours. Therefore, if one wishes to obtain colour harmonies she will combine warm colours with warm, and cool colours with cool. If contrasts are desired, some cool colour may be used in a warm colour scheme, or a warm colour note introduced into a cool scheme.

There are, however, degrees of warmth and coolness within the warm and cool groups. Blue of full intensity is colder than a somewhat neutralized blue, since the neutralising orange has also warmed it. Blue-green is warmer than blue because it contains yellow. Therefore, if it is desired to use a cool colour as an accent in a warm colour scheme, a tone of blue-green or some-what neutralized blue will be more harmonious with it than a clear, cold blue. Similarly, while introduc-

---

* Goldstein. H. and Goldstein, V., "Art in Everyday Life". MacMillan Co.

ing a large amount of warm colour into a cool colour scheme, it might be better to use the yellow, yellow-orange, and orange tones than a full intensity red-orange or red. If the warmest of the warm lines are used, they should be utilized in small amounts or else they should be neutralized, since the addition of the complement – which is a cool colour – tends to cool them.

The most unifying colours are the colours of the light-yellow, yellow-orange and orange. When these colours are dull enough, as in warm greys and tans, any hue looks well against them, and therefore, they make the most useful colours for background. It is also seen that the greyed warm hues, which are some what advancing, have a tendency to unify the colours placed against them. The cool hues, which recede, have a tendency to separate cool colours seen against them.

Rhythm

Rhythm refers to the arrangement of colours along which the eye can move easily from one colour to another. When colours are skillfully repeated in several places in a room, the eye travels rhythmically as it follows these colours. Rhythm also results from the use of gradations in hue, value and intensity.

Emphasis

Emphasis can be gained by contrasts of hue, light, dark and brightness. In any colour arrangement there should be one outstanding colour effect. Even if the colour scheme is quiet and simple or is complicated, one should be conscious of a major colour. Backgrounds should show less emphasis than the objects placed against them. Colours of background in rooms should be quiet, because only then the objects planned against them will be more effective.

GENERAL GUIDE LINES FOR PREPARING COLOUR PLANS FOR ROOMS

Most homes have at least a few good features which can be accentuated by skillful decorating. And almost every home has potential strong points which can be brought out by colour emphasis.

A cramped-looking room facing the Sun can be helped by painting atleast one wall in a cool blue or green. This area will seem to 'back away', giving the impression of spaciousness. But care should be taken to avoid the use of cold colours in a room on the shaded side of the house – otherwise it would be just adding to the iceberg effect. It is far better to use pastel tones of a warmer colour. The white in the pastel shades will give the desired illusion of space, while the warm colour will retain its warmth and help correct the room's coldness. Therefore, in a room not receiving much light, pure warm colours can create a sunnier feeling.

Long, narrow rooms can be given a wider look by painting the shorter walls in warm, advancing colour – tints of yellow, red or orange. Small, boxlike rooms can be made to appear larger by reversing this idea. Similarly, oppressively low ceilings can be 'raised' by painting, them with soft, receding colours, and high ceilings 'brought down' by giving them a bold, advancing colour. The badly – proportioned rooms can be corrected by using striped wall papering. Vertical stripes add height to a short room by carrying the eye upwards, unconsciously beyond the limits of the wall. This may over-correct, though, by making your ceiling seem 'miles away'. In this case painting the ceiling in a shade of a warmer colour would lower it to the right height. The purer this colour, the lower it will seem (Figure 6.7 – see colour plate).

Large picture windows should be framed with a warm colour; in winter, a blue window frame would give the impression of framing the glass with ice. As armies have discovered, the best way to be inconspicuous is to dress in the colour of the background they will be patrolling (shades of green in jungles, soft khaki in deserts). If there is a room with many doors which breakup the walls into small, ugly sections, camouflage is the best answer for them, too. Paint the doors, frames and all, precisely the same shade as the surrounding walls.

Equal tones will camouflage, even where the colours are different. Tones of the same 'weight' will neutralize bad features in a room, whereas a blatant contrast in tone (or intensity) will highlight. Any colour mixed with grey can be used in camouflaging because of its muting, receding effect. Seldom used doors – to store or cellos, say – can be hidden by painting them a deep grey. Alternatively, a shade of main colour, but deeper and greyer, will help fade them away. Under-stairs cupboards should, as a general rule, be made to disappear.

Dark colour will help disguise paneling which is broken up by small doors, or board's which are not well finished. Similarly, the margins beside a stair carpet are best painted in a 'retreating' tone chosen from the colours in the carpet. The following points can help a person to organise and decide the colour schemes for their rooms:

1. A safe scheme consists of tints and shades of one colour.
2. Definite schemes, such as complementary, adjacent, or trial are recommended.
3. An easy scheme consists of white or off-white plus one or two clear colours.
4. Three colours and their variations in value are sufficient for any scheme.
5. A scheme should be definitely dark or light.
6. Either warm or cool colours should predominate.
7. A colour scheme is often begun with a tertiary colour.
8. A neutralized colour is generally best for large areas.
9. A more definite colour is suitable for medium areas.
10. One or two brighter colours sometimes complementary to the dominant colour are often used in small areas for accents.
11. If a scheme lacks sparkle, a brilliant contrasting colour note or white may be added.
12. Equal areas of different colours are monotonous.
13. Every colour scheme should have a dominating colour and a secondary colour.

So, to summarize the general rule; use bright, vivid colours to pick out any surface that you like and that projects forward into the room and use deeper hues on surfaces you want to hide or which are recessed. This magnifies the natural light-and-shade effect of both daylight and artificial light. The projection catches the light, and recesses are naturally in shadow anyway.

TONAL AND CHROMATIC DISTRIBUTION

In developing a colour scheme for an interior space, one must consider carefully the chromatic and tonal key to be established and the distribution of the colours. The scheme must not only satisfy the purpose and use of the space but also take into account its architectural character.

Decisions must be made regarding the major planes of an interior space and how colour might be used to modify their apparent size, shape, scale and distance. Traditionally, the largest surfaces of a room, its floors, walls, and ceiling, have the most neutralized values. Against this background, secondary elements such as large pieces of furniture or area rugs can have greater chromatic intensity. Finally, accent pieces, accessories and other small-scale elements can have the strongest chroma for balance and to create interest.

Neutralized colour schemes are the most flexible. For a more dramatic effect, the main areas of a room can be given the more intense values while secondary elements have lesser intensity. Large areas of intense colour should be used with caution, particularly in a small room. They reduce apparent distance and can be visually demanding.

Of equal importance to chromatic distribution is tonal distribution, the pattern of lights and darks in a space. It is generally best to use varying amounts of light and dark values with a range of middle values to serve as transitional tones. Avoid using equal amounts of light and dark unless a fragmented effect is desired.

Typically, large areas of light value are offset by smaller areas of medium and dark values. The use of light values is particularly appropriate when the efficient use of available light is important. Dark colour schemes can absorb much of the light within a space, resulting in a significant loss of illumination. Another way of distributing values is to follow the pattern of nature. In this tonal sequence, the floor plane has the darkest value, surrounding wall are in the middle to light range, and the ceiling above is fairly light. Of course, the distribution of values and their degree of contrast will depend on the size, shape and scale of the space. Since light values tend to recede while dark values advance, their placement can modify our perception of these spatial dimensions.

Texture

Texture is often the last aspect that is considered when decorating a room; yet, its importance should not be overlooked, especially where pattern is absent. Texture not only adds variety to plain furnishings, it also modifies the effect of colour. While rough or matt textures absorb light and appear to add warmth, shiny ones reflect it and create a cooler atmosphere. A colour that's light-absorbing, like olive green, may seem black when used for carpet or velvet upholstery if light is limited, while a colour that's reflective, like white, may look bright when used on shiny surfaces like ceramic tiles.

Pattern

The selection of colours used in pattern on wallpaper, fabric or carpet and their tones with other furnishings extends the theme for complete coordination, linking rooms by the use of colour. For example, the apricot of the living room is repeated in the blue and apricot frieze in the hall and the blue in the hall is taken up by the cream and light blue in the kitchen. Tints rather than bold cold colours in large areas should be used where one wants to reflect light. Large patterns have the same effect as advancing colours; they reduce the impression of space. In contrast, abstract striped designs in related colours and delicate sprig patterns on a white ground may, like receding colours, increase the sense of space. The size of the pattern is not the only important factor, because a large pattern in pale, receding colours may blend into the background more than one with small, multi-coloured motifs which will tend to look busy. In addition, a design will stand out more on a flat surface (like wall paper or carpet) than on a gathered one like curtains, which is why coordinating ranges usually save bold designs for curtains and provide neater patterns for wallpaper.

Space

Colour is an essential ingredient in the creation of a scheme because it is possible to play decorating 'tricks'. Much of the most common problem these days is that of trying to make a room look larger, or of giving a spacious feel to a cluttered area. The cool colours of blue, green, lilac or grey are the receding ones and so this is the group to choose from if one wants to create a feeling of space.

Colours vary in intensity too, so a pale, dull (or subtle) colour will be much less dominant than a clear, bright one; so choose the paler, more subtle colours as opposed to the strong, vivid ones to create an illusion of increased space.

Beauty

The most beautiful colour schemes are those which give a single impression, an impression of warmth with, perhaps, a note of coolness for variation, or of coolness with an accent of warmth. It is particularly desirable to follow this order when one is planning colours for large rooms. For example, while preparing a colour scheme with blues, blue-greens and greens, it is desirable to bring in a little of orange and red-orange to make it more interesting. One should visualize a complete colour scheme for the room before finalising and executing them. Every piece of a room – walls, floor, wood work, curtains, accessories like pottery, pictures boxes, etc. should be considered for their contribution in creating beautiful colour combinations.

Keyed colours

A combination of several colours is said to be keyed when each colour has something in common with every other colour. Keying of colour is the secret of creating successful colour effect. Colours may be keyed to each other in the following ways:
- By neutralizing them
- By mixing them to introduce a colour in common
- By glazing, veiling, or topping them
- By tying them together by means of a neutral colour
- By the use of a rough texture

Since all the hues are present in a neutralized colour, it is apparent that all neutralized colours are keyed, because they have something in common. Many colours can, therefore, be combined, if they are somewhat neutralized. The mixture of a common colour with the two rather widely separated colours can also help to key a colour. Yellow and blue are brought together by the use of green, and green and purple are keyed by combining blue with them.

One colour placed over the top of a group of colours will key them. In painting, this method is called glazing, where a golden-coloured varnish or a flat wash of one colour over a picture, will unify all the colours in the composition. In dress, this result is achieved by placing a transparent or semi-transparent fabric over several colours. When this method is used in dyeing, it is called topping. A piece of multi-coloured embroidered material is dipped in a beautiful colour dye bath to give a keyed colour effect.

Colours that are not entirely agreeable, may be tied together or harmonized by the use of a neutral tone between the two colours. In addition to the neutrals, black, white and grey, an inter mixture of silver and gold can serve to tie the colour. It is a well-known fact that tan will bring the colours together, i.e. partly neutralized green and purple have a tendency to harmonize a group of colours when combined with them. A rough texture will have a tendency to key colours because of the variations in light and shadow over its surface. For example, bright colour prints on a terry cloth would appear softer than the same colours on glazed tile, for, the rough weave of the terry would tend to melt the colours together.

Thus colours combined in various ways can create wonderful effects. By using various means, as mentioned earlier one can balance the colours, harmonize them and please the eyes of the creator as well as that of the viewer.

Chapter 7

LIGHTS AND LIGHTING

L ight is the primary requisite of life. This statement is the foundation of the basic interest that man has for the phenomenon of light. It is not without reason that man is continually fascinated by the phenomenon of light. Without light, life on this earth would be scarcely possible. Light is the prime animator of interior space. Without light, there would be no visible form, colour or texture, nor any visible enclosure of interior space. Lighting a space is an important element in determining the beauty and comfort of the home. The first function of a lighting design, therefore, is to illuminate the forms and space of an interior environment, and allow its users to undertake activities and perform tasks with appropriate speed, accuracy and comfort.

Until well into the nineteenth century, one could speak of the continuous battle against the limitations that darkness imposed on human nature. Engineers have devoted a tremendous amount of effort in exploiting the aesthetics as well as the functional possibilities of artificial lighting. The development of artificial light through the ages led to the use of torches, oil lamps, candles and then the gas light. The invention of the incandescent lamp in 1879 by Thomas Alva Edison marked the final victory in the battle against darkness.

Light is an art element as well as an utilitarian element. Good lighting design has an elusive quality. When we walk into a room which is effectively lighted, our eyes sense only the things that are easily visible. This is because, our eyes do not see the light itself, of course, but only the objects on which it shines. Good lighting can also increase efficiency, relieve eye strain, cut down on accidents and help to set the mood of the room. Thus, light serves as a silent partner in enhancing our surroundings. Moreover, lighting is a form of decoration and creates an atmosphere of cheer and relaxation for the family. On festive occasions like marriage and Diwali and on national celebrations like Independence Day and Republic Day etc., lighting is used to create gaiety and excitement, reverence and dignity. However, the industrial light advancement and the residential lighting have not kept the same pace with each other. People generally decorate the house/ apartment first and lighting comes only as an after thought.

Light has a definite emotional effect. Light is stimulating, darkness is depressing. A sunny shiny day makes us sparkle, and a dark day makes us dull. On the other hand, light that is too brilliant exhausts us physically and aesthetically, and is as offensive aesthetically as loud noise. In our homes, we therefore, should have available, but under control, all the lights that we can use. The shadow element too is an important decorative factor both indoors and outdoors. Architects use shadows for emphasis in the front door, in over hanging caves, under porch roofs, and on uneven surfaces such as stone/brick. In interiors, shadows assist decoratively, serving as a foil for light.

SOURCES OF LIGHT

Natural lighting is when we harness the light given out by the natural elements in the environment such as that of sun and moon.

Natural light

Sun is the first and the oldest source of light known to us. Its radiation not only permits us to see, but also allows the life supporting processes to take place. Sight is one of our senses dependent solely on an outside agency i.e., light for its functioning. Natural light may be provided by wall windows, ceiling windows or roof windows. Natural lighting (Figure 7.1 – see colour plate 15) gives us a warm and cozy room and extends a personal feeling. Rooms that are illuminated by sunlight have vitality about them. Areas used early in the day, such as kitchens, breakfast areas and bedrooms, that face the east are warmed up by the awakening rays of the morning sun. The adequacy of this light would seem to be determined by the number of windows and skylights.

Glass fixed on a window, however, may absorb or reflect as much as 35% of the light that falls on it. Although the thickness of the glass does not seem to have much effect, it is the smoothness that does. A rough surface absorbs radiant energy more readily than a smooth one. Much light is absorbed by dirt on the glass and thus it must be kept clean. Glass helps in diffusing the light as it distributes the light and provides more illumination at the farther side of the room.

The use of shades and draperies also influence the amount of daylight that is available in a room. It has been found that drawing the roller shade so that the upper half of the window is covered, cuts off sixty percent of the daylight, but only fourteen percent is lost when the shade is restricted to the upper fifth. By contrast, the light transmitted through the lower panes is reflected from the ground, and surrounding areas which may absorb more light than what they reflect.

Opaque shades may cut out 98 percent of the light while translucent shades transmit 18 percent. Curtains and draperies may cut off as much as 75 percent of the light. Removing the valence may double the available day light at the farthest side of the room. Many heavy draperies and curtains increase the shadows and tend to produce a spotty condition that may result in glare, whereas a clear window and very thin curtain materials diffuse the light and soften the shadows. If draperies are desired for purposes of decoration, the windows must be more in number and larger to make up for the loss in light. A window area equivalent to one-fourth of the floor area is desirable.

It is important to know how much natural light enters the room, and for how long a period during the day. In sunny rooms with a southern exposure, cool colours will be more pleasing than warm ones. Hot yellows and reds will produce unpleasant glare during a large part of the day and are more effectively used in rooms that face north and need the warming effect of sunny colours during winter times. Darkening the room by covering the windows does nothing for it, except that it makes all the colours in it to look dark and gloomy; so the window treatment one uses will affect the amount and quality of natural light in a room and therefore the colours. Therefore, it can be said that windows provide ventilation, light and decoration and thus satisfy three principal needs.

Westside exposures can create an uncomfortable glare and heat in some parts of the country. Unless such lighting is carefully controlled by shading devices such as overhangs or blinds, problems arise from the harshness of the direct light of sun. Landscaping with deciduous trees may help solve the problem of heat and glare. A deciduous trees has its full foliage during the summer and sheds it during the winter, thus providing the house with shade during the hotter months of the year and allowing the warming rays of the sun to reach the windows during the colder months, resulting in savings on the heating bill. Conversely, coniferous trees on north side can block some of the piercing northern winds in the colder months of the year.

Thus, daylight is such an important factor in the appearance of a room that no plan of decoration should be made without considering the exposure, number of windows, amount of sunshine that enters the room, trees and vines that shut out light and what season of the year the room is used most. As much daylight as is desired can now be procured by means of glass walls and large windows extending from floor to ceiling. Besides, movable curtains should be provided, so that the quantity of light can easily be controlled to fit the needs of the occupants of the room.

Artificial Light

Artificial light is produced by manufactured elements. The quantity and quality of light produced differs according to the type of lamp used.

Artificial light had an important influence in the growth of civilization. The greatest single improvement in lighting was effected by *electricity,* since it has reduced simultaneously the labour, dirt, heat and the danger of fire that were connected with the other means of illumination. Modern artificial light, particularly electric light is not only a remarkable functional utility but also a marvelous flexible art medium. It should be used in a room as an artist who uses light in a picture. Artificial light may produce unity by its diffusion through the entire room. It may show contrast and emphasis by bringing bright light to important areas while subordinate areas remain in the shadows. It can bring rhythm and continuity to a room's furnishings by linking together the various points of emphasis.

Sources of Artificial Light

There are two common sources of artificial light – the incandescent bulb and the flourescent tube. Light bulbs and tubes can be grouped in general categories according to the way they produce light. They are installed in ceiling fixtures, floor and table lamps, and structural light designs that direct the light to where it is to be used.

Incandescent light source

The incandescent light is the kind used most frequently in our homes. The light here is *produced by a tungsten thread that burns slowly inside a glass bulb.* Due to the manner in which light is produced this source also produces high quantities of heat. The variety of incandescent bulbs available has been increasing at a rapid rate for some years and continues to do so. It includes colours from most of the spectrum, but has a large proportion of yellow and red. When dimmed, incandescent light becomes even more towards red colour.

It is possible to know the wattage, average light output, life, and efficiency of these bulbs from the wrapper in which they are available. They are available in a variety of rated lives. For instance, the 60-watt bulb and under, have the rate life of approximately 1000 hours, the 75-watt and over are generally rated at 750 hours. Some manufactures use different shapes, finishes and names to identify longer life ratings. It is possible to purchase bulbs that provide three levels of brightness. The size varies from three watts to 10,000 watts; light bulbs are available in colours such as pink, yellow, blue and green. All types of bulbs are made in a wide variety of shapes and finishes, including frosted coatings that soften and diffuse the light as well as minimize the glare. The most familiar incandescent bulb is the standard frosted one that gives a warm light, much warmer than the light produced by the most familiar fluorescent tubes. The finish on the incandescent light bulb is important to the quality of light. White-light sources may be finished with inside frost and soft white. The inside frost lamp gives off an identical quantity of light to that of the clear lamp; however, the soft white finish absorbs approximately two percent of the light that normally is generated within the source.

Another finish deals with colour. Colour is produced with incandescent lamp bulbs through the application of enamel or ceramics on the bulb wall. These finishes produce certain colours by absorbing all colours other than the one desired and therefore transmit the selected colour. Coloured light becomes less efficient in terms of light output but this should not be a basic concern as these sources are more appropriate for decorative or festive purposes. Some coloured sources are selected for very specific purposes. The yellow bulb is useful for controlling insects. Night-flying insects are nearly blind to yellow light and are not attracted to this light as they are to other colours and white light.

Bulbs are the most commonly used item for transmiting light. Reflector, Parabolic Aluminised Reflector (PAR) and Ellipsodal Reflector (ER) bulbs produce a more controlled beam. 'Silver bowl' types diffuse light. The bulbs have an advantage over the other type of light source especially fluorescent tubes in that, bulbs of different wattage can be used in the same socket. Because of the convenient size and shapes of the bulbs, incandescent lighting allows more flexibility than fluorescent lighting. Incandescent bulbs are generally less expensive to install, they light immediately when activated and produce more light from a small source than fluorescent light.

However, it takes more energy to use incandescent lighting than fluorescent lighting. They have a low efficiency rating. Only about 12 percent of the wattage used goes toward the production of light; the remainder is heat. They also have comparatively a short life.

Fluorescent Tubes

Fluorescent light is produced when electrical energy and mercury vapour creates an arc that stimulates the phosphorous coating on the inside of the tube. Because the light comes evenly from the entire surface of the tube, it spreads in all directions, creating a steady, shadowless light. Tubes require a ballast to ignite and maintain the electrical flow called a choke. The fluorescent light has more than 200 colours of fluorescent tubes available in the market. Generally speaking, they are low in red and high in green and blue light waves.

Fluorescent tubes are unrivalled for energy efficiency. They also last for longer period than incandescent bulbs. U-shaped fluorescent tubes now available in the market can be used in smaller, trimmer fixtures. One of the unique advantage of fluorescent sources is their linear quality, which is handy in lighting some architectural elements such as valances, coves, cornices and brackets and for installations along the length of a surface such as kitchen cabinets and shelves.

The fluorescent installation is more complicated and hence more costly. However, the tubes last longer, are cooler, require fewer fixtures, produce less glare, and use from one-third to one fifth less electricity than incandescent light bulbs. This makes up for the initial installation costs.

Other sources of artificial light

Apart from the incandescent and fluorescent electric light sources, there are other sources of light as mentioned below:

• Quartz Halogen

Quartz halogen bulbs contain a tiny quartz filament that produces a brighter, whiter beam than other sources. They are excellent for task lighting, pinpoint accenting and other dramatic effects. One major disadvantage of halogen is that it is very hot and requires only special halogen fixtures, beside its initial cost. They are popular for commercial display and museum lighting, as well as for some specific residential uses.

• High-intensity discharge (HID)

The high-intensity discharge bulbs produce light when electricity excites specific gases in pressurised bulbs. Requiring special fixtures and ballasts, these lights may take upto seven minutes to ignite after being switched on. But they offer long life, high efficiency and brighter light. Some examples are mercury vapour, metal halide and high-pressure sodium vapour bulbs.

• Neon-Light

It is generated when electricity passes through neon gas. It is technically known as cold-cathode. It is most commonly used for sign making. The greatest advantage of neon light is that it can be shaped into any form. It also has a long life and presents few maintenance problems. Neon lights could be used to provide extra sparkle and excitement in a room. Recently artists have used neon lights to create light sculpture. However, the low light output makes them undesirable as a functional light source and are expensive.

TYPES OF LIGHTINGS

Lighting of houses must be serviceable and should also be a decorative one. From the point of view of service they provide, three types of lighting are necessary, namely, general, task and accent lighting—

I. General/Ambient lighting

The lights which are used generally to light up the room are known as ambient or general lighting like the wall lights, chandeliers etc. General illumination by general lighting describes the total light available in any room. It must be sufficient to ensure safety of movement and should be of high level for simple tasks. In other words, it fills in the undefined areas of a room with a soft level of light – say, enough to watch TV or to navigate safely through the room. The general lighting should be provided in various intensities, so that pressing different buttons should produce bright, medium or subdued light, or dim light for watching television. Ambient lighting usually comes from indirect fixtures that provide a diffused spread of illumination. Directional fixtures can also be aimed at a wall to provide a wash of soft light. General lighting may emanate from troughs at the top of the walls or above doors or windows, from ceiling or wall fixtures, from ground glass panels flush with ceiling or walls or from indirect lamp. Cove lighting or other architectural lighting is a great improvement over fixtures as it is inconspicuous and conforms to the lines of the architecture.

II. Task lighting

It illuminates a particular area where a visual activity takes place. It is often achieved with individual fixtures that direct light onto a work surface. The location of task light is self-evident though its characters will differ according to the task in hand. For example, a good reading light may also be suitable for darning but not for sewing, it might serve for sketching but not for painting. Task lights offer high intensity lighting without high general illumination. Their inclusion in a lighting scheme provides the required practical lighting and leaves one free to vary the general and accent lighting according to the atmosphere desired (Figure 7.2 – see colour plate 15).

Task light is produced in particular places, usually by portable floor and table lamps, but also by straight or curved lighted rods and by lights behind ground glass that is flush with the wall. Lamps are necessary for reading, writing, working and card games. Every easy chair should be near a lamp; a desk, piano or table should have its own adjustable lamp. A major design factor, this type of lighting enables one to create moods, to emphasize important objects, and to bring the visual delights of variety and rhythm. Task lighting also helps to define a field of concentration; a bright light on a desk surface, for example, tends to fix our attention on work.

III. Accent lighting

Accent light chiefly fills an aesthetic need: a spotlight dramatizes or highlights an art object. Accent sources give local rather than general illumination. Their uses are many and varied, chiefly to make visual focal points of centres such as coffee table or dining area or special objects in the room. They are supplementary to the general light sources. The effect of accent lighting is instantly appreciated when one enters a room with much sparkling light.

Traditional accent sources are standard picture lights, the occasional wall bracket, strip lamps over bookshelves. Even a simple table lamp can provide accent if used independently of the general lighting scheme. Downlighters, uplighters and wall washers can also become accent lights.

Most accent sources are directional in their construction so that their output can be pointed exactly where it is needed. The range of such sources encompasses semi-recessed eyeball fittings, surface and track mounted spot lights, together with fully recessed directional fittings. While choosing a spotlight, it is essential to assess the exact characteristics of each fitting and specially the beam angle. For best results, it is wise to try to restrict the beam within the outline of the object to be featured rather than allowing too much light to be spilled over the edges. To accent a picture or any wall mounted feature, the general rule is to position the light between the viewer and the wall. The fitting must be placed at a suitable distance from its beam to cover the picture. The effective means of placing an accent light within a home would be approximately at a distance of 80-100 cm or 31-39 inches from the wall. Many other types of accent sources also exist such as special lighting for shelves and niches and uplighters for plants.

FACTORS AFFECTING ILLUMINATION

Light provides radiant energy. It radiates equally in all directions and spreads over a larger area as it emanates from its source. As it spreads, it also diminishes in intensity according to the square of its distance from the source. As it moves, light reveals to our eyes the surfaces and forms of objects in space. An object in its path will reflect, absorb, or allow the light striking the surface to pass through.

The sun, moon, stars and electric lamps are used by us because of the light they generate.

Most of what we see, however, is visible because of the light that is reflected from the surfaces of objects. Our ability to see well i.e. to discover shape, colour, and texture, and to differentiate one object from another, is affected not only by the amount of light available for illumination, but also by the following factors:

- brightness
- contrast
- glare
- diffusion
- colour

Brightness

Brightness refers to how much light energy is reflected by a surface. The degree of brightness of an object is, in turn, depends on the colour value and the texture of its surface. A shiny light-coloured surface will reflect more light than a dark matted or rough-textured surface, even though both surfaces are lit with the same amount of illumination. Generally speaking, visual activity increases with object brightness. Of equal importance is the relative brightness between the object being viewed and its surroundings. To discover its shape, form and texture, some degree of contrast to brightness ratio is required. For example, a white object on an equally bright white background would be difficult to see, as would a dark object seen against a dark background.

Contrast

Contrast between an object and its background is especially critical for visual tasks that require the discrimination of shape and contour. An obvious example of this need for contrast is the printed page where dark letters can best be read when printed on a white paper.

For performing tasks requiring discrimination of surface texture and details, less contrast between the surface and its background is desirable because our eyes adjust automatically to the average brightness of a scene. Someone seen against a brightly illuminated background would be silhoutted well, but it would be difficult to discern that person's facial features.

The surface brightness of a task area should be the same as its background or be just a bit brighter. A maximum brightness ratio of 3:1 between the task surface and its background is generally recommended. Between the task area and the darkest part of the surrounding room, the brightness ratio should not exceed the ratio of 5:1. Higher brightness ratio can lead to glare and associated problems of eye fatigue and loss in visual performance.

Glare

Even though our eyes prefer even lighting, particularly between a task surface and its background, our eyes are able to adopt to a wide range of background levels. We can respond to a minimum brightness ratio of 2:1 as well as to a maximum of 100:1 or more, but only over a period of time. Our eyes cannot respond immediately to extreme changes in lighting levels. Once our eyes have adjusted to a certain lighting level, any significant increase in brightness can lead to glare, eyestrain and impairment of visual performance.

Glare is the transmission or reflection of light in such a way that the eye gets an unpleasant sensation and may even result in serious health problems. Lighting can produce two kinds of glare, discomfort and disability glare. Discomfort glare may be caused by an over bright source or too much contrast between an object and its background and can result in eyestrain and headache. For example, when reading a book under a lamp in an otherwise dark room, the reflection from the white page in contrast with its surroundings may make reading uncomfortable. Similar is the case when we are driving at night and a car with a bright light approaches us. Disability glare can make seeing difficult or impossible especially when an object is placed too near a bright light. For example, when an unshielded bulb is hung low over a work bench, the eye adapts to the brightness of the bulb but cannot see the object on the bench distinctly.

Glare also results when a bright light falls on a polished or glossy surface and is reflected from that surface to the eye. The shiny covers of books and magazines, glossy table tops or a mirror often cause glare.

To avoid glare, following considerations may be kept in mind:

- Do not use unshielded lamps, and shade the lamp to screen it from the eye.
- Do not place lamps and objects to be viewed too close together.
- Provide a proper degree of illumination for the space to be seen.
- Make use of frosted bulbs, opaque glass shades which diffuse and soften the light.

Diffusion

Diffuseness is a measure of a light's direction and dispersion as it emanates from its source. The quality of light affects both the visual atmosphere of a room and the appearance of objects within it. A broad source of light such as a luminous ceiling produces diffused illumination that is flat, fairly uniform and generally glare-free. The soft light provided minimizes contrast and shadows, and can make the reading of surface textures difficult.

On the other hand, a point source of light such as an incandescent bulb produces a directional light with little diffusion. Directional lighting enhances our perception of shape, form and surface texture by producing shadows and brightness variations on the objects it illuminates.

While diffused lighting is useful for general vision, it can be monotonous. Some directional lighting can help relieve their dullness by providing visual accents, introducing brightness variations, and brightening task surfaces. A mix of both diffused and directional lighting is often desirable and beneficial, especially when a variety of tasks are to be performed in a room.

Colour

Another important quality of light is its colour and how it affects the colouration of objects and surfaces in a room. While we assume most light to be white, the spectral distribution of light varies according to the nature of its source. The most evenly balanced white light is noon daylight. But, in the early morning hours, daylight can range from purple to red. As the day progresses, it will cycle through a range of oranges and yellows to blue-white at noon, and then back again through the oranges and reds of sunlight.

The spectral distribution of artificial light sources varies with the type of lamp. For example, an incandescent bulb produces a yellow-white light while a cool-white fluorescent produces a blue-white light. The apparent colour of a surface is a result of its reflection of its predominant hue and its absorption of the other colours of the light illuminating it. The spectral distribution of a light source is important because if certain wavelengths of colour are missing, then those colours cannot be reflected and will appear to be missing or greyed in any surface illuminated by that light.

The interaction of light and colour is a subtle but powerful force in planning lighting for an interior. Fabric in a store can seem different in colour under fluorescent lighting, as compared to the incandescent lighting at home or to natural light. Thus, it is important to test a colour under proper lighting in the home. Colours containing a lot of white reflect a larger amount of light and darker colours absorb light. A white objects reflects 80% of light whereas black objects reflects only 5% or less. A room with dark coloured walls would require more light sources and higher wattage bulbs to get reflective light levels. The illumination in a room with light coloured walls is distributed farther and more evenly as the light is reflected from surface to surface, until it gradually diminishes.

Colour rendition

The colour of an object as we perceive it, is determined by two things, the surface colour of the object and the colour contained in the light shining on it. The colour of a blue vase under a blue light will be heightened as the colour of the light intensifies the colour of the vase. Under a red light, the same blue vase will appear dull and greyish, because the red light waves are absorbed and there are no blue waves to be reflected by the vase. This interaction between an object and light source is called colour rendition, and this may influence the colour of the objects.

Benefits of effective lighting in the house

Light as an element of design should be used for visual comfort and to achieve desirable emotional responses. Emotional responses are influenced by the manner in which light is used to communicate ideas about colour, texture, shape, form, and line. Structural aspects of the interior may be accented as light is used to interpret visual elements that define space, to denote which surfaces shall be lighted and which ones shall remain dark, and to convey means by which patterns of brightness may be merged with the structural patterns of the house design. The nature of a space is greatly dependent on distribution and patterns of illumination. Light, used as an integral part of the total environmental design, can give the house an atmosphere in which people may respond in the most favourable way to other members of the family and friends.

To understand further the benefits of effective lighting in the home, there is a need for an understanding of how people respond to light. Human seeing responses improve as lighting levels are increased. These responses bear directly on human performance and productivity. Mental and physical responses are slower and less precise when lighting is not suited to the seeing tasks being performed. As seeing conditions are improved, the following responses are more positive; seeing becomes more reliable and takes less time; texture, colour, and fine details are seen more clearly; human energy is conserved; and productive seeing is prolonged Figure 7.2 – see colour plate).

Eyes and light work together to provide humans with sight – through which approximately 85% of the responses to the environment are experienced. Each aspect of human growth, development, and performance may be influenced by the luminous aspect of the environment. More specifically, physical, psychological, psychophysical, and aesthetic responses are associated with this environmental factor. The physiological aspects of human response to light are the structure of the eye, perception, factors involved in the seeing task, recommended levels for seeing for task performance, neural and muscular reactions to environmental stimuli, and energy. Each factor involves an understanding of human physical response to light in the environment.

The elements of seeing are the eye, the source of light, and the object to be seen. Each element must be given careful consideration in terms of the part to be played in effective seeing. The age, health, and activity of the person may well determine the quality of the response of the eye to the environment. The source of light will influence the quantity of light available for seeing and the colour of the light. The nature of the object or surface to be seen combined with the colour of the light source will determine the colour to be seen, and the other factors related to perception of the object or the surface to be seen.

Human psychological responses are related to the quality and quantity of light in the seeing environment. The specific responses related to the lighting in the environment are stimulation, comprehension, motivation, and distraction. Psychophysical responses related to the lighting are visibility, visual performance, and evaluation of visual responses in terms of physical characteristics.

Many aesthetic aspects of the environment are influenced by the quantity and quality of light. Among those of significance are spatial effects. Bright attitudes and emotional moods are influenced by the relationship between the sources of illumination, the surfaces and objects to be seen, and the backgrounds against which they are viewed.

HEALTH AND LIGHTING

Sight is the only one of our senses dependent on an outside agency, light, for its functioning. Light, whether natural or artificial, must be adequate if eyesight is to be protected. Poor illumination will necessitate a still greater effort to have a clear picture of the things one sees.

Illumination is defined as the amount of light falling on a surface from a source of light. This is what actually determines comfort in seeing. The source and intensity of light must be adapted to requirements. When there is an adequate source of light, one can work with ease. There is less tension, and the heart beat is closer to normal. There is less blinking or staring, as the general sensitivity of the visual sense is greater. Improper lighting puts a strain on eyes and nerves. If such conditions are prolonged, serious damage to eyesight and to health in general, occurs. Nearsightedness, specially in children, may have been caused by reading material close to the eyes due to inadequate light. A small, concentrated source of light, like a very small reading lamp, produces sharp shadows, and harsh effects. It thus causes eye strain and indirectly, mental and nervous strain. Thus adequate lighting is very important not only from the point of seeing the objects clearly, but also to protect the general health of a person in many ways.

Acute eye discomfort is associated with an excessive feeling of fatigue, often ascribed to other causes but actually caused when the eye strains hard to see. Other symptoms are aches, redness, burning and scratchiness of the eye. Old people need more light than young people, since the size of the pupil decreases and the eye tissues become less translucent with age.

Eyes become defective through abuse. The pupils gradually dilate under work requiring continuous and great visual effort. Eye strain resulting from the attempt of the eye to adapt itself to unfavorable conditions may not always reveal itself in eye irritability. Among school children, lack of interest, failure to concentrate, sleepiness and apparent laziness may all be caused by eye-strain. The objective in home lighting is, therefore, to provide adequate light for all the tasks that have to be performed in the home. Lighting is also a form of protection or security, for prowlers are wary of entering illuminated premises.

Decorative aspect of Lighting

Lighting is one of the basic requirements in decorating a room. It is as important as the various objects and the colour scheme. It can be said that the quantity and the quality of light in any space

PLATE 15

Figure 7.1. Natural light not only brightens a room, but also creates interesting designs shadow

Figure 7.2. Local (task) lighting in dining area

PLATE 16

Figure 7.3. Movable Light Fixtures

Figure 7.4. Speciality Lamps

determines our essential perception of these elements. The quality of light itself as well as the appearance and placement of lighting fixtures determines the use of lighting system as a decorating tool. Along with the functional purpose of light, aesthetic considerations, need to be weighed to make lighting as an integrated whole.

Establishing character of room with lighting

Most people react happily to sunshine and take another sort of pleasure in moonlight but respond with depression to an underlighted room. On the other hand, highly dramatic lighting created by theatrical designers has limited adaptability to residential purposes. Lighting can evoke any number of moods, from hectic gaiety to funeral gloom, even to terror in horror shows. Therefore, one can establish character of the room with lighting. For a warm, serene character, use of shaded lamps is recommended but for liveliness, unshaded incandescent bulbs and brightly illuminated white walls may provide the effect. The choice of furnishings, accessories and the subsequent choice of lighting all together, modify the character of a room. Thus, a large number of items alongwith lighting in a room determine its character.

Lighting for colours of the room

In the absence of light, there is darkness and therefore, no colour is made visible. Thus colour and light are closely related. Incandescent lighting has a yellowish tint whereas the fluorescent tubes generally impart a bluish cast. Both the intensity of artificial light and its source, whether incandescent or fluorescent tend to distort colours. When one needs to select some fabric of a specific colour, one should try it under natural light as well as in home lighting conditions since most shops use bright commercial fluorescent lighting as a rule.

Lighting fixtures

The lighting fixtures in a room must relate in scale, colour, and texture to the room's furnishings. An imaginative combination of various types of lighting fixtures can add zest to a room. The many styles, shapes, their placement, etc. are the various factors that can enhance the appeal of the decorative element of lighting and are discussed later in this chapter.

It can be said that lighting recognizes and fulfills following needs:

- function
- safety, and
- beauty

A good guide to assess the amount of light needed is to allow 20 watts per square metre of floor space for filament lamps and 10 watts per square metre for fluorescent lamps. For example, for a room 13.5 sq.m, the total lamp wattage for general lighting should be atleast 270 watts for filament lamps or 135 watts for fluorescent lamps.

More light is however, needed for activities such as reading or sewing. It is found that a 60 watt bulb is likely to be suitable for a bedside lamp, a 150 watt bulb for a floor standard lamp, and a 100 watt bulb for a table lamp. The older people need more light than younger ones. But the amount of light that falls on a surface depends not only on the power of the bulb, but also on the degree of reflection from the surface and its distance from the bulb, as well as the way the light is distributed by the fitting. The best way to find out how much direct light is needed is to experiment with bulbs of various wattages and then come to a final conclusion.

Measurement of light

Several methods of measuring light are practised. In one system the standard unit is a foot-candle, which is the illumination on a surface one-foot away from, and perpendicular to, the rays of a standard-size candle. Ten thousand foot-candles of light occur in sunshine; about one thousand in the shade of a tree on a sunny day. About 5 foot-candles of light fall on a book directly underneath a 40-watt lamp. About 50, ten foot candles are considered suitable for occupations like kitchen-work/card playing, 50 - 70 foot-candles for close work like studying, newspaper reading/sewing, and 50 to 100 foot-candles for close work over a long period of time.

Table 7.1: Recommended Footcandles for residential activities

Activities	Recommended Minimum Footcandles
• Kitchen Activities	
– Sink	70
– Range and work surfaces	50
• Laundry Activities	
– Trays, ironing board, ironer	50
• Reading at Any Location	
– Prolonged periods	
Small type-low contrast	70
Large type-high contrast	50
– Casual periods	
Small type-low contrast	50
Large type-high contrast	30
• Sewing	
– Dark fabric (fine detail, low contrast)	200
– Prolonged periods (light to medium fabric)	100
– Occasional periods (light fabric)	50
– Occasional periods (coarse thread, large stitches, high contrast thread to fabric)	30
• Facial Grooming-Shaving-Make-up	50
• Study-Concentrated and prolonged	70
• Casual letter writing	30
• Table Games	30
• Work Shop	
– Bench work	70
• General illumination*	
– Entrances, hallways, stairways, stair landings	10
– Living room, dining room, family room, bedroom, common room	10
– Library, game or recreation room	10
– Kitchen, laundry, bathroom	30

* *Note:* General lighting in residential areas need not be uniform in character.

Understanding and using electricity

Electricity flows along wires in the same way as the water flows in a pipe – except that electricity cannot flow until a circuit is completed. The amount of electricity flowing along the wire is measured in amperes (A). The pressure that pushes electricity along the wires is measured in volts (V). When an appliance is switched on, the amount of power consumed, or watts (W), is calculated by multiplying the volts by the amperes.

Every appliance has atleast two wires; one live (brown) carrying the electricity in, and one neutral (blue), returning to the circuit. In addition, there is usually a third wire – the earth wire (green and yellow). This is a safety device which carries electricity harmlessly to earth if there is any failure in the appliance. The previous colour-coding used was live (red), neutral (black) and earth (green). Old appliances still in use may be wired in these colours.

The wattage used by an appliance gives a guide to the running costs – 1,000 watts used for one hour is one kilowatt-hour (kwh) – the unit by which electricity is measured on the electric meter. The price of electricity varies from region to region and there may be different kinds of off-peak tariff rates. The electric meter may be the digital or the dial meter. To check consumption, the previous reading is subtracted from the new reading.

Separate circuits of cable distribute power from the consumer's unit to the various appliances, via power points or socket outlets but power is carried from plug to appliance through flexible cable – called flex. Flex must be connected correctly, otherwise an appliance or circuit may be damaged. The flex must also be the correct type and the amperage to match the appliance. It is better not to extend flex by taking on an extra length, which is unsafe. A qualified electrician can fit a new length of flex to the appliance without causing any damage to the appliance.

Light Switches

Dolly switches – flicked up and down – are most commonly used in the home. Pull cord switches are the only type of switch that can be used in a bathroom to comply with electrical safety regulations and are cheaper to install than wall switches as they do not need wiring down the wall to a switch plate. Wall-mounted push-button switches need only a very light touch to operate, and are more expensive than dolly switches but as they can be operated with any part of the body, they are useful for handicapped people too.

Dimmer switches, to vary the amount of light given out by a lamp, can be used, for example, in a child's bedroom to produce the right level of illumination. Since they decrease the power consumption, they can save appreciable amount on the cost of electricity too.

Door switches that turn on the light when the door is opened and switch off when the door is closed can be used for cupboards or refrigerators. Automatic time switches can be used for outside lighting, night circuits, or as security lighting to deter intruders; they can be set to turn lights on and off at certain hours. Photo-electric switches are sensitive to daylight, and turn the lights on at dusk and off at daylight, can also be used for automatic switching, but are expensive.

The modern markets are providing a kind of switch called rotary switches, specially meant for the use of electrical fans. These switches have a rotary system which alters the speed of a fan by rotating the switch to the desired level. A rotary switch helps to dispense with a speed regulator for the fan. The switch, thus, plays the role of a speed regulator too for the fans.

PLANNING FOR LIGHTING

It is worth noting that a specific level of illumination can be supplied by various combinations of luminaries. The choice of what type of luminaries are used and how they are laid out should be based not only on visibility requirements but also on the nature of the space being illuminated and the activities of the users. The lighting design should address not only the quantity of light required but also its quality. The layout of luminaries and pattern of light they radiate, should be coordinated with the architectural features of a space and the pattern of its use. Since our eyes seek the brightest objects and the strongest tonal contrasts in their field of vision, this coordination is particularly important in the planning of task lighting.

Planning the lighting system of the house, whether in a new house or in a remodelled one need to satisfy all essential functional and aesthetic requirements resulting in efficient lighting for safety, comfort and beauty of the house. The most important consideration is the people who will use it – their habits, hobbies, work methods and style of living. A mental or written catalogue should be made of various requirements for general, local and accent lighting based on all the activities that take place in a room.

Whether one wants restful pools of light or more dramatic effects need to be resolved at the beginning because that would determine the choice of various types of lighting fixtures resulting in a particular character of the room. A floor plan indicating all pieces of furniture, areas of circulation and the lighting requirements, is very important in thorough planning for lighting of the house. If need be, one may also solicit professional help from architects, interior designers and lighting consultants.

For the purpose of planning the visual composition of a lighting design, a light source can be considered to have the form of a point, a line, a plane or a volume. If the light source is shielded from view, then the form of its light and the shape of its surface illumination should be considered. Whether the pattern of light sources is regular or varied, a lighting design should be balanced in its composition, provide an appropriate sense of rhythm, and give emphasis to what is important.

Art Principles in lighting designs

The use of lighting is visually pleasing when it follows the principles of design. As principles of design are mutually reinforcing, it is difficult to speak of one without considering the other.

Balance is achieved by placing light sources throughout the room, avoiding a concentration of light on any side or area of the space. Balance can be symmetrical or asymmetrical, depending upon the kind of shade used in the lighting system. **Harmony or Unity** can be created by duplicating the fixture as well as by repeating the materials, finishes, colours, relationships or lampshades. Using similar, but not identical, fixtures will create variety. If the fixtures are the same throughout the room, it will be monotonous, but if they are totally different, it will be chaotic. To avoid monotony and to incorporate variety, in lighting systems, both general and local lighting can be used. Changing the level of illumination with dimmers can also provide variety in lighting. If the contrast between accent lighting and general lighting is too strong, there will not be enough luminous transition, and the lighting composition will lack harmony.

As the entire space is illuminated, there will be a need to highlight certain objects or areas, thus creating **emphasis**. Lack of emphasis in an interior causes boredom, confusion and uncertainty as to the design intent. Accent lights can be used on the points where one would like to focus or to create emphasis. The sequence of lighting will create a luminous **rhythm**, giving the space a dynamic quality. The light fixtures should be in **proportion** and scale to the room and to the related objects in the room. In the case of table lamps, both the holder and the shape should be in proportion to each other.

SIX CONSIDERATIONS IN CHOOSING LIGHT

Imagine that you are setting out to plan lighting for your room. The considerations would be:

1. *Space and Atmosphere:* Objectively assess the space to be lit. Is it a kitchen or a bedroom, a bathroom or a corridor? The garden patio or dining room? How is the space to be used? What type of *ambience* is the most appropriate?

2. *Form and Design:* What are your plans for the room? What are the main pieces of furniture and where will they be sited? What will the colour scheme be? Will there be carpets or bare floorboards, wallpaper or paint, curtains or blinds?

3. *Focus and Detail:* Which are the main features to be highlighted in the room? These might be architectural or decorative, and could include cornices, fireplaces, niches, fine plasterwork or panelling, paintings, sculptures, furniture, bay windows or plants. You might equally consider what would be best left unlit.

4. *Tasks:* What *activities* to be performed in the room will require special lighting consideration? Where will you read, sew, chop vegetables, shave or make up?

5. *Practicalities:* Are there any restrictions to the installation of new wiring? For example, do the walls or ceilings consists of concrete, or is there access to the ceiling from the floor above?

6. *Control:* Finally *technical details* must be considered. Where do you want your control? How many switches and power points do you need in the room? Do you want dimmers, and how many different circuits are needed?

REQUIREMENTS OF AN IDEAL LIGHTING INSTALLATION

1. Steadiness of the source of light

There should be no appreciable fluctuation or flickering of light which overstrains the eye so that the source of light remains stable.

2. Elimination of glare

Glare of the vision is likely to occur if gas or incandescent electric lamps are used. It may be remedied by placing the source of light high above the level of the eye, so that it is not ordinarily seen, and screening the light by means of a suitable shade, or interposing frosted or opaque glass silk, celluloid, etc., to diffuse and soften the light, so that glare is eliminated.

3. Avoidance of shadows

Inconvenient shadows can be avoided by proper shading of the source of light, using light colours on walls and ceilings which reflect and diffuse light in all directions, and providing a general mild light to illuminate the entire room, and one or more stronger lights, in addition, in proper places for specific purposes such as reading, sewing, etc. All these will avoid shadows so that the objects are seen clearly in light.

4. Sufficient illumination to suit the nature of the visual task

For the comfort of the eye and efficiency of the particular visual task, proper degree of illumination is required. Thus, reading requires more light than playing cards, and for sewing more light is required than for reading. These requirements are already discussed earlier in this chapter.

5. Non-production of excessive heat

Production of excessive heat is a great disadvantage in a tropical country like India, particularly in the summer. Therefore the light source should be properly shaded whether by natural or artificial means.

6. Minimum consumption of oxygen from the air

Except the incandescent electric light, every other source necessarily consumes oxygen from the air. This factor needs to be kept in mind while choosing the kind of light for a room.

Electric light (except the arc-light which is now rarely used) is the best, as with proper shades and reflectors, it fulfils all the above conditions.

LIGHTING FOR VARIOUS RESIDENTIAL AREAS

The amount of light required, varies according to the needs or the functions of a room. In a residential area, the need for lighting and the amount of lighting for a bed room will be different from the main entrance. These differences for the various areas are discussed in detail in the following paragraphs.

Entrance and Passage Lighting

With good front door and porch lighting, the door itself, the keyhole, the ground immediately in front of door and the steps or step leading upto it will clearly be visible. It is a safety measure too, to identify the unexpected callers or intruders by illuminating the front entrance brightly.

One can create a welcoming atmosphere by brightly lighting the entrance hall. One can use ceiling lights to highlight a painting or sculpture with spotlights. The entrance hallway should be well lighted by a ceiling fixture and other supplementary light as needed. A hall may have a direct or indirect ceiling light that can give brilliant, medium or subdued light as required. A pair of table lamps or a floor lamp may constitute a part of a group of furniture in a hall.

Living Room Lighting

A living room is exactly what it says. It is a place to be lived in than a work room, study or bedroom. And living covers a multitude of different moods and occupations, both social and solitary. The main criterion of any living room is that it should be a relaxing, inviting space, flexible for all that goes on, but nevertheless with a definite style.

One's living room like most others, is probably the beehive of household activity – reading, watching television, talking and entertaining. The living room is the very heart of one's home, which is sufficient reason to light it well and requires a variety of lighting techniques. Precisely because of its different uses, lighting must be flexibile, combined with, aesthetics. A balance of all three types of lighting – general, task and accent – gives the desired effect. General and accent lighting should be decided first; task light can be added afterwards.

Living and family rooms need general illumination, preferably both direct and indirect, to bring walls and furniture, floors and ceilings into soft visibility. In addition, some scintillating light adds to the animation. Each group of furniture especially in a large living room should have its own lighting subsystem, which gives the room a degree of visual organization. Direct light is a requisite where people read or sew, play games, or do homework.

When entering the living room at night, general orientation lighting must be present. A single luminaire from the ceiling will provide the ambient lighting needed for this purpose. Accent lighting is particularly important as it may impress a visitor by directing his attention to a favourite painting. It gives the finishing touch and needs to be planned as carefully as general lighting.

The lighting in one's living room would not be completed without some highlights on paintings or on the crystal in one's cabinet. Apart from its stylish appeal, accent lighting also provides a certain amount of ambient light. Spotlights meet the need very well when fitted with incandescent/halogen reflector lamps.

Dining room lighting

The dining table is the spot where the family gathers to enjoy a meal together. Dining spaces deserve primary emphasis on what is most important – the table, the people around it, and the food that is going to be served. Not only should the table itself and the dishes on it receive direct lighting but the faces of the diners should be included as well. In many families, a first-class row can spring up over the question of dining room lights. One faction demands no-nonsense visibility whereas the other craves romantic candlelight. Both options, of course, have some justification. If a fish is on the menu, mere safety requires a fairly high level of illumination. Most people agree, however, that the evening meal, at least, should be a restful and special occasion which calls for soft lights.

A ceiling light fixture is sometimes the most practical solution for lighting a dining room. The use of low-level general illumination will create a serene atmosphere. Bright overhead lighting tends to have too harsh a character of a pleasant dining-room-lighting. Under no circumstances should coloured bulbs be used in the dining room. Their effect on the appearance of food can be, literally, nauseating. A luminaire suspended above the table is a good solution. Chandeliers should shed a soft light and should hang fairly low over the table. However, keep the brightness low in the direction of the faces. A dimmer control allows lighting to be adjusted to varying conditions and to family preference, which can help to establish the moods for the room. The positioning of the luminaire is very important. If too high, there is the risk of seeing the lamp. If too low, the luminaire will obstruct eye contact. The correct height is usually about 60 cm. from the table top.

Because most dining rooms have fixtures that hang over the table, the bottom of the fixture should be approximately 30 inches (or approx. 76.20 cm) above the table if the room has an 8 feet ceiling. In rooms with higher ceilings, the fixture will need to be raised for better proportion. Chandeliers of 21 to 29 inches (or approx. 54cm to 74cm) in diameter should be chosen unless the dining area is less than ten feet wide, in which case the diameter should be less than 24 inches (or approx. 60.96cm). These guidelines need to be flexible, however, if the chandelier is massive in design and colour, it may appear heavier than its actual measurement.

Ceiling lamps are available that can be moved along an installed track; this allows a change in the position of the dining table. Ceiling lamps/fixtures used should be well designed and inconspicuous. However, other lighting should also be used in the room; since a single dimmed source of light will not provide adequate lighting for dining, and one light source shining down on those sitting at the table can be unflattering. The light could come from fluorescent tubes concealed in coves along the top of the walls/window frames/from one or more ground glass panels set flush in the ceiling. Special lighting effects can be obtained from two indirect floor lamps or from two lighted urns on the buffet. A hidden spot light can be placed to throw extra light on the table. In lieu of a light fixture hanging over the table, built-in recessed lighting may be used to frame the table with incandescent light and bring out the sparkle of silver, china and glassware.

Candlelight is pleasant for the dining table because of its softness to the diners and to the table appointments. Its flickering, uneven quality gives it additional interest and suggests a friendly, festive, and mysterious mood. At least six candles are needed on a small table; they should be dripless. White candles always look well with white dishes. However, more dramatic colour effects are also possible. Tall candelabra and tall candles are necessary to keep the light above the eye level.

Bedroom Lighting

Every bedroom, since it needs to answer only to one or two people, has a personality of its own. The amount of light acceptable for sleeping will, of course, differ from person to person. Those who need complete darkness for sleeping, or who dislike morning light, will want dark window shades or thick draperies. Those who are apprehensive in the dark/disoriented on waking will want some kind of night-light. For any number of reasons, bedroom lighting must be planned with more attention to the needs of a specific user than lighting for general living areas. Bedrooms merit light for the multiplicity of purposes served by the bedroom. Apart from sleeping, the bed room introduces a multiplicity of lighting requirements such as for reading, either in bed or in an armchair, dressing in front of a long mirror, studying or writing letters, eating at a breakfast table or off a sick room bay; viewing television, sewing etc.

In a large bedroom, general lighting can best be provided by a centrally located lamp with a luminaire that spreads the light around comfortably. As in the other rooms, general lighting is needed in a bedroom too for orientation purposes and in order to find things in cupboards. White fluorescent lamps are best suited for general lighting. This ambient lighting should be switched off when reading in bed with the aid of a reading light installed at the bed head.

The dressing table, for any women needs at least two small lamps, one on each side of the mirror and should be 36 inches apart. They should have translucent shades and should carry a 100-watt bulb. The centre of the shades should be at the height of the face. An excellent dressing-table light is one that hangs over the table not far above the head of the seated person. The important point in placing the mirror light is to be sure that the light falls on the person rather than on the mirror. An overhead fluorescent fixture is also satisfactory if the light is properly diffused. Deep shadows must be avoided, however, and colour distortion must be accounted for. The dressing table may be placed directly in front of a window, to make use of the natural light.

Lamps for reading may be on stands or attached to the wall or ceiling if space is at a premium. High-intensity lighting will allow reading without disturbing the partner; however, this is not recommended, for, a glare is created if the rest of the room is dark. If two people use the room, each should have his own reading lamp with individual switch. Reading in bed is an indispensable luxury for many, but too often basic lighting requirements are ignored. The lamp should provide a good spread of light, and the bottom of the shade should be no higher than the reader's eyes. The recommended height for a table lamp is roughly 20 inches (or approx. 51 cm) above the mattress, that for a wall lamp hung in back of the reader about 30 inches (approx. 76.20 cm) above the mattress. In at least one bedroom in the house there should be a special master switch to light the whole house, in case of an emergency.

Child's room Lighting

If space permits, it is a good idea to furnish a child's bedroom with a well-lighted desk so that he will have a private and quiet place to do his homework. As children's eagerness to explore extends over all objects and areas within their reach, it is best that their room is brightly and uniformly lit with a general lighting. Long, harsh shadows as well as glaring, unscreened lamps should be avoided. Similarly, narrow-beam spotlights, which are often used to provide decorative effects in other areas of the home, are not recommended in the children's room. Children like a lot of light, although they may need less than their parents. Dim lights, often considered relaxing by older people, might be frightening for young children. Most of the young children like some kind of night lamp in the room.

Study lamp should be carefully selected. A good study lamp should have a diffusing bowl and translucent shade to provide good shielding from direct glare. It should carry a 200-watt bulb. If homework involves a great deal of drawing, a pair of lamps, one on each side of the working surface, will provide more satisfactory shadowless light. If fluorescent lighting is built in above the desk, the working surface should receive at least 70 foot-candles; this level is produced by two white 40-watt tubes. A study desk should not be placed immediately in front of a window where sunlight might produce too much glare. However, the study table placed in close proximity to a window, can provide good natural light while working on the table.

Lighting for study

A study is by definition, a work space. The priority is thus to ensure that adequate illumination is provided for reading and writing.

Practical illumination for a study can be provided in one or two ways. Either the entire space is lit to a very high level, with halogen uplighters/an array of powerful downlighters or the general lighting may be kept to a lower level similar to that for a living room and a number of task lights are used in addition.

A basic scheme for an average-sized study, say around 3 m or 10 ft. Square is to use a single freestanding halogen uplighter with a 300 or 500 watt bulb, plus one or two task lights. If fitted with a dimmer, the uplighter can provide strong overall illumination for daytime activities on overcast days and softer glow for late night reading. In a room where books are dominant with an expanse of bookshelves across one wall a more appropriate effect could be wall washing again supplemented with task lights.

Task lighting for writing should be positioned so as to avoid both glare as well as a shadow. An ideal position should be a little to the left for a right-handed person and a little to the right for a left handed person as the hand will not cast a shadow over the page. The general lighting should also be arranged to cast light evenly over the entire working surface. The same principles also apply to natural light. Thus a desk should not be placed immediately in front of or under a window where sunlight will stream in and produce too much glare.

Workroom Lighting

The lighting needs of a studio, home office or other workroom depend on the work for which they are used, from drawing and painting to needlework or carpentry, often requiring a higher level of an overall illumination. Halogen uplighters can be ideal, provided the ceiling is over 2.5 m or 8 ft high. Should this not be so, even light distribution will be difficult to achieve from uplighters. Where the ceiling is too low for uplighting, ceiling-mounted halogen wall-washers could provide a similar degree of indirect light but with a vertical emphasis. Ceiling mounted fluorescent sources would also be effective, and less expensive to run. Fluorescent battens now exist in many attractive casings, and most are fitted with carefully designed anti-glare lourdes or diffusers.

Light for sewing

Women who do a great deal of sewing or needlework, should remember that the recommended lighting level for these activities is extremely high. It is suggested 100 foot-candles for prolonged sewing of any kind and 200 foot-candles for work on dark fabrics.

Lighting for an artist's room

Lighting is any artist's first observation when he plans a studio. For some, techniques like fine ink drawing, or etching, there is no real substitute for daylight. Windows should be large and located high up on the wall; they need curtains or blinds for fine light control. Painters also often prefer daylight because it does not distort colour. The fabled 'north light' is not mere artistic preference but because this exposure never admits the sun directly, light is less variable in colour. Since paintings today, however, are usually exhibited under artificial light, most artists no longer insist on natural light for working. They generally prefer fluorescent light to incandescent containing both cool white and warm white tubes for an acceptable balance of colour. Fluorescent lamp must be carefully maintained since prolonged work under flickering tubes can produce serious eye fatigue.

Bathroom Light

A bathroom lighting should have safety and utility requirements. Bathroom is commonly a small room and its floors will often be wet, electric cords and electric fixtures should be out of the way. It is desirable to have both general and task lighting in the bathrooms. Overall bathroom lighting is best provided by a ceiling fixture. Both incandescent and fluorescent lamps are satisfactory for lighting the bathroom mirror. A single 100 watt incandescent fixture or a fluorescent tube of 40 watts for a fixture above the mirror is recommended. A vapourproof fixture is needed if lighting is provided over a bathtub or shower stall. Under no circumstances should a light switch or outlet be located inside the shower.

Lighting the Laundry Room

The lighting in the laundry room should have at least one source of strong light to facilitate spot removal. Good overall illumination should be provided for ironing.

Lighting the recreation room

A recreation room needs a strong ceiling light for a ping-pong table or billiards. For cards and board games, a bridge lamp or pull-down fixture focused directly on the playing surface will, in most cases be more pleasing than general overhead lighting.

Lighting the strairways

The lighting of the staircases in one's home is most important in order to prevent accidents. Stairways should always be well-lighted, especially at the head and foot. Avoid the mistake of lighting the steps from below, since a person descending the stairs will not be able to properly pick out the edge of each step. If on the other hand, the fixture is positioned above and somewhat in advance of the topmost tread, shadows will be created and the edges of the steps clearly defined. When the fixture does not light all of the steps adequately, additional lights should be provided on the edges of the steps or by means of a post light. Recessed lighting may be secured to the risers on the stairs and give a feeling of elegance to the room while adding to the safety of the stairs.

Lighting the garage

Extra lighting in the garage can transform this space into a part-time workshop. At the least, an outlet should be provided for an occasional light to be used for minor repairs to automobile or garden machinery.

Outdoor and Garden Lighting

A lighted garden can be arranged for viewing from inside the house, or for effect when viewed from outside or for garden parties. Moreover, an extra advantage of lighting the house and the garden is that it also discourages trespassing. The entrance and the house number need to be clearly illuminated and both the visitor and the host should be able to see each other in good light. Weatherproof fixtures mounted on exterior walls or over sunshades provide typical solutions. Use of spot and floodlights concealed in the landscape can be made for more elaborate installation.

LIGHTING FIXTURES

Light fixtures are integral parts of a building's electrical system, transforming energy into usable illumination. Light fixtures normally require an electrical connection or power supply, a housing assembly and a lamp. While choosing a lighting fixture, the concern is not only with its shape and form but also with the form of the illumination it provides. Point sources give focus to a place since the area of greatest brightness in a space tends to attract our attention. They can be used to highlight an area or an object of interest. A number of point sources can be arranged to describe rhythm and sequence. Small point sources, when grouped, can provide glitter and sparkle.

Linear sources can be used to give direction, emphasize the edges of planes or outline an area. A parallel series of linear sources can form a plane of illumination which is effective for the general diffused illumination of an area. Volumetric sources are point sources expanded by the use of translucent materials into spheres, globes, or other three-dimensional forms.

TYPES OF FIXTURES

There are mainly two types of light fixtures, namely, movable and surface-mounted fixtures.

1. Movable light fixtures

Table lamps, floor lamps, and small speciality lamps are easy to buy, easy to take along when one moves, and are the movable light fixtures (Figure 7.3 – see colour plate 16).

Table lamps show individuality and style, at the same time they serve as sources of light. Variety, mobility, and ease of installation add to the appeal of such lamps. The choice of lampshade is very crucial. Diameter of the lower edge dictates the spread of light below. The height of the bulb within the shade also affects the circle of illumination.

Floor lamps offer great flexibility. The traditional floor lamp provides a continuation of levels, serving either as a reading light or as a source of soft ambient light.

Pharmacy lamps offer options for tasks such as reading and sewing. Lamps with adjustable arms provides greater range and flexibility. Bright Torches available in halogen and incandescent versions, bounce light onto the ceiling for a dramatic form or indirect lighting specially for high ceiling rooms.

Speciality lamps in new varieties are constantly appearing in the market. These lamps can fill a definite need while remaining movable and requires standard wiring. A lamp designed for an artist's table can be called a speciality lamp Fig. 7.4 – see colour plate 16).

Table 7.2: Types of Light Fixtures

Movable	Surface mounted
• Table lamps	• Ceiling and wall fixtures
• Floor lamps	• Chandeliers and pendant fixtures
• Pharmacy lamps	• Mini & strip lights
• Speciality lamps	• Track lighting
• Recessed ceiling fixtures	• Built-in-indirect lighting

2. Surface-mounted fixtures

Installed either on walls or on ceilings, surface-mounted fixtures are integral to most home lighting designs. They are also a common sight in all show rooms in the market.

Ceiling and wall fixtures provide general illumination in traffic areas such as landings, entries and hall ways where safety is a consideration. Kitchen, bathrooms and workshops benefit from the added light from ceiling fixtures used in conjunction with task lighting on work surfaces.

Chandeliers and pendant fixtures add sparkle and style in high ceiling entries and above dining and game tables. These decorative fixtures can give direct or diffused light or a combination of the two. If a fixture is used over a table, the width should be at least 12 inches (or approx. 30 cm) less than the width of the table to prevent collisions with diners or passers-by. Hanging it about 30 inches (or approx. 76 cm) above the table surface helps to avoid glare.

Mini-lights and strip-lights are partly for fun and partly for effective task lighting. They add a splash of light and colour and may be used to provide very good effect in highlighting high windows and other architectural features.

Track Lighting offers great versatility of installation. Available in varying lengths, tracks are extended electrical lines from the outlets they hold. Fixtures can be mounted any where along the line. A track can be 'lush-mounted' or suspended on ceilings and walls. Tracks come in one-and-multi-circuit and they allow one to extend systems in a straight-line angle, or even in a rectangle. Tracks can accommodate pendant fixtures, clip on low spot lights, as well as large lamps.

Recessed Ceiling Fixtures are used to provide illumination in downlight offices without the intrusion of a visible fixture. For this reason, they are effective in rooms with low ceilings and sleek lines. Recessed fixtures can be added in existing areas provided there is enough space between the ceiling and the roof above.

Built-in Indirect lighting are the ones like coves, cornices, valances, brackets and soffits which can be used to provide indirect light. Simple and in architectural designs, these devices and light sources are shielded from view, allowing spill-out around the shields. Coves direct light upward, while cornices spead light used over windows. Valances spread lights both upto and down the draperies. Wall brackets, mounted direct into interior walls, spread light, both up and down and can be used to highlight artwork or to provide ambient light in living rooms. Soffits, used over work areas, throw a stronger light below.

LIGHTING FIXTURES PROVIDING DIFFERENT TYPES OF LIGHTING

Lighting fixtures are used to provide different types of lighting. Earlier in this chapter various types of lighting were discussed. Though the type of lighting required is different, it is the kind of fixture or the shades of the lamps that provide these. The different kinds of fixtures provide different types of lighting.

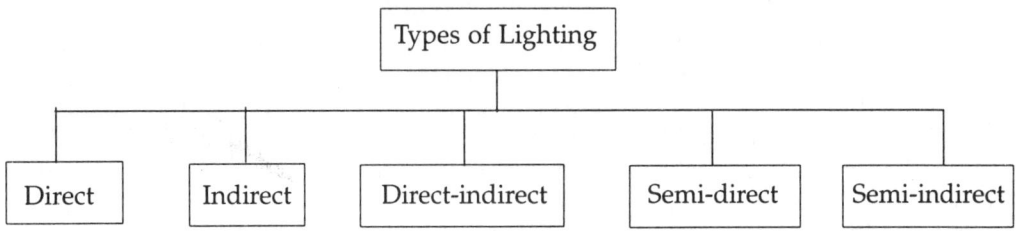

I. Direct Lighting

The greatest quantity of light is obtained by a direct fixture. It concentrates the light at the desired surface and a minimum amount of lighting is lost in transmission. The quality, however, is inferior, for the space covered by the light is comparatively small and forms an objectionable contrast with the darker surroundings so that glare is produced, unless with reflector bulb which makes the most dramatic shadows. Gooseneck lamp and track lighting are examples of direct lighting.

(i) *'Down-lighters'* are fittings designed to direct light straight downwards, the effect is to evenly illuminate the horizontal surface below. Wall-mounted downlighters are available for use over counters and sideboards. The downlighter fitting itself is only a means of housing the bulb, and/or in most cases providing glare control.

Moreover, in-built downligher reflectors are usually designed carefully, often by a computer, to make the most efficient use of the bulbs which restrict glare and produce a defined beam of light. Reflectors are available with either silver or gold finishes, silver leaving the colour of the light unchanged, gold producing a warmer coloured output. The gold effect is particularly suitable for interiors with a brown or beige colour scheme; the yellow light intensifies the depth of these colours and helps to create a very warm and inviting atmosphere.

One single downlighter placed centrally over a coffee table or a work centre, will provide a very effective local illumination but will not illuminate an entire room. For general lighting, it is necessary to have an array of downlighters across the ceiling. The number and spacing of the array will depend on the beam angle of the selected fitting and also on how much surface area it will illuminate. Downlighting may be used to highlight a group of plants, a piece of sculpture, as well as to provide light for reading and other tasks if a selector bulb is used. A downlight can be placed where the chandelier is ordinarily hung, thus illuminating the table or the arrangement on the table.

(ii) *Wall Washers:* Because track lighting does not require elaborate electrical wiring, this form of direct lighting may be used to throw light on an empty textured wall. This process is called wall washing. Wall-washers are ceiling-mounted fittings designed to cast light evenly across and down a wall. They place emphasis on the vertical surface rather than the floor or the ceiling. They can also contribute to the general lighting of the space by using the wall as a reflector. The paler the colour of the reflecting surface the more light is reflected into the room. However, its use is limited because of the following reasons. Firstly, the effect requires a large stretch of clear wall space in order to produce adequate general lighting on its own. Secondly, by emphasizing a wall, wallwashing may make it seem to advance. This is inadvisable to use in small rooms.

The wall-washers are similar in construction to downlighters, with both a reflector bulb or a GLS bulb and a separate reflector within a curved housing. Often a broadly curved 'scoop' reflector is also added to ensure that the light reaches the very top of the wall. Mostly all wall-washers are ceiling mounted. Majority of the wall-washers take 60, 100 or 150 watt bulbs.

Positioning is critical in this case, since the spacing between one wall washer and the next must ensure even distribution of light. If they are too far or towards the centre of the ceiling, there is a risk that people may cross into the path of the beam, being dazzled and creating shadows at the same time. On an average, the fittings should be placed 80-100 cm from the wall and the space between each fitting and the next should not exceed twice the distance from the wall. Wall washers which take linear tungsten-halogen bulbs lit up double the width of the wall than the GLS/reflector bulb fittings; as a result, only half as many fittings are required.

In all, there are three types of wall-washers. They are:

(a) Track-mounted 'scoop' wall-washer

(b) Recessed cone wall-washer

(c) Track-mounted halogen wall-washer.

Track lighting can also be used to spotlight a collection of pictures or for task lighting if a reflector bulb is used. If track lighting is not mounted correctly, the light may cause disturbing shadows or glare. To achieve the best lighting effect, the track should be mounted on a specific distance from the wall for the following wall heights:

Ceiling height*	Track distance from wall*
7½ - 9 feet	24 - 26 inches
9 - 11 feet	36 - 48 inches
11 - 13 feet	48 - 60 inches

* 1 inch = 2.54 cm. (approx.) and 12" = 1 foot

II. Indirect light

Indirect lighting is produced by a light source that is hidden. The light is directed to a ceiling, to a cove, or to another surface from which it is reflected back into the room. Indirect lighting creates almost no shadows when it is used next to the ceiling, and it is ideal for general illumination. However, when it is used alone, it can be flat and uninteresting. Indirect lighting reflected down from the ceiling tends to raise the ceiling height by creating a visual illusion. Totally indirect light loses most in transmission and absorption, on account of the length of path the rays traverse, but the light is highest in quality. The almost perfect diffusion practically eliminates shadows.

(i) *Uplighters:* Uplighting is form of indirect lighting. The purpose of uplighters, as their name suggests, is to direct the light upwards towards the ceiling. The effect is to create space and height at the same time providing general light reflected from the ceiling surface. Uplighters offer greater flexibility than down lighters in that the fitting of different bulb types and wattage are available for fixing almost anywhere: wall-mounted, suspended from the ceiling on a stem or freestanding.

In terms of the bulbs they use, uplighters for general lighting fall into two categories; those designed for the new tungsten-halogen and those using the more traditional GLS or tungsten reflector bulb. A tungsten-halogen bulb can be made much smaller, yet produce the same amount of light as a tungsten bulb of higher wattage. The output of a tungsten-halogen bulb is also much whiter than that of a tungsten bulb, extremely crisp and sharp, and giving very faithful colour rendering. For this reason halogen bulbs were first widely used in the galleries, museum and photographers' studios. The type of halogen bulb most often used in uplighters is the linear one which runs on normal mains voltage. The standard-wattage are 150, 200, 300 and 500 watts. Because of their extreme brightness it is better to fix a wall-mounted uplighters above eye level and avoid placing them near the foot of a staircase or on a wall below a flight of stairs where they may produce glare in the eyes of someone coming down.

Even more efficient of the tungsten-halogen are the metal halide bulbs. There is a growing need of uplighters in the offices rather than the homes. Besides, the blue output of metal halide sources and the fact that they cannot be dimmed somewhat restrict their decorative possibilities. Uplighters that take standard GLS and reflector bulbs tend to be far less powerful. With most such fittings just one or two would scarcely be enough to illuminate a ceiling.

Canister shapes are used in a variety of locations on the floor in the corners of a room, at either end of a sofa, under the sofa, or behind plants. Uplighting can be aimed at a piece of sculpture or a plant to create interesting shadows on the wall. The technique of uplighting can add a dramatic mood to the room as well as introducing softness to the room's rectilinear forms.

III. Direct-Indirect Lighting

Direct-indirect light uses both direct and indirect light distributed evenly in all directions. A fixture that has bulbs both inside and outside the reflector, as some table and floor lamps do, produce a direct-indirect diffused light.

IV. Semi-Direct Lighting

In semi-direct fixtures sixty per cent of the light is directed down to the work surface and the remaining amount is directed upward.

V. Semi-Indirect Lighting

Conversely, semi-indirect lighting directs 60-90% of the light towards the ceiling and upper walls, using the ceiling as the main reflective source, the other 40% to 10% is directed toward the work place. A lamp providing semi-direct lighting with a diffuser is preferred for reading and study.

Factors for Choosing lighting fixtures

Once the quality and quantity of light needed is worked out, fixtures are to be chosen that would provide exactly the type of lighting as planned. The various factors that influence their choice are as follows:

(1) *Function:* One of the primary considerations is how the fixture would direct the light. Will it direct the light to the area where you want it? Make sure that directional fixtures have enough maximum bulb-wattage to allow the use of bulbs that are strong enough to 'throw' the light from the fixtures to task or display areas.

(2) *Size:* Fixtures on display will often look smaller in the store than they will in the homes. Make sure the right size is chosen.

(3) *Design:* Here personal preference and taste would guide the choice. Manufactures offer 'families' of fixtures available as spotlights, pendants, track lights and ceiling fixtures. This offers a sense of decorative continuity by using similar fixtures.

(4) *Flexibility:* Because tastes, habits and technology often change, flexibility is one important consideration. Change in display artwork would need lighting adjustments. Movable or adjustable lamps are longtime favourites. With track systems, location of fixtures can be altered along the track as well as the way each fixture is aimed.

(5) *Cost:* Purchase price and operating costs are very important in fixture selections. When a fixture is to be burning for several hours at a stretch, it may be wise to invest in a more costly low-energy unit than a less expensive kilowatt-eater.

(6) *Maintenance:* To operate efficiently, all fixtures should be cleaned regularly. Consider using a fixture which is easy to clean and provides better accessibility towards changing light bulbs.

Balancing and Layering Light

One rule of efficient lighting is to put light where you want it. But to ensure an attractive, comfortable lighting scheme, you also need to think about balancing light, that is, creating an effective spread of dim and strong light throughout the room.

The key to balancing is in the way the light is layered. To do this one should first determine the focal point or point of the room (having two or three focal points is usually best). This is where the brightest layer of light is to be directed. Next, a middle layer is added to provide interest specific areas without detracting from the focal points. The last layer fills in the background.

The first two layers are usually done with task or accent lighting. The 'fill light is usually indirect. The ratio between the brightest light in the room and the "fill light" should be about 3 to 1, or at the most 5 to 1. Ratios of 10 or even 100 to 1 are great for creating high drama, but they are too uncomfortable for everyday living.

KINDS OF LIGHTING FIXTURES PROVIDED IN THE LIVING SPACES

Several methods may be used to provide illumination; frequently more than one type of lighting is used in the same room.

Lamps and Fixtures

The most commonly used sources of light are portable lamps and fixtures that are mounted on walls or ceilings. Lamps provide a certain amount of flexibility in that they can be moved from one place to another. Both lamps and mounted fixtures may be useful in emphasizing a particular decorative theme or in adding some special note of individuality. Fixed lighting makes it possible to conserve table and floor space and to light areas where portable lamps seen impractical. There are no rules about where to use lamps or mounted fixtures. The ultimate aim is to provide the type of light needed and to choose designs that are keeping with the decorative scheme of the room (Figure 7.4 – see colour plate).

Portable Lighting

Portable lighting is simply fixture or lamps that one can unplug and move to a new house or a different area of the room. While there is an infinite variety of portable fixtures in the market, the most common types are floor lamps, table or desk lamps, pole lamps, and swag lamps, as well as some types of wall lamps. In addition to providing a good local or accent light, a portable lamp is also a vital decorating accessory. So, when you are selecting one, be sure that it relates to the colour, texture, and style of the room's furnishings. Choose a design that reflects the mood of the room.

GUIDELINES FOR THE SELECTION OF A LAMP STYLE

While there are no strict rules about choosing lamp style and sizes, here are a few guidelines to assist you in your selection:

- Try to scale the lamp to the furniture around it. Do not flank a massive piece of furniture with dainty feminine lamps.
- Create contrast between the tabletop and the material of the lamp base. Contrast will set each off to its best advantage.
- Check the balance or weight of the fixture to ensure that it will not be tripped over easily. Look for lamps with weighted bases for children's bedrooms or play areas.
- Fragile floor lamps should not be used in heavy traffic areas.
- If you use any antique lamps, be sure that you place them in a safe location to avoid any accidents.

Style of Lamps

The lamps you choose need not conform to a single furniture period. Many of the popular lamp styles mingle well with one another, and an artful combination of periods can add zest to a room. Lamps with simple, classic bases – cylinders, columns, urns, blusters, and classic oriental porcelain designs – are available in a wide spectrum of colours and finishes to complement the various decors.

Traditional lamps take their inspiration from period candleholder and oil lamp designs. The traditional lamps you choose may be either authentic antiques or convincing reproductions. Designs that are appropriate for period or formal traditional rooms are eighteenth and nineteenth century – French, English, and American colonial lamps, Greek urns, and the classic Oriental shapes. Traditional European and American lamps are often ornate with rococo embellishments. These are made of materials that reflect the opulence of their heritage – crystal, silver, porcelain, pottery, bronze, and glass.

Informal country lamps, appropriate for using with French Provincial, Early American, and Spanish or Mediterranean decors, are much more rustic in flavor and design. Common materials for these lamps include pewter, copper, wood, wrought iron, brass, and handcrafted pottery. Finishes have a hand-rubbed look, and colours are apt to be the natural earth tones. The popular traditional contemporary fashion in lamps of Indian origin expresses a natural or country motifs. These designs emphasize the use of natural materials – wrought iron, pottery, rope, wood, and wicker. They are particularly attractive in rooms using similar materials for other furnishings.

Contemporary fixtures have taken on new and fascinating shapes. Fluid and free form, the designs are trimly tailored, sleek, and imaginative. Some resemble futuristic sculpture, others have exposed bulbs with no shade at all, and still others feature sound-activated switches. Improvements in the quality of molded translucent plastics have made possible fanciful new shapes in lamps. Other common materials in contemporary lamps are chrome, aluminum, and glass.

Lampshades

Lampshades vary widely in size, colour, and translucency. You should, of course, choose a size in scale with your lamp base and the general decorating scheme. Also decide what texture you want. A nubby texture may be perfect for a country or contemporary room, while a traditional room demands a smooth silk shade. Translucent shades yield more light than any other type. White or cream-coloured shades are the most translucent, but colours such as grey or beige also give optimum light. The inside surface of the shade should be either white or near white for effective reflection.

The relative translucency of the lampshade should vary, depending on its prescribed use. For example, at a bedroom dressing table, where good light is essential and eyestrain is not an issue, use highly translucent shades. On a desk or study table, however, where the reader's eyes will be about bulb level, pick a shade that is only slightly translucent.

Opaque shades are used chiefly as accents. Tole lamps, for example, with their decorative metal shades, are attractive in themselves and provide a pleasing light for highlighting tabletop accessories. But they are highly inadequate as reading lamps because the total field of light is too small. On the other hand, opaque shades are preferred for hanging lamps. Because hanging lamps are located above the illuminated surface, the light should be directed down, not dissipated near the ceiling.

Diffusers prevent direct glare from the bulbs, and as such, are vital for reading and study lamps. These are widely available in either clear glass or plastic. Shapes range from a simple bowl to the more elaborate prismatic diffusers, made as a series of concentric plastic cylinders. You also can obtain a large white bulb, classified as R-40, which is made of a diffusing glass and needs no additional diffusing. Paper lanterns are an economical, appealing solution to the lamp and lampshade problem. This compact, folded paper lamps are available in various geometric shapes – round, oval, and cylindrical. Most of them are made in Japan, although some of the more sophisticated expensive versions are from the Scandinavian countries. Handsomely used as permanent but replaceable fixtures, the lanterns are most appropriate in contemporary room settings.

The safety of these shades is covered by government regulations governing the sale of flammable fabrics. A very large paper globe can safely use a bulb of upto 100 watts, but do not equip any lantern or lamp with no more than its recommended wattage.

GUIDELINES FOR THE CHOICE OF LAMPS

In choosing lamps the following points might be considered:

1. A sturdy base or one that is heavily weighted at the bottom prevents tripping. With the tall slender bases currently in vogue, this is an important factor.
2. A lamp that has a diffusing bowl will give less glare.
3. A harp makes it possible to adjust the height of the shade or to tilt the shade if necessary.
4. A table lamp intended for reading purposes should have the lower edge of the shade about 40 to 42 inches above the floor. The lower edge of the shade on a floor lamp should be from 47 to 49 inches from the floor (1 inch = 2.54 cm approx.)
5. Lampshades should be similar in colour and texture or else they should contrast. One beige shade, one white shade, and one pink shade would probably be unattractive. On the other hand, two identical beige shades and one gold metallic-paper shade might provide an interesting combination.

6. For some areas, swing-arm or adjustable gooseneck lamps may be practical.

7. A floor lamp should be placed so that the light comes from behind the shoulder of the reader, near the rear of the chair, at either the right or the left side, but not from directly behind the chair.

In choosing fixtures, these points may be important.

1. Adjustability of position often increases functionalism. Chandeliers that may be raised or lowered and wall units that swing provide a variety of lighting effects.

2. Diffused light is more pleasant. In many fixtures the bulbs may be exposed and present an irritating glare.

3. The design of the fixture should be in harmony with the character of the room.

A FEW PRACTICAL HINTS ON ECONOMY IN LIGHTING

1. It pays in the long run to use a costlier wire than spending on decorative fittings. Besides, ornamental fittings are difficult to clean.

2. It is advisable to use lamps manufactured by a firm of long standing and repute, though they may cost more in the beginning. They will pay their extra cost in two or three months only, by effecting a saving in electrical energy.

3. Instead of using lamps of high wattage, it is possible to derive the same illumination with lamps of much lower wattage by (a) providing a scientifically designed reflector, (b) adjusting the distance or height of the lamp, (c) fixing them just at places of maximum advantage, and (d) keeping the reflector and the lamp free from dust and smoke.

4. The shades or reflectors should be cleaned from time to time with soap-water. Dust, if allowed to be deposited on them, absorbs light.

5. If a switch is found to get warm, it should be immediately replaced by a new one.

6. The flexible wires should be occasionally examined. If at any place the cover is worn out, exposing the copper wire inside, it should be replaced by a new piece. The flexible wires should be renewed every three or four years; by the action of the atmosphere they deteriorate. They are often the cause of break-out of fire.

Lighting is an important aspect of good living. The owner's taste is reflected through the kind and the amount of lighting provided in her home. More than this, the health and good eye sight are the key factors which are promoted by the right kind of lighting used for the various purposes in the home. A good knowledge and an awareness on the various factors related to lighting is a must to achieve the above mentioned goals.

Chapter 8

WALLS, WALL FINISHES AND CEILINGS

Just as grass and sky are the background for the subject in a landscape painting, or music in a film set the background mood of serenity or excitement, so are the background of walls, floors and ceilings to the furnishings of an interior. The basic functions of walls and ceilings are to provide protection and privacy. They determine spatial relationships within the home and they have a tremendous effect on heat, light, sound and odour in a room. In addition, they make a major aesthetic contribution. Their colour and texture become an integral part of each room, since they provide the background for the other additional decorations in the home.

Man's compulsions to decorate the walls that surround him asserted itself very early in his existence. The remarkable wall paintings and carvings that are seen in many architectural buildings, forts, caves, temples and the other monuments of India offer an idea on wall decorations that were practised in ancient India. As indicators of a given historical period, wall decorations can also be made as a reliable source as the styles of furniture or clothing. Brilliant tapestries, linenfold panelling, brocades, gilded or white plaster, applied ornamentation with paintings, clear and colourful glass and stones were all used to cover the walls and ceilings in different ways at different times in the past.

Most of the early wall treatments had functional as well as aesthetic purposes. They provided insulation in drafty, damp stone buildings. They also proved to be the status symbols as the elaborate decorations on the walls of the palaces and buildings occupied by the kings and ministers of the yester years. They were also the indicators of the ways these places were used, as can be observed in the case of temples as the places of worship, which are painted with the pictures of gods and goddesses. They were also the indicators of the skills of the craftsmen, painters and architects of those periods and their levels of prosperity. The more prosperous a kingdom was, more prevalent were the architectural marvels and wall decorations on their walls. These wall decorations are not only the indicators of the talents and skills of the people of the past, but are also the proof of the way the walls and ceilings were used to exhibit their ideas and followings.

148

ARCHITECTURAL ELEMENTS IN THE INTERIORS

While designing a room, one has to consider all the elements of the architecture of the interiors. A room encompasses three dimensions – wall, floor and ceiling, surfacing of which combine with their own colours, textures and forms to give the total effect as an unit. For all these reasons, it is essential to study each one of them in detail, and they are explained in detail in this chapter.

Walls

Walls are the essential architectural element of any building. They traditionally served as structural supports for floors above, ceilings and roofs. They form the facades of buildings. They provide protection and privacy for the interior spaces they create. Walls enclose space, establish a room's dimensions. Walls occupy the room's largest area and set its tone. They can be inconspicuous or dramatic, formal or casual depending on how they are treated. Walls are the primary elements with which the interior space is defined. Together with the floor and ceiling planes which complete the enclosures, walls govern the size and shape of a room. They can also be seen as barriers that limit our movement, separate one space from the next, and provide the occupants of a space with visual and acoustic privacy.

A distinction should be drawn between the wall itself and wall coverings, which are the materials attached to the wall. Contemporary designers are well aware of the 'honesty' of genuine materials and are concerned with maintaining it. Some wall materials, if properly used, need no covering at all. Many interior designers, for example, use concrete in its natural texture, showing the pattern left by the wooden frames used to pour the concrete or brick walls that reveal an interesting pattern in their natural state as a conscious expression of material.

The vertical orientation of the walls make them visually active in our field of vision. In defining the boundaries of a room, they give form and shape to the space and play a major role in determining its character. Stable, precise, symmetrical walls convey a feeling of formality, one which can be considerably enhanced with the use of smooth textures. Irregularly shaped walls, on the other hand, are more dynamic. When combined with a rough texture, they can import an informal character to a space.

Walls provide a background for a room's furnishings and its occupants. If smooth and neutral in colour, they serve as passive backdrop for foreground elements. When irregular in shape, or given texture, pattern or a vigorous colour, the walls become more active and compete for attention. Light coloured walls reflect the light effectively and serve as efficient backdrop for elements placed in front of them. Bright, warm colours on a wall exude warmth while white light, cool colours increase a room's spaciousness. Dark-coloured walls absorb light, make a room more difficult to illuminate, and convey an enclosed, intimate feeling. A wall's texture also affects the amount of light it will reflect or absorb. Smooth walls reflect more light than the textured ones, which tend to diffuse the light striking their surfaces. In a similar manner, smooth, hard wall surfaces will reflect more sound back into a space than porous or soft-textured walls.

Wall Coverings

Today many wall-coverings are available to be installed. These coverings are placed over the original wall surfaces. Different kinds of materials are at the disposal of a homemaker – prepasted surfaces for wallpapers, self-adhesive wall tiles, wooden planks with interlocking grooves, stick-quick plastic mouldings and modular ceiling tiles. These are available with easy and practical installation methods. There is a considerable standardization in the sizes and shapes of interior coverings for walls, ceilings and floors.

In applying colour, designs and materials to walls, one should always be aware that a large span of the room's surface is being used up. Wall coverings define the mood and become the principal setting against which we place all other objects in the room. According to design experts, the design of the room's walls should be the first consideration in planning for space. One should decide on the type of wall covering to be used according to how colourful, active, textural and functional the room is intended to be. Depending on the selection of the wall covering, the room can look larger/smaller, noisy/quiet, formal/informal, light/dark, cluttered/empty, festive/serious or bright/dull.

Choice of wall coverings

The choice of wall covering will depend upon the role the walls are likely to play in the overall decorating scheme of the room. Because of their extensive area and high visibility, walls constitute a most important decorating element in any room. This major element of the walls can certainly be used to establish the basic colour scheme or to provide a subtle foil for more dramatic, smaller components. If bold or bright prints for draperies and upholsteries are chosen, the neutral back-ground of plain painted walls will give them the needed visual space in which to flourish. Similarly, when a space is desired to display an art collection, it is better to keep the walls unobstrusive and non-competitive. A room furnished with handsome antiques, on the other hand, seems to call for a more elaborate and traditional frame in which the furniture can be properly displayed.

A family's style of living, as well as the style of the room's furnishings, will govern the style of wall treatment – whether plaster is rustically rough or urbanely smooth, whether wall paper is boldly striped or delicately patterned, whether wood panelling is highly polished and trimmed formally with mouldings or knotty and laid casually in random widths.

Wall coverings may make a major decorating statement or even provide the entire decoration for a room. A small room occupied only for brief periods can have an extravagant wall decoration which may prove too strong for a bed room. A forceful red-and-gold wall paper, for instance, might overwhelm one's senses in a living room, but would produce a feeling of pleasurable astonishment in a front room. In an entrance hall, a large mural painted on one wall might provide interest and even a subject for conversation, while in a study it would be an irritating distraction.

Wall treatments can be designed to cure defects of dimension or to camouflage awkward structural necessities like protruding columns. If the ceiling of the room is too high, it can be brought down by artful wall decoration. The other way of lowering the ceiling visually is to paint the walls in pale colour and the ceiling in a darker shade or to cover the ceiling with a bright boldly patterned wall paper.

FACTORS FOR SELECTION

As seen earlier, walls play an important role in the appearance of a room. Since they can change the appearance of a room, and also because a wide variety of materials are available in the market, it is important to consider and understand the factors that influence their choice.

Scale of walls and room dimensions

Scale of walls in relation to room dimensions and other items placed in that room must be considered in the selection of wall coverings. Walls with bold patterns such as large rough stone or concrete textured surfaces can be overpowering in a small room with furnishings that are light and delicate in appearance. On the other hand, wall papers with a small pattern in a light single cover gives a room the appearance of a larger scale. As a general rule, it is appropriate to use light colours or small scale patterns in a small room and large scale patterns in large rooms. An outsized wall paper pattern or a very dark coloured wood panelling will tend to overpower a small room whereas a dainty pattern will appear inconsequential in a more generously proportioned space. Thus the scale of walls and the room dimensions are the factors to be considered before making a final decision on the choice of the kind of wall coverings for a room.

Open/enclosed looking walls

The appearance of the walls in terms of colours, textures and materials that are used to cover them are the determining factors whether a room will appear open and light, or massive and heavy. Wherever the space is a constraint, there is an inclination towards providing spacious and open areas. This effect can be achieved through the use of simple, inconspicuous wall coverings and flat textureless finishes, which are thin, light and unobstrusive. Rooms that open into one another should share a common colour relationship. Picking up a colour from one to use in the other – the rug colour from one room for the walls of the room it adjoins or a wall paper that echoes the paint colour of a neighbouring wall tends to give a feeling of space and continuity.

The complexity/simplicity of the wall

The wall can become a centre of interest – a focal point and as the main determinant of the appearance and mood created in a room. Patterned wall papers, textured wood grains and colourful tiles suggest a sense of movement that becomes an active part of the room, competing for attention with the furnishing. It can be said that active wall surfaces should require less furniture and simpler arrangements, for, the wall itself acts as a busy space filler and creates visual interest. A wall of a single neutral or off-white colour having a smooth surface may recede into the background, playing a passive role in the room's appearance. It gives a sense of cohesiveness to the room as the neutral colour may bring diverse elements and furnishings together. Variations in wall treatment may create contrasts and accentuate the different areas of a room. Therefore, the wall treatment depends upon this factor – whether a person wants to segregate the different areas of a room by using variations in wall treatments, or to have an uniform, continuous inter-linked activity areas in one room through the use of single subdued colour treatment for the entire walls of a room.

The texture of walls

The room comprises of other textures as well in addition to the texture of the wall covering. Thus all the textures should be kept in mind to create the overall harmonious impact. Similar to the textures of the other items in a room, the texture of room's walls can also suggest formality or informality; the smooth white wall reflecting a great deal of light is associated with formal, whereas textured walls of wood or of cork suggest more intimacy and less formality. The various types of textures of wall coverings can be glossy, smooth surfaces of ceramic tiles, laminated wall papers plastics or even glass or mirrors to rough stony/brick surfaces or textured painted areas (Figure 8.1 – see colour plate 17).

Durability of the wall covering

The wallcovering is one thing that is generally changed more often as compared to other things in the room. Thus a flexibility to change the room's appearance through new types of wall papers, paints/colours/textures should be there.

Also, it is important to know who will use the room and for what purpose. For example, in a child's room, one may prefer wall tiles, plastic laminates which are not only durable but also easy to clean whereas such things may not matter when elderly people occupy the room. A shopper should be able to judge the durability of a wall covering if a factual label was provided with it. She can get definite information about the wear resistance materials from the Bureau of Indian Standards and the standardization/certification marks mentioned on the label.

Sound and light

Owing to the noise pollution, the sound seems to be a problem of many of today's homes because of the use of sound system and the greater amount of sound that children make, we depend on walls and ceilings for the insulative qualities. Generally speaking, soft coverings such as cork, padded-fabrics and acoustical material absorb sound whereas smooth, hard surfaces such as plastics or wood are not likely to lessen sound and bounce it around the room.

Light must also be considered in the treatment of wall surfaces. For rooms that lack natural lighting, it would look better if appropriate wall covering is provided. Light-coloured walls increase the apparent size of a room and make it easier to illuminate. Smooth and glossy surfaces on walls make the best light reflectors for the rooms.

Functional considerations

The selection of wall finishes will also depend in part on the physical impacts they are likely to undergo. The accumulation of moisture in a bath room, for instance, requires that the wall covering be resistant to damage by water. For this reason, ceramic tiles and porcelain enamel are conventional and highly satisfactory choices for bathroom walls. On the other hand in the kitchen, where spattering grease and water are inevitable, washable wall surfaces such as tiles, glossy enamel or plastic laminates are preferred. Vinyl wall papers not only introduce colour and pattern in the kitchen, but also provide washable and grease or water-proof surfaces. They are also durable and easy to clean and maintain.

Decorative accents

Walls need not be restricted to the purely passive role of providing background. They can also be utilized to punctuate a decorating scheme as in the case of a kitchen where a wall of colourful or patterned tiles can prompt such an effect. They may also serve to unify a decorating scheme by covering one wall of a bed room with a fabric that matches the bedspread and the curtains. They can also be used to establish visual organisation in a multi-purpose room, with the use of a wall paper at one end of a room to identify it as a dining area. Or they may be used as to furnish entertainment by decorating a child's room with a story book mural.

Use of Colour

In the distribution of colour, the nature is taken as the source of inspiration. Just as the earth is darkest, trees are medium, and the sky is light, so the floors might well be the darkest, walls medium and ceilings light. However, interesting variations can be the opposite of this. One colour idea throughout the entire background is usually desirable – for example, beige walls, ceiling, wood trim doors, floors and rugs, with some areas darker than the others. Various tints and shades of grey-green for the walls, ceiling, woodwork, carpet and draperies would make a cool, delightful background that could be relieved by dull yellow and salmon in the furnishings. However, two colours are often effective in backgrounds, such as greyed lime walls (somewhat off-white) with muted orange floor covering.

Although the ceiling is now considered a part of the room, not a separate item, and is usually treated exactly like the walls, many different solutions are possible where a more novel effect is desired. For instance, the ceilings colour can be like the background/foreground colours of the wall paper, or the colour in the drapery (rug/floor) or colour of woodwork, doors, or absolute white when a tall effect is desired.

Now-a-days, people choose to create dramatic effects by painting, one-wall, two-walls, three walls or all the four walls in a different colour. This may be an interesting way to create emphasis or for the sake of variety. The principle of proportion can also be applied to colouring. For instance, an increasingly large room may be made to appear a bit small by applying a warm dark colour on all the four walls, or a long-narrow room can be made to appear some-what wide and small, by using warm and cool colours alternatively on the narrow and long walls, respectively. For creating more illusions in a room, refer the chapter on colours, its uses and application.

The imperfect walls in the rooms of a house can be improved upon by the use of colours like light green, light blue, flesh colour or ivory. The colours like lemon-yellow, and white exaggerate irregularities. The dark rooms should have white or yellow or peach colours for its reflecting qualities. A large room should have warm, advancing colours, whereas a small room should have cool coloured walls. The vertical stripes improve too low ceilings whereas the horizontal stripes improve too high and small rooms.

Neutral colours like white, off-white, grey, beige, brown and natural colours are good for backgrounds. It is recommended to have white in enormous quantity to contrast with brown/dark gray. The most preferred colour for the ceiling is white because of its highest receding effect.

The colours like grayed lime, grayed pink, grayed peach, light slate blue, light turquoise-blue, sky blue, sage-green, pale lemon-yellow and golden yellow can be combined with other colours. Beautiful effects can be created in an otherwise well-balanced room by colours like blue-green, clear medium blue, cherry-red/chartreuse or ivory white.

Whenever a colour arrangement is planned, it is desirable to consider the backgrounds and furnishings together. This concept, if kept in mind, will create an interior which unifies and creates a total effect. Unifying effect for the entire house can also be created by planning the colour schemes for all the rooms together. The planning should start with the colour scheme for the living room, then the passage/corridors, then the bed rooms and so on. If one colour is made to run in the colour schemes for all the rooms it would help to create unity in the entire house.

WALL TREATMENTS

It is important to finish the interior as well as the exterior walls for functional considerations. Since walls are the largest area seen in the interior of any room, it is necessary to pay enough attention to their finishes. Today's markets provide a large number of options as wall coverings to create colour effects, illusionary space or to provide theme, mood or style of a room. The wall coverings are, therefore, used not merely for functional purposes, but also for aesthetic and other reasons. Information on a few popular materials for wall treatments are discussed in this chapter. They are:

- Wood Panels
- Paints
- Cork
- Leather
- Masonry

- Fabrics
- Wall Papers
- Tiles
- Murals
- Mirrors

Wood paneling: The charm of wood as a wall covering arises from its warm colour and beautiful grain, which provide a rich but restrained background for all styles of decoration. Wood offers, in addition to warmth, texture and vibrancy of a living material and is thus a valuable complement to materials such as gold, glass, stone or metal.

Plywood panels are known for their finest decorative paneling effects for a long time. There are, however, modern substitutes like a veneered wood panel. The stiff, non-warping character of plywood, a result of its laminated cross-grain construction, eliminates many difficulties of installation and uncertainties of stability. Because veneer are real wood, they can be finished exactly like any other lumber, with varnish, strain, wax, or paint. Plywoods are available as pre-finished sheets as well. They, therefore, require no further treatment.

As a low-cost substitute for veneered plywood, manufacturers of wall panelling make wood grain hard-board. The authentic wood-grain pattern is reproduced on a vinyl film by a combination of photographic and modern lithographic print methods, and the vinyl film is then fused to the hardboard base to provide an almost undestructible surface. Besides reducing the cost of real wood panelling, this process also eliminates unsightly irregularities like knots.

Paint: The major reason for the continued popularity of paint as a wall covering undoubtedly is its low cost. Besides being inexpensive, there is also the scope for changing the decorating scheme frequently. Paints also permit a precision in matching colours that is virtually impossible with any other wall covering. While replacing or renovating or repairing, it becomes necessary to take a sample to the market for colour matching, and it may not always be available at the time of need. Paint, however, can be got easily, and if not, can also be matched with existing colours or mixed to the prescribed colour specifications.

Application of paint is also easy. Its application is fast, unlike wall paper, wood paper, wood panelling, fabric etc., which must be cut to measure and matched. The application of paints presents fewer problems for an amateur. The required equipments are also not complicated, and do not demand extra skill in application. Any blemishes that do appear, can simply be repainted. Thus paint is ideally suited to "do-it-yourself' projects.

Cork: This resilient material is valued for its acoustical properties and because of its surface, that can be punctured repeatedly, without noticeable ill effect. In a study or library, where silence is required, cork provides an efficient sound absorbtion. In a children's or an individual's room, it can be ideally used on the walls to exhibit the changing art work.

Made from the inner bark of the cork oak, the wall covering is made of small particles of cork held together with a composition binder. Its colour is generally a combination of tan and mustard yellow, but tinted cork panels are also available. The material is available in flexible sheets, in square tiles, and in plywood-backed panels.

The recent development in cork wall coverings is the use of large, irregular particles of cork fused to a heavy, burlap like fabric. This fabric may either be matched or contrasted with the cork particles so that, for instance, a bright red burlap may occasionally show between particles of dark brown cork. Cork's chief disadvantage is its vulnerability to stains, particularly to those caused by grease. To overcome this problem, a clear vinyl coating provides sufficient protection.

Leather: A luxurious wall covering, leather was popular in seventeenth century. In Holland, effects of great splendour were achieved with heavily embossed "Spanish" leather. Treasured examples of this material are still carefully preserved in museum collections. At present, less expensive vinyl sheets designed to resemble leather are often used where a masculine setting of a particular richness is desired. The surface of this material, like that of real leather, may vary in colour, and in texture, from suede through soft glove leather to shiny patent leather. Leather-covered walls are often ornamented with moulding strips or with brass-headed studs.

Masonry: Most true masonry walls are part of the structure, and they are often regarded as old buildings. Masonry works such as brown stones, brick-walls, rock-stones, stones, pebble-finishes etc. are all popular for their natural finishes. In most cases, these walls require very little maintenance except occasional cleanings and the application of a sealant to protect the porous surface of the masonry against a build up of soot and dust.

At times, stone and brick walls are simulated with moulded polyester sheets. This relatively light weight wall covering reproduces the texture of a brick wall both in its larger scale and in its smaller scale. An alternative use of these also produces a patterned effect. These sheets are popular since their installation is made simple by nailing and finishing the nail heads with a matching point.

Mirrors: Mirrors make the walls glittery and therefore, are currently fashionable as the major component of the super shiny interiors. Besides providing a sparkle of their own, mirrors reflect the dazzle of the bright metals and sharp spot-lights that are also inseparable component of this style. It is no wonder that shops selling gold and silver jewellery, artificial jewellery and cosmetics are made up of mirror walls!

A major function of mirrors in interior decoration is, of course, to create illusionary space. A single mirrored wall can double the apparent size of a small room. Facing mirrors carry space out to infinity. Mirrors are relatively expensive to install, costing three times as much as medium – priced wall paper. In compensation, however, the material is for all practical purposes, permanent. The threat of breakage is quite slight, if the mirrors have been properly framed and braced.

Murals: Wall pictures, the oldest known interior decoration, remain a favourite way to give a room invividuality. Murals provide a forceful and exciting presence, although other decorations must be restrained if the total effect is not to appear excessive. A mural, properly speaking, is necessarily a specially commissioned work of art. At a considerably more modest level, a poster may be considered as a substitute. Colour posters, in all sorts of antique and modern designs, are obtainable in sizes ranging from large to enormous. Costing very little, these disposable murals allow a lot to experiment with (Figure 8.2 – see colour plate 17).

Tiles: Two special properties of ceramic tiles – its imperviousness to moisture and its wash-ability – make them a customary choice for bathroom and toilet room walls. Despite its high initial cost, it may prove a long range bargain, if family members habitually take long, hot showers that would shorten the life of a less durable material. The only maintenance required for ceramic tile is the occasional recalking of the wall joints, and with the development of the plastic calking compound, this problem is comparatively reduced at present.

Bathroom wall tiles are also available in solid vinyl and in porcelain enamel (a metal tile with a vitreous surface). All of these materials come in a range of colours, and all are suitable for installation in bathrooms, water closets, kitchens, etc. Ornamental tiles used near the fireplace, add an air of special distinction to any interior. Most of the retail stores have catalogues of the available patterns, and one can choose and order any of them. The tiles are also available in standard sizes and therefore it is easier to replace broken, cracked or otherwise damaged tiles.

Wall Papers: They have become so attractive and exciting in recent years that it's tempting to go all out for them. The romantic history of wallpaper dates back to ancient times, but its popular use can be traced back only to about the seventeenth century. Wall paper was first made to take the place of high-priced tapestry and textiles. At the outset Italian booklining papers in small sheets were used; later they were followed by Domino papers, which were small marbleized squares. Subsequently, France produced the present type of wall paper of continuous design but printed by hand and also picture papers and scenic wall papers. Machine-printed wall papers such as we have today were the next development. Today the diversity of wallpapers presents a real challenge to the imagination and creative ability of interior decorators.

Wall papers may be used to create not only an atmosphere, they may also be used in new and different ways to create an infinite variety of novel effects. The wide range of textures, colours, patterns and interesting special designs almost defies any attempt to classification. Some papers introduce architectural features into a room, and still others provide scenic effects. The traditional all over patterns are ever present, but panels and 'spot' motifs have become extremely popular (Figure 8.3 – see colour plate 18).

Wall paper is no longer confined to walls. It may be used on folding screens, furniture and decorative accessories. Closets and other storage areas lined with wall paper are often more attractive and may even be more useful. For example, a cedar wallpaper may make a closet repellant to moth damage in the walls.

Advantages of Wallpaper

The use of wall paper as a wall finish has been on the increasing trend owing to the following advantages:

1. It can be applied to walls by amateurs, especially now that the pasting medium is already on the paper.

2. It is very useful in covering imperfect surfaces. Defects and blemishes in the wall surface can be easily covered or camouflaged.

3. A problem area can be made interesting and attractive. A small foyer, a long, narrow passageway can become a dramatic center of interest without the use of furniture. Difficult, or uninteresting alcoves can be given importance with the use of wall papers.

4. Colour, texture and pattern of wall papers lend a distinct individuality to the character of the room. It can establish the colour scheme for the entire room and introduce the element of design into it – a design as everyone knows is exceptional.

5. Colour and pattern can create illusions. A large, sweeping design can make a room look smaller; a small room can be given unity with a tiny-patterned paper and the matching furnishing fabrics.

6. Vertical stripes on wall papers will make the room seem higher. To make the ceiling seem lower, the papering should be only as far as a molding or border placed below the ceiling.

7. A boldly patterned paper calls for comparatively plain floor covering and fabrics in the room. Thus the choice of patterns and colour plans have become increasingly interesting.

8. Patterned wall paper effectively contributes to the theme of a room. Polka dots or gingham plaids suggests intimate cottages; broad five inch stripes/stylized motifs add to Modern effects; Greek motifs contribute to the formality of an Empire room; western, military, or nautical motifs usually suit a boy's room or floral patterns in pastel shades can help to create a feminine effect.

9. Wall-paper may be used to emphasize or to minimize architectural features that are either pleasing or unattractive. A large room may be made to appear smaller and vice versa. The design of the paper may change the apparent proportions of the room; for example, an emphatic treatment of one wall will make it advance, or a natural scenic patterns like a garden to recede.

10. Wall-paper can be used both to separate and to coordinate areas when other means are impractical. One end of a small living room can become the dining area without a room divider simply by the use of a different wall-covering.

11. Wall papers can emphasize a furniture arrangement or make a center of interest more dramatic. A popular application is a panel or spot design at the head board of the bed in an otherwise plain bedroom.

12. A wide variety of designs are available in the market to select from. Thus the wall-papers offer a great variety of choices according to one's likes and requirements.

Disadvantages of wall-papers

The major disadvantage of wallpaper is that after two or three layers have been applied to a wall, they must be removed before a new finish is applied. With professional steam equipment, old wall paper can be removed quite easily, but sometimes the amateur without equipment must soak the paper with hot water and then scrub it off. This can be very time-consuming and an expensive affair as this can damage the final plastering of the walls.

The wrong choice of pattern of a wall paper can become tiresome and irritating. This may call for either a patient waiting period till the time a family budget finds enough funds to replace them or an immediate demand for extra expenditure to replace the existing one with a dissatisfying note.

TYPES OF WALL-PAPERS

The wall-paper industry has offered to the consumer various kinds of wall-paper at different price levels. The prepasted paper that can be simply moistened and applied to the wall has made the use of wall-paper more popular as it is not time-consuming. The wide variety of materials that are available now almost defies classification.

1. Non-washable coverings

These papers have been printed with water-soluble dyes and so, any application of water may quickly damage the design. Many of the foil (or metallic looking) wall-paper fall into this category.

2. Washable covering

Various degrees of imperviousness are produced by special finishes that set the colours and make it possible to wipe the paper with a damp cloth to remove surface soil. A thin coating of a finish on the paper is applied to make it washable. More vigorous treatment of spots and stains will remove the coating and possibly damage the design.

3. Scrubbable coverings

Some wall papers have a design that is impregnated or coated with durable protective finishes that make them quite resistant to spots and stains. This finish is of a highly protective nature, and therefore it can usually be washed with soap and water. In addition to the coated papers, there are various coated-fabric wall coverings. In general, these tend to be more expensive than the wallpapers, but they are durable and easier to maintain.

Paper may be given a plastic finish that is quite tough and resistant to scrubbing; however, the more durable scrubbable covering have a fabric backing. A light weight canvas may be impregnated with a coating that has the design applied in such a way that will withstand repeated washings. Both vinyl coatings on cloth and nonwoven vinyl coverings are available. These are resistant to spots and stains and can withstand constant washing with soap and water. They are virtually indestructible, and a light gauge vinyl wall covering may be a worthwhile investment in an area that receives punishing use, such as the kitchen or the children's room.

Olefin, a spun-bonded fabric, is another plastic rapidly gaining favour as a wall covering. Its structure consists of a single filament of olefin plastic randomly laid to cover an area, then treated with heat and pressure to fix its texture. This material has remarkable dimensional stability and toughness.

KINDS OF WALL-PAPERS

If anything fits the phrase "an abundance of riches", it has to be today's wall-papers. The choice in every respect-pattern colour, design, texture is overwhelming. Some wall papers to consider are those with floral patterns of big and small sizes. Traditional and stylized patterns may include:

- Grass cloth surfaces
- Thin wood surfaces
- Solid colours
- Marbled patterns
- Polka dots
- Plaids and checks
- Stripes

- Diamonds/stars
- Leaf forms
- Copies of botany prints
- Copies of period prints
- French trite flower patterns
- Damask
- Geometrics.

Scenic paper and murals: Scenic wall-papers, usually landscapes, but sometimes including figures and buildings, are produced in both traditional and contemporary styles.

Scenic paper and murals give depth to a wall, and as the perspective 'pushes' the wall further back, it makes the room seem larger. These centres of interest are particularly impressive above a stained wood or painted dado. It is probably the most difficult kind to select and use, but it is sometimes hung as a background for period furniture where it is historically authentic. Halls and dining rooms, where people do not remain long, seem to be favorite places for it. The person who has scenic paper should treat it with proper respect. There should be no high furniture to cut the figures in the wrong places. An effort should be made to fit the motif to the wall places adequately.

Since there is little restraint in the pictorial aspect of scenic paper, it is usually advisable to select a one-colour effect, preferably brown or grey. The wood work and doors in the same room should usually repeat the chief light colour in the scenic paper.

Grass cloth paper: Textured wall coverings include such genuinely textured materials as gross cloth, a lustrous material woven in Orient, and shiki, a coarse silk resembling shantung.

Grass cloth paper from Japan is a straw-like material that comes in many colours and variations of texture. Its textural effect is interesting but subtle, and does not compete with patterned or textured materials used elsewhere in the room. The appearance of these materials, is often reproduced photographically on wall papers and dimensional textures are added where appropriate, by embossing.

Flocked wall-paper: It has the texture and feel of cut velvet. It belongs to a very formal and elegant atmosphere and is excelled at creating one, as for example in a dining alcove.

Flock papers are made more practical now-a-days and so are also popularly used. Flocking is made of chopped synthetic fibres instead of silk or wool fibres. When this flocking is applied to a vinyl ground, the paper can be washed with soap and water. When flocking is applied to foil, it produces a contrasting shiny background and velvety figures, which produce the extravagantly dramatic wall-paper patterns. Both flock papers and foils require special care during installation. Flocking highly attracts paste. For this reason, selvage is ordinarily not trimmed until the strip is stuck on the wall, and the paste is removed immediately using a sponge.

Metallic wall-papers: These are the latest in wall coverings. Dramatic and rich-looking in themselves, they have the room-deepening effect of mirrors, but reflect light much more softly. Their shimmering colours and designs respond instantly to light changes. Particularly effective in modern rooms, foyers, and dressing rooms, they can provide a glowing and appropriate touch in formal rooms.

Metallic foils: Metal foil wall papers follow the current decorating vogue for slick, smooth, and shiny surfaces. Gold and silver leaf have been used in the past when interior decorators wanted to create an effect of special splendor. These metal pieces were not only expensive, the intensive labour involved in their installations also added to their costs. However, the modern technology requires only a thin sheet of metal – usually, gold, silver or aluminium – bonded to a sheet of paper to toughen the material and to facilitate hanging. Patterned foils are also available. They can be printed with transparent inks with a gleam of metal, or with opaque inks, producing a highly effective contrast of visual textures. Designs are very often geometric prints, in two or more colours, or luxurious imitations of marble or tortoise shell.

SELECTING A WALL-PAPER

It is difficult to select wall papers, because so few good patterns are available, and because those that appear satisfactory as samples may become overpowering on large area. Many wall-papers are now water-proof and washable.

A general thumb rule is to choose a wall-paper design according to the size of the room – small scaled patterns for small rooms, medium-scaled patterns for average rooms, large patterns for large rooms. But many other factors must also be considered. In a small room, for example, a dramatic effect can be created with a large-scaled pattern on one wall, or a scenic design can lend a perspective that can increase the apparent size of the room area. A large room that has many doors, windows or other architectural features may have a chopped-off appearance if papered with a large pattern. The theme of design and the colour tones must, of course, blend with the decoration of the room. Small, delicate/quaint patterns seem appropriate for less formal rooms; large motifs and scenic designs lend themselves better to formal and elegant themes.

Some wall-papers to avoid are those with:

- Weak, tailing or scattered flowers or leaves.
- Pressed or embossed surfaces.
- A spotty effect with motifs of dark colours.
- Figures that are out of scale with the rooms.
- Mixed motifs combining both delicate and bold figures.
- Strong diagonals (opposed to architectural unity).
- Papers that have three-dimensional effects.
- Contrasting colour schemes.
- Brocade and satin stripes and medallions.
- Restless busy patterns suggesting movement
- Silver or gilt.
- Pictorial patterns, like hunting scenes
- Imitations of draped or puffed materials
- Imitation plaster effects.

Installing Wall-papers

The techniques for hanging wall-papers, includes measuring the wall space, preparation, cutting, matching and gluing. Special circumstances, however, will require special handling.

Fine wall-papers – antique scenic wall-papers, for example – are customarily mounted on an underlay of muslin to ensure a smooth surface and to protect the paper against cracks. More importantly, the muslin lining facilitates removal so that the paper can be hung elsewhere. The removal of these fine papers, is, however, a difficult job that should be entrusted only to a top-notch professional. It is a good idea to use a lining under flock papers and foils. Both these wall papers depend on a smooth surface to display their best qualities, and foil, in particular, will reflect any irregularities in the surface it covers.

When the wall-papers are bought, the dealer should be asked to provide the installation instructions. Some papers may require special adhesives. Directions generally accompany the papers and should be made available by the dealers. When these instructions are followed with care, one can achieve best results. If anyone wishes to install the wall-papers at home, the general procedure to be followed is given in the following paragraphs.

Taking measurements

The dealer can figure the required number of wall paper rolls, if he is given the dimensions of the walls, windows, and doors. Therefore, it is necessary to take the measurements of the doors and windows, besides taking the measurements of the walls.

Preparation

Any electrical fixtures and switch plates should be removed before fixing the wall papers. All materials and equipments/tools such as table, steel straight edges, roller knife, seam roller, smoothening brush, etc. should be gathered. Other materials required are sponge, scissors, yard-stick, paste, brush and a string chalk line. Keep the wall paper roll ready for measuring and cutting.

PLATE 17

Figure 8.1. Various Textural Effects can be created on the walls using different materials

Figure 8.2. Murals on walls provide variety as well as decoration

PLATE 18

Figure 8.3. Wall papers with floral and geometrical patterns

Figure 8.4. Wood Panels for textural effects on ceilings

Surface Preparation

The surface needs to be free from oil, grease, dirt and foreign matter. Plaster of Paris is used to repair undulated surfaces and bridge cracks. If walls are freshly plastered or painted, treat with wall "size". Then a thin coating of glue is applied with a brush.

Cutting

The wall paper is unrolled for measuring, and cutting. Extra allowance for trimming at baseboard ceiling and for matching pattern should be provided. After measuring the first strip, the wall paper is cut to the desired size. Then the next strip is unrolled and the pattern is matched with the first piece before cutting. The subsequent pieces are cut using the same procedure.

Application

Phenol-resin or its equivalent mixed with a hardener in powder form is applied on the walls and then on the paper. This is then struck to the walls. Creases and air bubbles, if any, are removed by applying pressure in the right direction. Then the second strip is hung by matching the pattern. The wall paper should be stuck, by beginning at the top and then rolled towards the floor. A wet sponge is rubbed against the wall paper to remove excess paste. If the paper is not washable, a clean cloth is used to wipe the paste.

Care and Maintenance

Most wall-papers, if printed with insoluble oil-based or vinyl inks, can be cleaned with mild soap and warm water. To make sure that the wall-paper is washable, washing can be tried in a small corner that is ordinarily hidden by a furniture. Detergents are harmful to wall-papers and should, therefore, be avoided.

The cleaning process starts with simple washing in sections. After soaping, the wall-paper should be rinsed with clear water and then wiped with a clean cloth. Walls should be washed from bottom to prevent stains caused by dripping water. Dirty streaks are easily sponged off clean walls, but may form ineradicable stains, if surface soil is allowed to accumulate for a long period of time.

For non-washable wall-papers, commercial cleaners are available. A gum cleaner, resembling plastic wax, is used as an eraser to remove surface dirt. The cleaner should be re-kneaded after some use, to work with only the clean area. Wall-papers can also be cleaned with a cloth dipped in dry borax. There are also products that may be applied to newly hung wall papers to make them spot-resistant. Grease spots can often be removed with dry cleaning fluid or with a hot iron and blotting paper. Vinly wall-papers are grease-resistant, but grease spots should be sponged off immediately.

Although vinly flock papers can be washed, this treatment should not repeated very often, as the flocking will eventually shed. Flock papers can easily be cleaned with a vacuum cleaner and a soft brush. Tough vinyl-coated and vinyl-impregnated wall coverings can be scrubbed fairly but should be scoured with abrasives. Whatever may be the kind of wall-paper, all of them require regular attention. They should be dusted and wiped often with a clean cloth. Occasionally they should be washed by the process mentioned above. Any kind of damage—torn, scratched, loosened seam edges, appearance of air bubbles etc. – should be attended to, immediately. If minor, the damage can be repaired at home, but a professional can be called to repair heavy damages.

FABRIC COVERED WALLS

Many types of cloth have been produced and used to add interest and warmth to walls. When they originated back in the Middle Ages, rug, tapestries, velvets and brocades were used as hangings, not so much for decoration as for insulation. Today, too, there are more than decorative reasons for fabric walls. Fabrics are now being used frequently to cover one, two, three or all four walls in a room. Fabrics may be tacked on frames, hung loosely, stretched on frames so that they can be removed for cleaning, pasted directly to the wall or glued to heavy paper and applied as wallpaper. Some fabrics for wall covering are backed with a plastic coating so that they can be cleaned with a damp cloth. In such a case only solid colours or unobtrusive patterns should be considered.

The fabric covers are used to completely conceal badly cracked walls that cannot take paper, for example, and they are still useful as good insulation cover. It can be attached with staples or small tacks which can then be hidden by molding or a decorative panel. They can be pleated, gathered or just eased on, depending partly on the stiffness of the fabric. Gingham, toile, corduroy, raw silk, chintz, brocade, canvas, denim or in fact almost any firm material is usable. This method is ideal for persons who live in rented places and wish to take their improvements with them when they move.

If walls are relatively smooth, fabric can also be fixed to them with an adhesive that will not stain or discolor the fabric. Fabrics to be attached to a wall should be firmly woven, such as canvas, linen or burlap. Where texture is to be emphasized, grass cloth is the best material. Where pattern is designed, ginghams and percales in checks, plaids and small floral patterns are attractive. Leatherette makes an appropriate finish in a library, or a man's room. Dull-finished oilcloth for walls comes in acceptable designs and is practical, for it cleans well and wears well.

Felt, burlap, chintz, velvet, damask canvas and grass clothes are popular wall fabrics. Damasks and brocades are also used for rich elegance. For a modern look, a contemporary cotton or linen print will do. In recent years, a popular trend has been to use bed sheets as wall fabrics. Sheets are generally less expensive than traditional wall-covering fabrics. There is almost no limit to the variety of colour and texture that fabrics can provide. Many of them can be treated or impregnated with plastics so that they will resist soil and can be easily cleaned.

It is sometimes desirable to cover only a part of each wall with a fabric. The upper two-thirds can be a fabric, with a moulding and a dado below. Fabric-covered panels can be placed on the wall behind a sofa or a bed; generally the same material should appear elsewhere, possibly all over the opposite wall, or as a bed-spread or draperies. Panels are easily made by tacking a narrow wooden piece around the space and stretching or pleating the fabric over this, using small tacks. Sometimes the edge is covered with a narrow wood moulding.

Use of fabric for overhead drama

Ceiling drama can be the solution to make an ordinary room extraordinary. It is particularly apt where badly placed doors and windows prevent the centering of interest on a wall – the more conventional way to focus on a decor.

Fabric on the ceiling is not really a new idea. Castles and palaces set styles in such treatments long ago. In those days, fabric served as insulation against cold, and also as an acoustical device to prevent sound from echoing against the stone or marble walls of vast, high ceiling rooms.

The Styles – sophisticated, romatic, formal, informal – can be very different, depending upon the fabric, its colour, its pattern and the manner in which it is mounted. It can be as correct in an elegant living room as in a dashing bedroom or bath for the very young. Dining rooms also can benefit from rejuvenation that focuses on overhead drama. Basement recreation rooms whose overhead pipes and beams offer handicaps in remodeling projects can be stylishly reclaimed with the right ceiling installation. Fabric can be pasted on the ceiling or mounted on rods, depending upon the kind of fabric chosen and the mood to be established. For sheer fabric to initiate a canopy, a series of curtain rods can be mounted on the wall just below the ceiling. Each panel is shirred down to almost nothing at the center of the ceiling. The effect secured is distinctive and is also practical.

A point to remember when using fabric on the ceiling is that cool colours of pale value will make the ceiling seem higher; warm colours and deep hues reduce the apparent height. Carpeting is becoming more popular as a covering of both ceilings and walls. As well as being decorative from the standpoint of colour and texture, it has great sound-absorbing properties. If the effect of a fabric ceiling are desired with even less maintenance, the vinyl-coated wall-papers available on today's market can be used for the purpose. Overhead, they'll look like fabric and the trickery will be even harder to detect if a matching fabric is chosen for draperies and slip covers.

DESIGNING INTERIOR WALLS

In designing a house, tentative interior elevations are developed along with the floor plans. These cannot be made definite, however, until the exterior elevations of the house have been designed.

'**Function**' directs much of the designing of interior walls, for space must be devised to suit the furniture, and good circulation determines the location of doors. Moldings, which are dust-catchers, are minimized for the sake of function. '**Beauty**' can be attained along with good functioning by proper designing. The five principles of design, proportion, balance, emphasis, rhythm, repetition and harmony are excellent guides to pleasing appearance.

'**Proportion**' applies to the walls and to all the shapes that occur in them. Height is an important factor in the proportion of walls and in their expressiveness. High walls are necessary in large rooms; low walls are desirable in small rooms, especially those of cottage types. Proportioning the other architectural features that give character and interest to rooms require study. Fireplaces, built-in-bookshelves, corner cupboards, and seats may be ugly or beautiful. A fireplace must be in scale with the room and have good proportion. A large fire-place is suitable only in a large room or a rough-textured room; a small fireplace is necessary in a small room or in a room of refined type. It scarcely need to be said that gas or electric-heated logs or coals do not require built-in fireplaces or chimneys; such imitations are insincere and should be avoided. '**Balance**' of openings desired in an interior wall may conflict with the location desired on the exterior of the house. In a compromise between them, usually the exterior balance is considered more necessary, for, interiors can be balanced by the location of furniture. Balance in the amount of colour used is also an important factor to be considered.

Emphasis on some part of a room may be obtained by means of an important architectural feature. Emphasis on one direction of line often creates a desired effect and focuses attention on the desired objects. **Rhythm** occurs in the relation of doors, windows and other architectural features. The use of similar-looking objects and colours would carry the eye movement from one object to the other, thereby creating rhythmic effect. **Repetition** is a factor conducive to unity. For example, built-in-bookshelves, cupboards, and seats should continue the lines of the doors and windows. The upper line of doors and windows should be at the same level if possible. **Harmony** of ideas, texture, colour, lines and shapes are also important factors. When harmony is created, it brings in an "unifying" effect. When harmony is ignored, it results in discord.

CEILINGS

The third major architectural element of interior space is the ceiling. Although out of our reach and not used in the sense that floors and walls are, the ceiling plays an important visual role in shaping interior space and limiting its vertical dimension. Ceiling is the sheltering element of interior decoration offering both physical and psychological protection for those beneath its canopy. No one can go wrong by painting the ceilings white. That's the general belief and it is true. But if it leads one to assume that nothing else could be right, that is the farthest thing from the truth. There is no reason why a ceiling cannot be as decorative as any other part of a room.

Whether we're aware of it or not, a ceiling does affect us. A soaring ceiling in a specious room raises our spirits; floating above in a small room, it can be disconcerting. Reactions vary with the individual, a low ceiling seeming heavy and restricting to some, cozy and inviting to others. The height of a ceiling has a major impact on the scale of a space. While a ceiling's height should be considered relative to a room's other dimensions, and to its occupancy and use, some generalizations can still be made about the vertical dimension of space.

High ceilings tend to give space an open, airy, lofty feeling. They can also provide an air of dignity or formality, especially when regular in shape and form. Instead of merely hovering over a space, they can soar. Low ceilings, on the other hand, emphasize their sheltering quality and tend to create intimate, cozy spaces. Changing the ceiling height within a space, or from one space to the next, helps to define spatial boundaries and to differentiate between the adjacent areas. Each ceiling height emphasizes, by contrast, the lowness or height of the other.

When a flat ceiling is formed by a floor above, the height is fixed by the floor-to-floor height and the depth of the floor construction. Given this dimension, the apparent height of a ceiling can be altered in several ways. Since light values appear to recede, smooth, light-colored ceilings that reflect light convey a feeling of spaciousness. Carrying the wall material or finish onto the ceiling plane can also make a ceiling appear higher than what it is, especially when a cove is used to make the transition between wall and ceiling. The apparent height of a ceiling can be lowered by using a dark, bright colour that contrasts with the wall colour, or by carrying the ceiling material or the finish down onto the walls.

TREATMENTS FOR CEILINGS

Taking their inspiration from the handsomely decorated ceilings of chateaus, castles and stately country mansions, some of today's leading designers use out-of-the-ordinary ceiling treatments to call attention to the usually neglected 'fifth wall'. Wall-paper, fabrics, carpeting, cork, paneling, floor planking, luminous panels and beams are used imaginatively to spotlight ceilings. Architectural lines that include dormers, arches, slanting angles and exposed beams lend themselves naturally to unique decorative ceiling treatments (Figure 8.4 – see colour plate 18). But if the room lacks such features, it can be just as exciting to transform an ordinary plastered ceiling with pattern, colour and texture to make it the stand-out feature of the decorating scheme.

At one time ceilings were strictly functional considerations. Now, with the aid of new materials, ceilings can have decorative functions as well, and be suited to a particular furniture style. Beams made of plastic, styrofoam or plywood are lightweight and simple to install. Ornamented, they go well in a classic room; untreated, they fit into either a country-type or a contemporary room. Floor planking used to cover ceilings can be tinted to carry out a basic colour scheme and finished in a natural wood tone. This type requires no maintenance and goes well with provincial styles of furnishings. Luminous ceiling panels are effective with modern furnishings. They can be installed over the entire ceiling or over just a part of it to highlight decorative accessories. Concealed lighting behind coves can go with modern or traditional furnishings, depending on the design of the coving.

FALSE CEILING

Additional ceiling created after the original ceiling is called false ceiling. In a room with a high ceiling, all or a portion of the ceiling can be dropped to lower the scale of the space around it. Because a dropped ceiling is usually suspended from the floor or roof structure above, its form can either echo or contrast with the shape and geometry of the space.

A suspended ceiling creates a concealed space that serves the following purposes:

(1) The false ceiling helps to change or decrease the room height.

(2) Such a ceiling conceals light wires. After false ceiling, light fixtures can be put anywhere and the wires can be concealed behind it.

(3) Air conditioner ducts can be provided and the wiring can be concealed behind the false ceiling.

(4) False ceiling is also provided to make the room sound proof. This can be done by putting glasswool behind it.

(5) False ceilings also have heat absorbent and cooling quality. False ceiling is therefore provided to make a room cool or hot. For this, we can use glass wool pads or thermostat.

(6) Most importantly false ceilings are used for aesthetic purposes. Attractive coloured and beautifully patterned false ceilings can contribute to the beauty of a room.

In commercial spaces, a modular suspended ceiling system is often used to integrate and provide flexibility in the layout of lighting fixtures and air distribution outlets. The typical system consists of modular ceiling tiles supported by a metal grid suspended from the overhead structure. The tiles are usually removable for easy access to the ceiling space.

Walls and ceilings are the important parts of the structure of a house. They also provide ideal places for expressing various ideas and decorative aspects. They are also considered to be the indicators of one's personality and social status.

INTERIOR FURNISHINGS

> *Floors and floor coverings*
> *Soft Furnishings*
> *Furniture*
> *Art Objects: Selection and use*

Floor coverings, furniture, furnishings and art objects are an integral part of any room in the home. The appearance of a room depends upon the way it is furnished. The furnishings contribute to the general atmosphere and add to the personality of a home. They act as the indices to judge the taste and culture of the family members living in that home. At times they also serve as status symbols – the more sophisticated and elegant, the higher the status of a family.

Next to walls, the larger colour area in a room is its floor. It binds all the elements of a room together. Floors are the flat horizontal surfaces meant to be walked upon, besides holding furniture items and the other such household objects. It is therefore expected to give long service without frequent and extensive repairs. A soft floor covering makes a room comfortable and gives it a furnished appearance. Since they cover a large area of the house, their design, colour and texture have considerable importance. Besides being used to fulfil aesthetic needs, the use of carpets and rugs as floor coverings take care of a number of practical functions such as insulation against sound, earth's coolness, slippery, dampness etc. Being an important aspect of a home, a chapter on floors and floor coverings is included in this unit.

Any study of interior furnishing is incomplete without tracing the use and application of the various furnishing materials. Soft furnishings are an integral part of any decorative idea for the interiors. There is a wide range of fabrics available in the market for use as furnishings. However, to choose wisely from them, a thorough knowledge of the terminologies used in textiles would considerably help an interior decorator. An interior decorator can benefit not only in the selection of textile products, but also in their use to their best advantage and to get maximum satisfaction from their purchases. This unit, therefore, also deals exclusively with the various aspects of soft furnishings.

It is generally seen that all other items such as accessories, soft furnishings, wall hangings, foot mats and floor coverings are planned and selected only after finalising the plan for the furniture items for a room. Any furnishing plan for an interior starts with the furniture plan, as they provide character to the room. The function and comfort of a room, to a large extent, are determined by the way the furniture is placed in a room. Careful planning for furniture arrangement, is, therefore, necessary which determine the quality and style of a room. As stated earlier, the furniture silently express the personality and living style of a family. The chapter on furniture covers all these aspects in this unit.

In giving an aesthetic look and a beautiful environment, art objects play a significant role. Provision of such an environment calls for the selection, acquisition, and placement of art objects. In interior furnishing and decoration, one has to select and use art objects which reflect the aesthetic taste and personality of the people who live in the home. While some may have the natural ability to do so, others may find it difficult to create an aesthetic effect while arranging art objects. However, good taste can be acquired by understanding the factors in selection and use, which has been discussed in detail in a separate chapter in this unit.

Chapter 9

FLOORS AND FLOOR COVERINGS

The beauty of a room begins with its floor, since this is the foundation for comfort as well as for a decorative background. Floors, along with walls and ceilings, form the shell of the room. Next to walls, Floors are the largest colour area in the home interior. Generally, the floor is expected to give a long service without extensive repair.

Floors are the flat horizontal surfaces meant to be walked upon. It binds all the elements of a room together. It holds furniture items, imparts insulation against the earth's coolness and darkness. It gets the greatest wear and the most dust in a room.

A soft floor covering makes a room comfortable and gives it a finished appearance. In winter, it is usually good to have a carpet or a large rug in all the rooms in the house except the kitchen and the bathroom. However, in the summertime, or at any other time in tropics, bare floors appear clean and cool but the soft floor covering gives an aesthetic appeal.

CARPETS AND RUGS

Every person in the world dreams for a beautiful interior, a part of which is compulsorily composed of floors and the floor coverings. Carpets as the beautiful floor coverings softens foot-steps, subdues sound and enchants the eye. Small wonder it is the all-time, world-wide favourite covering for the floor!

The soft floor covering includes the carpet and rugs. Since they contribute to a greater area of the house, their design, colour and texture have considerable importance. The word carpet comes from the latin world 'carpere' meaning to card wool. Since carpets were originally made with wool, they were appropriately termed as carpets. Carpets have been used all the time as the most important, preferable and elegant floor coverings. Besides, being used to fulfil aesthetic needs, the use of carpets and rugs as floor coverings, reflect a number of practical functions as well. They bring warmth and comfort under the feet, and soften and muffle sound more effectively than any other surface material. They offer safety against slippery on high glossy surfaces. At the same time, they are the most expensive options as floor coverings and require, therefore, considerable maintenance as they tend to attract termites, insects, bugs, dirt and dust.

The foremost thing that comes to our mind while selecting a floor covering is that whether we should have a rug or a wall-to-wall carpet. Rugs are soft floor coverings mostly fastened to the floor. They do not cover the entire floor and they are usually manufactured or cut into standard sizes. The larger sizes are considered room-size rugs.

Some of the advantages of a rug are that it can be reversed to prevent wearing out in spots and is more easily handled and cleaned. However, rugs and carpets can now be dry-cleaned on the spot without being removed. Rugs are adaptable for use in different rooms and also in different houses. Those who move often should have rugs of standard size. These rugs can be lifted, moved and changed as and when one feels like, so as to give a different look to the interiors, or to have variety in arrangements or colour schemes in a room.

A carpet is generally a soft floor covering fastened to the entire floor and sold by the roll in widths. It is pieced, if necessary, to fit the room size. Wall-to-wall, stairs and hall carpets are examples of this kind. Carpeting from wall-to-wall makes a room seem larger than a rug does. It can be made to fit irregularities such as the bay windows so that they will not appear to be separated from the room. Wall-to-wall carpeting permits having inexpensive floors in a new house, and an old one serves to conceal worn out floors.

The first carpet looms produced fabrics 27 inches wide which were cut into length that had to be sewed together. In the early 20th century, wider looms were developed to permit carpets to be installed with little or no joining. These were called 'broadlooms'. Broadloom carpet means any carpet made on a wider loom. It refers to carpeting fifty-four inches wider and describes no other qualities or characteristics except the width. The standard widths today are 9, 12 and 15 feet. Broadloom may be used as wall-to-wall carpeting, or as room size rugs or area spot rugs.

Rooms-Sized rugs can be cut to fit any room. Any length of carpet can be purchased in standard widths. They cover a room within about 3-9 inches of its border. For example, to cover the floor of a room of 11' × 16', a standard finished 9' × 12' rug would be small; a 12' × 15' would be too wide. A wise size would be 10' × 15' rug. The rug would then have a 6" border or floor area seen around the edges. Because rugs usually stretch on the floor, it is always desirable, to allow a small leeway when ordering such a rug. They offer the same decorative advantages as wall-to-wall carpet without the expense or bother of installation, and are of course, easier to move (Figure 9.2 – see colour plate 19).

Area Spot-rugs cover only a small area of the floor. They may be of a solid colour, often fringed, but usually they are boldly patterned. They are useful to change the proportions of a room, to define areas within a room or to give a room a splash of colour, warmth and design (Figure 9.1 – see colour plate 19).

There are two kinds of area rugs – **accent and scatter**. Accent rugs resemble area rugs but are usually more dramatic in pattern, shape, texture or colour. They are often used for decorative effects. Oriental, hooked and braided rugs are usually included as accent rugs even though some of them are of room sizes. These are small and vivid, and meant to be, as the name suggests, points of contrast and emphasis. Some of them are the prettiest, inexpensive and exquisite.

Scatter rugs are smaller rugs that add spots of colour and texture to make a bare floor look interesting. They are often used to protect areas that are subject to heavy wear or soiling, they are usually inexpensive and can easily be washed. In many new apartments and houses, a carpet is laid directly over the concrete flooring to deaden sound and to cut expenses. Since this is done with costs and not decoration in mind, the carpets are apt to be fairly neutral and plain, flatened uninteresting in construction. Accent and scatter rugs are ideal here, bringing a colour and texture to such a situation that considerably relieves its dullness.

Carpet tiles are the squares of a carpet. When laid collectively, they form a wall-to-wall carpet. They are self-backed squares that can be installed by an amateur. Carpet tiles are available in solid colours, patterns, and various fiber contents. Damaged tiles can be removed and replaced, leaving the spot looking like new, provided the older tiles have not faded or worn out (Figure 9.3 – see colour plate). They offer a number of advantages, such as:

- They can be easily cut to fit odd-shaped contours with a minimum of waste.
- Individual tiles can be replaced if worn out or damaged.
- Carpet tiles can be moved easily and re-used.
- In commercial installations, the tiles can be removed for access to underfloor utilities.
- Different patterns can be created using different coloured or patterned tiles according to one's taste and imagination.

The recent innovation in floor coverings has been the indoor-outdoor type of carpeting. This carpeting often has different textural qualities from the more conventional soft floor coverings, but it has opened up new concepts of decorating in various areas of the home. Terraces, swimming pools, playrooms, hallways and kitchens are all achieving an entirely new look with the introduction of a carpet that can withstand weather, stains, and wear. Several synthetic fibers, particularly olefin and acrylic, are being used in either felted or tufted constructions that produce pile effects. These are more practical, washable, and therefore, easy to maintain.

SALIENT FEATURES OF A CARPET

Since carpets are here to play a significant role in a home, it is important to understand their salient features. These features not only speak of their style and utility, but also of their life, serviceability, maintenance and appearance. The features of a carpet are discussed in the following paragraphs.

Character

A rug/carpet must agree in expressiveness with the room where it is to be used. It cannot be expected to set the mood of a room, but it must not be in contrast with it, for example, grass or rush rugs carry on the idea of a tropical room or a sun room whereas, rag rugs have a cottage flavour. Deep pile suggest luxury. Some of the oriental rugs represent primitiveness. Still older rugs, appear to be "antiques".

Style

Carpets must conform to the style of furnishings in a room. A plain carpet is a desirable background for any style. However, the various period styles have their own designs.

The *Early American rugs* and carpets are excellent in design and colour, and have textural effects which make them appear homespun. In the *eighteenth-century* style (Georgian in England), plain colours were very desirable, but does not call for any specific floor coverings. Genuine oriental rugs are usable. The French styles of Louis XV and XVI suggest the use of reproductions of the French Aubusson and Savonnerie rugs with their scrolls and floral motifs and subtle colours. Solid pastel colours in broadlooms are most desirable for French furnishings. Bleached oriental rugs are also appropriate.

Directoire, Empire and English Regency need solid-coloured broad loom, light or dark, with or without borders. Reproductions of rugs with a classic feeling are appropriate. *Victorian* rooms require wall to wall carpeting with plain dark-coloured broadloom, usually red, or with subdued reproductions or adaptations of the large floral and scroll patterns of the original period. *The Modern/twentieth century style* usually requires plain solid colours in carpets reaching from wall to wall. Light colours are emphasized. Texture interest and texture patterns are employed if variety is needed.

Colour

Since the floor is relatively a large area of any room, the colour decision is indeed an important one. Besides, the rug will probably remain longer than other colour areas. It is wise, therefore, to choose a versatile colour that will lend itself to a variety of colour combinations. The colours preferred in floor coverings are beige or pale grey-green, off-green, off-white, grey, tan, brown, rust, olive green, grayed yellow, grayed rose and grayed blue. These colour form the background providing interest to allow changes of the drapery and upholstery. Dramatic effects may be obtained with carpets of pure, brilliant colours like vermilion, raspberry red, ultramarine blue or emerald green.

Light colours are the colours of the modern age. They are pleasing because they make rooms seem spacious and cheerful, and they do not show dust or footprints as darker colour do. Tweed effects which are mixtures of several similar natural colours are practical where there are children. The practical aspects of colour also influence choice. Very light and very dark colours will show foot marks and soil more quickly than the medium hues.

Pattern

Within the framework of personal likes and dislikes, pattern and texture should be considered in relation to the purpose and theme of the room. Occasionally patterned rugs are usable and even desirable. For example – a cottage bedroom with plain furnishings gain interest by the addition of a carpet in a hooked-rug pattern. Small patterns or tweed effects are serviceable in halls that are subject to hard wear in dining rooms or bedrooms used by children. However, patterned floor coverings or rug patterns are usually limited in their designs and patterns, and thus, they generally provide less choice. Texture patterns that are exactly the same colour as the rug itself are usually good looking, and so are more popular than patterned floor coverings.

Texture

A lot of interest has been laid on textural variations in home furnishings, recently. These are also evident in the rugs and carpets made these days. Texture variations are needed for contrast with the large areas of slide glares that are used today. They include the cup-loop, cut and uncut loop, straight and twisted yarn or by several heights of pile. These textural variations can easily be felt while using them for various purposes. Texture in rugs and carpets are of higher significance than any other item of furnishings in the home, since these can be used to create patterns also.

Durability

For checking the durability certain information is required like the kind of fibres used, particularly in the pile, the weight of each kind of fibre, the ply in the yarn, the pitch, the wires and the shot, height of pile, etc. Synthetic multi-ply yarns woven with close piles are more durable. Some good woolen fibres are also used to make durable carpets.

The weight/quantity of each kind of fiber woven into a rug provides information about the durability. The rug with the largest percentage of the most durable fibers will wear the best. The ply of yarn is determined by the number of small wool yarns that are twisted together to form larger yarn. Six-ply is a thick yarn which makes large tufts when used as pile. It is well to recognize the ply of the warp and weft yarns, for the larger ones are stronger. The pitch refers to the number of tufts of pile per inch crosswise of the rug. Strength is indicated in a rug by a large pitch number and close tufts. The wire refer to the number of rows of tufts per inch lengthwise. The shot is the number of crosswire yarns which attach the pile to the backing material of the carpet/rug. The height of the pile should also be considered. It is not so important as thickness. High pile is luxurious, but it crushes somewhat more easily than short pile. Thin pile crushes more than thick ones.

Cost

Usually it is well to buy the best quality of floor coverings in whatever way the budget allows. For example – a first-class velvet rug is the type most suitable for the average home. However, ample size of rug is more important than quality in producing an effective room. The modern market provides a lot of variety in quality at varied costs. Thus a consumer has a wider choice, but a wise consumer will analyse the factors like quality, size, colour effect, texture etc. along with cost while making a final decision. For the very low budget home, felt, rag, hooked, string, jute, fibre, and porch rugs are also good selections.

Size

The floor covering includes both wall-to-wall carpets and area rugs. It is important that a rug fits the size and shape of a room. The standard sizes available are 6'× 9', 9' × 12' and 11' × 15'. However, other sizes are also procurable on order. A rug should leave about 6-12 inches of border all around it. A rug without a border looks larger than one with a border.

The placement of rugs is an important aspect in floor covering. Small scatter rugs are sometimes desirable in halls, bedrooms, or living rooms. In the same room, if more than one is required they must be identical/nearly as similar as possible. When used on a plain carpet, they should be like their background coloring. Placing the small rugs diagonally violates the architectural lines of the room. Small rugs must be placed before the most important pieces of furniture or be combined as bases for the various groups and should not stray into the middle of the floor. It is better to have too few than too many small rugs because they must not suggest a store display (Figures 9.4 and 9.5 – see colour plates 21).

Fibres

These may be classed as natural, like wool and cotton, or man-made like nylon and rayon. Sometimes two or more different fibres are combined in one rug or a carpet. Wool is generally considered to be a superior fibre for carpets, because of their warmth, pile effect, softness etc. Wool is also resilient, durable and colour-fast. All domestic wool carpets are permanently mothproofed, but the foreign products are not. Carpet wools come from cold mountainous countries like New Zealand, Persia, etc.

Cotton fibres are soft and crushable. They soil easily and are likely to fade. However, they wear fairly well. Cotton is used mostly in cheaper rugs now-a-days, Nylon fibre is resilient, very durable, easily cleaned and resists mildew and moths. It is used alone or added to other fibres to give strength. Rayon is crushable and generally undesirable. So it is used in the less costly rugs. It may look well for a brief period. The acrylic fibres are long wearing and easily cleaned. Jute and hemp help to make firm backs for carpets and rugs. Jute is strong and inexpensive, but will decay under moisture.

Selecting the right fibre: Most people are confused about carpet fibres, not only because there are several of them, but also because they are sold under a number of trade names which are, with a few exceptions, meaningless to most consumers. The fact about fibres helps to choose the fibre that serves the best as each fibre has certain characteristics that suit it to a particular range of uses. The fibres used in floor coverings must be either carefully selected/specially manufactured to withstand the rigours of wear. Resilience, the ability to retain colour, and ease of maintenance are also extremely important factors for fibre options.

NATURAL FIBRES

Wool and cotton are the natural fibres used in carpet.

Cotton

Cotton is seldom used in carpeting today because it has only fair resistance to wear. It soils easily and loses its appearance rapidly, but the only advantage is that it washes easily.

Wool

Wool is the ultimate carpet fibre, the one that sets the standards for all others. It dyes exceedingly well. Wool yarn is generally a blend of various types of wool to achieve a luxurious natural texture. A good wool carpet, with sufficient density, is expensive, but worth it if price is not the main concern. A good wool carpet is an excellent buy for long hard wear, and good appearance.

Man-made fibres

Although, man-made fibres differ greatly, they have certain desirable characteristics in common – they resist moth, mildew and mold, and they are all non-allergenic. They are also available in a variety of colours and patterns (Fig. 9.3 – see colour plate 20). However, they are not as superior as woollen carpets.

Acrylic

This resembles wool more closely than any other man-made fibre. It is more durable than wool, less durable than nylon. It resists sunfading and water, which makes easy cleaning and rapid drying. It has excellent resilience and crush recovery. They wear well and have good stain and spot removal properties. Acrylic's colours are brighter than wool's and are obtainable in all popular carpet styles. They are also less expensive as compared to woollen carpets. Usually modacrylic is added to increase flame resistance. Acrylic carpet can be used wherever wool has always been customary – living rooms, bedrooms, dining rooms, etc.

Nylon

Nylon is the largest selling of all fibers because of its extra ordinary strength and abrasion resistance, and its economy. In wearability, nylon outlasts wool by at least three times. Less ounce weight is required of nylon than of wool because of its high resistance to abrasion. Nylon does not show signs of wear rapidly. It may pill and has the tendency to build up static electricity charges if no anti-static finish has been applied. Its high bulk (thickness), great elasticity, durability, and dyeability make it practical all around. Added qualities like washability and looks, and nylon becomes suitable for every room in the house.

Polyester

The look and feel of wool can be achieved in polyester. Its soft, luxurious feel and a blend of beautiful colours have contributed to its desirability. It has poor recovery from crushing. However, its soil and stain resistance cannot compare with that of acrylic or nylon. It gives good wear, but lacks the resilience when used for carpeting. It is durable and also machine washable, and moth resistant – both fine attributes for bathrooms and for area rugs in children's rooms. Its 'ice-cream' colours are perfect for bedrooms, where its softness is welcome, too.

Polypropylene

This is notable for its high strength and exceptional abrasion resistance, mainly because of its low pile. It resists soil, stains and water. It is used for indoor/outdoor and kitchen carpet but, as the least expensive of all synthetic carpet fibres, its greatest contribution is undoubtedly economy. Each carpet manufacturer offers blends of the generic face fibers which improve on specific characteristics such as durability, soil-resistance, cleanability, colour and lustre.

Table 9.1: Qualities of widely used soft flooring fibres

Fiber	Wear/ Abrasion Resistance	Texture Retention	Resilience (ability to bounce back to life)	Moisture Absorbancy	Soil Resistance	Cleanability	Flammability Resistance
Wool	Good	Good	Very Good	Little	Very Good	Very Good	Good
Cotton	Fair	Fair	Poor	Most	Poor	Very Good	Poor
Acrylics Modacrylics	Medium	Good	Very Good	Little	Very Good	Very Good	Poor
Nylons	Very Good	Exceptionally high	Very Good	Little	Good	Excellent	Fair
Polyesters	Very Good	Good	Medium	Little	Medium	Good	Fair
Olefins	Excellent	Medium good	Medium	Least	Very Good	Very Good	Fair

FACTORS FOR SELECTING FLOOR COVERINGS

The consumer today is provided with a wide range of possible floor coverings. Floor coverings fulfill many needs in the home. Most floor coverings are relatively expensive and represent an investment of a fairly large proportion of the furnishing budget. In all probability, colour, texture, and price are important in influencing one's choice, but there are other factors too, as enumerated below:

Economy

Although the cost of the floor covering is important, at the same time the economy of upkeep is equally important. It is lessened when the floor material resists stains. Neutral colours are the best ones. Floor areas without any cracks or crevices are easy to clean and therefore they are durable.

Resilience or cushion impact

It is the degree of springiness with which a carpet returns to its original condition after weight is removed. It reduces the foot fatigue. There is less breakage of things. It is especially good for children because they do not get hurt.

Warmth

This feature depends on the climatic conditions of the area. Usually it is desired that all the carpets and rugs provide warmth to the foot. The colours used should be between middle to dark shades.

Reflection

It is usually associated with ceilings and walls, but, much more light hits the floor day and night. The more light the floor reflects, the brighter the home will be.

Sound absorption

Rough, porous material lessens noise. Carpeting helps in insulation against sound.

Safety

Carpets and rugs provide safety from falls, slips, skidding, as well as flame retardancy etc. Kitchens and bathrooms should be given special considerations.

Anti-Static

Anti-static is the ability of floor coverings to disperse electrostatic charges and prevent a buildup of surface static electricity. Nylon will build up a great deal of static electricity unless an anti-static finish is applied.

Aesthetics

The principles and elements of design should be used to create the effect one desires. Floor coverings should harmonize with the furnishings and architecture of the home. Some of the traditional floor coverings, such as braided, hooked, rugs, and needle punched rugs, harmonize well in contemporary home as well as in a colonial home.

Carpet Construction

Carpets are primarily of two types; hand-knotted and machine-made. Hand-knotted carpets are available in silk and wool while mechanically manufactured carpets are either machine tufted or of the power loom variety. They can be of natural or of man-made fibres.

Whether hand-knotted, power loomed or mechanically produced, there are many similarities in the production methods. The side-to-side progression in hand-knotted is accelerated in a loom as the shuttle propels the welt yarn track and forth. This is missing in tufted and later methods. The direction of lay of the finished face fibres is always in the opposite direction to width. However, in

PLATE 19

Figure 9.1. Area-Spot rugs with patterns and animal figures

Figure 9.2. Animal Figures in Room-sized Rugs and Carpets

PLATE 20

Figure 9.3. Vinyl Floor Coverings

PLATE 21

Figure 9.4. Area Rugs

Figure 9.5. Finished carpet in rolls to be cut for use as wall-to-wall carpet

PLATE 22

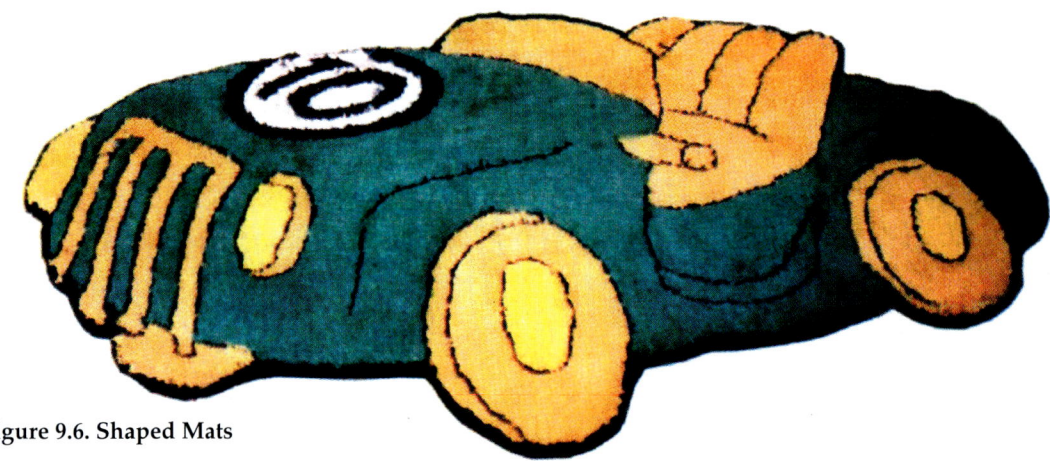

Figure 9.6. Shaped Mats

Figure 9.7. Flat Woven Rug

Figure 9.8. Floral design (pile weave) and geometric pattern (Flat-woven) in carpets

Figure 9.9. Geometrical designs in hand-made Dhuries

hand-knotted ones, the direction of lay of the face fibers falls one side or the other, depending on the style of the knot. In all tufted and woven broadlooms, it is imperative that the direction of lay be made to run in the same direction on all components of every installation.

TYPES OF PILES

1. *Loop pile:* They are made from uncut tufts in looped form, having all tufts the same pile height. The loop pile consists of a continuous strand of yarn. If one yarn is stretched the whole row comes out. It has jute or polypropylene backing. It is also called cording pile. 'Level loop pile' is a hardwiring surface formed by continuous loops of yarn of uniform length. Multi-level loop pile uses different lengths of yarn and therefore has of more sculptured appearance. Loop pile is tougher and more easily maintained but is less versatile in colours and patterns. It also lacks the softness of cut pile since light tends to be reflected off the carpet surface. It can be produced in tufted, woven or knitted constructions. It basically has a commercial purpose. It gives an aesthetic appeal. Its weight is 300 to 1000 gms per square metre of polyprophylene.

2. *Cut Pile:* Cut pile is a smooth, classic finish, often known as 'velour' or 'velvet pile', ideal for bringing a luxurious and sensual feel to rooms. The tops of the loops of wool are cut and the yarn is slightly twisted to produce tufts of yarn that stand upright and produce a smooth, even surface. It can be produced in tufted, woven or braided construction.

    (a) *Hand-tufted:* These carpets do not show the design at the back. They have a cotton or polyprophyene mesh at the back. It is the cheapest way of carpeting. They are available in both plain and designed patterns. The weight is 2000 to 3500 gms of wool per square meter.

    (b) *Machine-tufted:* These carpets also do not show the design at the back. It is a good quality carpet. Its cost increases with the interaction of the design. The weight is 1000 to 1200 gms of wool per square metre. The Transasia Group in India is popular for this kind of carpets.

    (c) *Handloom:* This carpet is not of a very good but of a medium quality. It is a cheap way of carpeting. They are thin as compared to other carpets. The weight is 1000 to 1500 gms of wool per square metre. At times, an acrylic base is provided which increases the cost. They usually have a jute backing.

3. *Hard Twist:* This style of carpet, also called 'faise cut pile' is where the yarn is twisted to a high degree and then set. The hardwiring texture minimises tracking (footprints), shading (irregular light and dark areas in the pile) and fluffing, making it a very popular and practical carpet.

4. *Cut and loop variations:* These carpets have a sculptured appearance created by cutting some loops and leaving others uncut at the same or a lower level. There are also multi-level loop piles where the higher loops have been cut to give a combination of cut and loop pile. This irregular appearance minimizes tread marks and shading. This combination adds a degree of warmth to all loop pile. It can be produced in tufted and woven constructions.

5. *Berber loop pile:* This carpet looks and feels extremely natural, thanks to the blend of many different natural shades of wool in one yarn. They are also very practical, hardwearing and versatile – ideal for areas that get plenty of wear.

6. *Random shear:* Random Shear is created by having two or more levels of loops, with the highest loop area sheared.

TYPES OF CARPETS

As far back as human history is recorded, people have yearned for the comfort and warmth of covered floors. First they used animal skins, then coarse fabrics of primitive weave. These were followed by tapestries, and then pile fabrics of which the hand-knotted Orientals from the East were the costliest and the most highly prized. From France, came tapestry weaves revered to this day – Aubusson, Savonnerie, Gobelin. These were exquisite and also expensive (Figure 9.6 – see colour plate).

It was not until the 18th century that Brussels carpets from Belgium and others made in England on wilton looms brought this desirable home furnishing within average reach. Both were machine made imitations of the precious handmade carpets. Tufting, the most recent of the construction process, has revolutionized the carpet industry in the period since World War II. As it is a much faster and economical method than the earlier weaving and knitting methods, the tufted carpets have eclipsed other types of carpets in today's world.

TYPES OF CARPETS AND RUGS ON THE BASIS OF CONSTRUCTION MODES

1. *Flat Woven rugs:* are made on looms. They have no pile. They are usually inexpensive and intended for casual use.
2. *Woven pile rugs:* are made on looms by interweaving weft threads and pile yarns crosswise through lengthwise warp threads.
3. *Tufted rugs:* are made by hooking pile yarn into a ready-made backing cloth that is later given a coating of latex.

Flat weaves

Grass, fibre and other flat rugs have been improved so much in appearance that they are now suitable for indoors as well as on porches. They are the most attractive without any decoration. Sometimes enormous ugly stenciled designs make then unfit for use (Figures 9.7 and 9.8 – see colour plate 22). Stripes and plaids are usually satisfactory.

(a) *Fibre rugs:* are made from spruce-wood paper twists or yarns woven in the basket, twill or Jacquard weave. They are durable when dry but disintegrate when wet.

(b) *Wool and Fibre rugs:* are woven on cotton warp with alternate strands of wool and fibre. The wool adds an interesting colour element and also pleasing softness.

(c) *Grass rugs:* are woven in the simplest over-and-under weave from continuous undyed grass strands. Grass mats of Southern India are an interesting material in the interiors. They are made from a special variety of grass, called 'kora' grass. These mats are soft textured; though many of them are coarse and have a simple striped pattern. They are of low cost and are available in different widths.

Grass mats are manufactured in a very traditional manner. The 'kora' grass is cut in lengths of five to three feet and left to dry in the sun till they get their fine golden colour. It is split along each length to make thin strips. Weaving also takes time. The warp is made of cotton thread that is sized in rich paste. Each thin piece of grass is lightly mois-tened and passed through the eye of a long stick that acts as a needle. Different colours require different needles. As each line is carefully navigated through the warp, the weaver presses it down to give a smooth finish. Eventually, the surface is given the final polish.

Thus, grass rugs and mats can be used imaginatively to give a different look to the floors.

(d) *Coir and Jute matting:* An alternative to high cost quality furnishing is the economical coir and jute matting as floor coverings. Made from coconut husk, the fibrous center covering of the coconut husk, the coir fibre is natural, bio-degradable and environment friendly. It is tough and durable, versatile and resilient, resistant to flame and fungi, provides insulation and helps sound modulation. The same applies to jute but coir is long lasting, while jute has a tendency to fade and wear out towards the edges. Coir, however, is a coarser material.

Traditionally, the coconut husks are wetted in salt water for a period of 6-9 months, then beaten with wooden mallets to extract the fibre from the coir pith, which is dried thoroughly, then dyed and bleached as required, for the weaving of carpets and other products.

The traditional natural shades have now yielded an exotic array of colours which makes coir and jute an exciting and affordable alternative for covering floors.

Coir and jute floor products range from doormats to wall-to-wall matting rolls, matting tiles and rubberised coir.

Doormats: The ubiquitous doormats are made in a variety of ways: low pile, brush mats, mesh mats, rope mats, loop mats, matting of different colours and designs made or cut to mat sizes.

Matting Tiles: There are rubber or late backed coir matting tiles. Non-skid they are, ideal for wall-to-wall carpeting. A combination of dark and light tiles make an interesting flooring.

Coir Mourzonks: The wearing surface of mourzonks is formed by weft yarn. Intricate designs not withstanding, these are heavy and more durable than matting, manufactured in natural and solid shades, in stencilled, floral and geometrical designs.

Maintenance: Coir and jute as floor coverings require relatively very low maintenance, and so are inexpensive in the long run. Stains should be removed immediately. It should be dry-mopped regularly to clear off dust.

Usage: Coir and jute mattings are not ideal for kitchens and bathrooms since they dry slowly when wet; and neither for children's bedrooms as it feels prickly to their feet. Inspite of limitations, coir and jute matting may be used anywhere on interior floors.

Sisal rugs: The new sisal-look wool carpets are proving popular. The fashionable raw, woven texture sisal (coir and seagrass) is stylishly translated into the warmth, softness and durability of wool. Sisal rugs are made from a tough, heavy fibre obtained from the leaves of a plant. Some strands are dyed and then used for making rugs.

Linen rugs: These are medium in cost, wear very well, look well and come in a large variety of plain or mixed colours. They are useful in dining rooms, sun rooms, porches and halls.

Rag rugs: Of the manufactured variety these are usually made of cotton rags, in fairly light colours. They are washable and durable but light in weight.

Thread and thrum carpets: These are made of a thick cotton warp and a thin wool weft, and resemble tapestry weaves.

The dhurries are an easy substitute for expensive carpets. With its range of colours and affordable price tag makes it possible for budget-conscious designers to incorporate it as floor coverings. Being a small investment, they can be disposed of so as to change the whole colour scheme of any interiors. They range from the traditional geometrics to bright continental designs blended with any style. Also, they are reversible. A dhurrie, thus, lasts longer and is cheaper and easier to maintain than a carpet. Also, they are lighter and absorb less dust (Figure 9.9 – see colour plate 22).

(i) The manufacturing process of a dhurrie basically involves a flat weave with a warp and weft like cloth, quite different from a carpet, which is knotted or tufted. The thickness of a dhurrie depends on the yarn. A fine yarn will give a light, fine dhurrie and a thick yarn, a more textured look. There are basically five types of dhurries, made out of wool, cotton, flex or silk.

(ii) Weft and warp are both of cotton. This is the cheapest and is normally available in plain colour or multi-coloured stripes.

(iii) Warp and weft are of wool.

(iv) Warp and weft both of cotton are made with shortest cotton pile on it.

(v) Woollen pile with warp and weft of flex wool or cotton is another combination.

(vi) Silk pile with flex wool is used for warp and weft.

Maintenance: Dhurries are easy to clean as surface dust can be brushed off without much effort. Dhurries can also be shaken or beaten while hanging to remove dust. Cotton dhurries can be scrubbed at home with soap. A quick remedy for stubborn stains is acetone.

Wiltons: They are named after the town in England from where it originated. They are usually patterned, and their texture, produced by sculpture/embossing. It uses only one colour at a time on the surface; the other yarn remains buried. The pile is woven into the backing. Thus, they tend to give an extra 'hand' or feel to the carpet. They are woven with the Jacquard mechanism which controls colour by means of perforated cards. The best carpets and rugs generally available in the shops are the Wiltons. They are available in a variety of values, depending on the thickness of the pile. Wilton is commonly known as the weave with "hidden value".

Chenille Carpets: They are widely recommended for any elegant interiors because of their glossy texture. They are woven on special looms. They are the most luxurious and also very expensive and are usually made-on-order. They may be plain or patterned.

There are actually two weaving processes required for the construction of chenille carpet. The first weaving produces a 'blanket' made with strong warp yarns widely spaced and soften yarns in the filling. The blanket is then cut into lengthwise strips, which are pressed into a v-shape. These furry strips are responsible for the name chenille – the French word for 'cater pillar'. In the second weaving operation, the strips of chenille are placed over the background warp yarns of the carpet, and each strip is held in place with several shots (filling yarns).

WOVEN PILE RUGS AND CARPETS

Velvet rugs and carpets: Velvet is the simplest form of weave. The pile loops of velvet carpet are woven over long wires – one wire for each row of loops – which are withdrawn as each row is completed. They are made by woven construction on a rather simple loom that does not require the use of yarn underneath the pile; therefore, the velvets are less resilient and less expensive than other woven pile rugs. The velvet weave is used for carpet at all price levels. Because the method of construction is relatively inexpensive, it is possible to produce carpet that will be durable, serviceable, and attractive at a comparatively low cost.

The velvet method is also called a tapestry weave when it is used with a low uncut pile for special effect. Velvets come in several qualities and in different effects, the frieze twist weave being one of the best. Traditionally, velvet has been a solid-colour, smooth surfaced, pile – woven carpet, but now it comes also in tweeds and stripes.

Aroministers: It derives its name from a town in England. Whenever many colours (more than 5 or 6) are used in an intricately patterned carpet, one can be quite sure it is an Arominister.

The texture of an Arominister carpet is usually even-level cut pile. Almost all of the stretch is lengthwise. Special and very complex carpet loom is used to weave these. They have a smooth cut pile surface with almost all of the yarn appearing on the surface. They are made on special looms that can handle many colours. Thus, the colours and patterns are limited only by the number of tufts in the carpets. Sometimes, the rugs and carpets are too colourful and too strongly patterned. The identifying feature is the heavy ribbed stiff jute ridges backing that only allows the carpet to be rolled lengthwise.

The initial setting up of a pattern is a slow and expensive process, but once this is done the design may be repeated over and over again with relative ease.

TUFTED RUGS AND CARPETS

Tufted carpets are gaining popularity in today's world. The process involved in construction of these is cheaper than weaving on a loom. In this technique, spacing of as many as 2000 needles on a huge sewing machine (12'-15' wide) determines the carpet gauge. Face yarn is stitched through the primary backing where it is bonded to a secondary backing with later before curing in a drying oven. The tufted carpets do not ravel when they are cut in any direction.

HAND-MADE RUGS

Rug making as a handicraft has been practiced throughout the world by many groups of people with different materials, methods and designs. As with any other handicraft, the design represent the artistic development of the people as well as their customs, beliefs and values.

European handmade rugs have constituted some of the finest products. Scandinavia, central Europe, France (Aubusson and Savonnerie rugs), balkan countries have been places from where carpets have been imported for a great many years.

American handmade rugs: Some rug carpeting is still made on hand-and-foot looms, and round or oval rag rugs are braided and sewed or crocheted.

American and Mexican Indian rugs and blankets are the most famous. They have geometric designs, neutral colours sometimes combined with positive colours, the patterns having the design and colour broken up and distributed over the rug rather than concentrated in a few large spaces.

American and Canadian handmade hooked rugs: Hooked rugs are made of rugs/yarns pulled up through burlap in loops which may be cut or left uncut. They look delightful, native, individual, and expressive of the lives of thrifty, sturdy people.

KNITTED CARPETS

They resemble weaving. Such a carpet is a warp-knitted fabric comprising of warp chains, weft forming yarns and face yarns. They usually have solid, tweed colours with level loop textures because a single yarn is used. By bending the edges of a carpet to expose the backing on the face side, one can see the continuous looping of pile yarns held in place by stitching yarn. They are made in one operation by using three sets of needles to loop together the backing, stitching and pile yarns. Latex is then applied to the backing for body. The pile can be cut or uncut. Knitted carpets represent only a small percentage of current carpet production.

Braided rugs: Strips on cloth were twisted and braided to form a thick strand that was stitched in a round or oval shape. Today, handmade, braided rugs are also made from strips of felt. Design is introduced by the use of contrasting bands of colour.

Hooked Rugs: A coarse fabric such as burlap or canvas is stretched on a frame and strips of cloth or yarn are drawn through it to form loops. Hooked rugs have become more associated with New England, where they became extremely popular during the 19th century. The patterns varied considerably and included geometric, scenic and floral motifs as well as mottoes and proverbs. In some rugs the looped pile was cut for a varied textural effect.

Navajo Rugs: Blankets, mats and rugs woven by Navajo Indians are characterized by geometrical designs and symbolic figures in grey, white and black, with accents in bright reds. Later, other brilliant colours were introduced.

Needle-point Rugs: The process of embroidering on heavy net or canvas with thick wool yarn has been used for some of the finest handmade rugs. Simple cross-stitches are used as the basis of construction but the designs are often complex floral motifs with beautiful delicate shading.

Rag Rugs: Strips of twisted cloth were used as the filling yarn on a loom. Simple rag rugs were made in a plain weave with cotton or linen warp yarns.

FLOCKED CARPETS

They are made by propelling short strands of pile fibre (nylon) electrostatically against an adhesie-coated prefabricated backing sheet. The finished surface can be printed after fabrication. It has a single-level, cut pile surface which resembles velour.

NEEDLE-PUNCHED CARPETS

Earlier, they were made of polypropylene fibers in solid colours; they are now made using wool, nylon, acrylic in variegated colours and designs. They are made by impinging loose layers of random, staple carpet fibres into a sheet of polypropylene, from both sides by means of barbed needles until the entire mass is compressed to a solid bonded fiber mass of carpet. They have no pile and are usually made of olefin fibres. The carpets made are harsh and stiff. They are good for heavy traffic areas and they resist moisture, and so they are a good choice for pool side. These carpets are used mainly as indoor or outdoor carpets, and for artificial grass surfaces. Its surface appearance is reminiscent of felt. It is also used in kitchens, children's rooms and family rooms where an inexpensive, easily maintained carpet is required.

FUSION BONDED

This process produces dense cut pile or level loop carpet in solid colours. It is especially suited to making carpet tiles.

Rya: Rya is the swedish word for 'rug'. It is now a popular type of Finnish rug with linen or wool yarns of various lengths and thickness. Yarns are knotted and secured into a backing to produce deep shaggy rugs in which the patterns are usually expressionistic or bold in design and colour. In Finland, they provide the warmth and colour needed for the long winters; in other countries they are usually used with modern furniture.

ORIENTAL RUGS

The historical symbol of the finest in floor coverings, come from Iran, China, India, Turkey and Russia. Their jewel colours and intricate patterning are equally handsome in modern or traditional settings. The symbolism and mysticism of oriental floor coverings hold an allure for many people who associate romance and tradition with the exquisite patterns and superb colours of Oriental rugs. No two Oriental rugs are ever exactly alike. Each weaver has his own style, and each rug to him is an unique work of art (Figures 9.8 and 9.9 – see colour plates).

Although oriental rugs are still made by hand, machine-made yarns, synthetic dyes, and a flaunting of traditional designs have impaired the intrinsic beauty of the ancient rugs. The older rugs were usually made in the homes of families, with designs and methods passed down from one generation to the next.

The designs in these oriental rugs are not just patterns nor are colours chosen for harmony along. Almost every element has a meaning.

- Yellow is the colour of royalty;
- blue stands for truth;
- white signifies purity and the holy life;
- orange is for devotion;
- rose, for divine wisdom.
- the serpent represents desire;
- the lion and the dog mean victory;
- the phoenix is the bearer of good tidings;
- the stork, emblem of longevity;
- the fish, sign of domestic and married tranquility.

'Antique Oriental rugs' are those that are more than fifty years old. These rugs were made by the nomad tribes of south-west Asia who wandered about seeking-pastures for their sheep, which provided them with food, clothing and rug materials. The wool was washed by hand and dyed with vegetable and animal dyes, and the rugs were woven by hand. The designs were often symbolical of the history of the fabric or tribe. Family pride, care and patience helped them to produce rugs that were the excellent works of art. The older ones are now in museums and in private collections.

'Semi-antique oriental rugs' are newer used rugs made by commercial methods, but not chemically bleached and glazed. Some of these rugs have pleasing design and colour.

'Modern oriental rugs' comprise most of those that are available for purchase. These rugs are made in the orient by weavers who have been gathered into factory centers to work on a commercial basis. The wools are now chemically washed and dyed with aniline dyes, and much of the rug is made by machinery. The designs and colours have been adapted to suit the taste of American and European rug merchants. Importers subdue the garnish aniline colours by a process of bleaching with chemicals and then retouch them and add an artificial gloss by means of glycerin and hot rollers. These products are not works of art and should not be confused with beautiful old rugs.

Oriental rugs are not fashionable now. They are difficult to use because they attract too much attention to the floor and because all other furnishings must be subordinated to them. Some owners of oriental rugs have had them bleached to very pale tints to accompany Modern furnishings.

There are some features that need to be considered while buying oriental rugs. The condition of a rug is important – damage results from age, wear and moths, the workmanship and materials are deterring factors in the price; long, even, erect pile is the most desirable; the number of knots in a square inch is significant, 100 to 200 knots denoting good quality. Some other factors are, small compact patterns and many borders are desirable, central medallions are not; strong colour contrasts and sharp value contrasts should be avoided.

Orientals are a sound investment because there is a constant increase in their values. Wear gives them a mellow patina, so that the more they are used, the better they look. Since they were originally made for use in the desert, they are exceptionally durable and do not show dirt and dust.

PERSIAN RUGS

These rugs have been the most popular of the oriental rugs. They are outstanding for expert workmanship, subtle colouring, and finer design. Their rather small conventionalized designs are based on natural forms such as flowers, trees, vines, birds, rivers, and clouds. The rugs are completely covered with a profusion of these graceful motifs. The names of cities and provinces where they are made are usually given to oriental rugs. Some of the best known include:

- *Feraghan*   : Small stylized flowers in rows on a deep blue ground.
- *Isaphan*   : Intricate all-over design on deep red background.
- *Kerman*   : Light background of cream, rose, blue with designs in other pastels.
- *Saraband*   : Palm leaves on a rose or blue back-ground.
- *Sarouk* (Saruk, Sarook) : Dark reds and blues with floral designs in lighter colours.

TURKISH RUGS

They are bolder in design than Persian rugs. Patterns consist of more highly conventionalized floral and geometric forms. The pinks, tulips and hyacinths are the favorite motifs. The colour is less varied and suave than in the Persian rugs but not so limited as the Caucasian. In the small sizes, used as prayer rugs by the Mohammedan, the design includes a 'mihrab' or niche, which is always printed toward Mecca when the owner kneels on the rug. Among the Turkish rugs are the Anatolian, American, Bergama, Ghirdes and Yuruk.

CAUCASIAN RUGS

These rugs are made by the tribes living on the mountainous isthmus between the Black and the Caspian seas. These rugs are even bolder in design and colour than the Turkish. Among the design motifs are geometric forms of animals and humans, snow crystals, the eight-pointed stars, crosses and itch hooks which are combined into mosaic effects. These are done in brilliant yellows, blues, greens and reds, with some black. Representative rugs are the Cabistan, Aghestan, Razak, Shirman.

TURKOMAN RUGS

These rugs are made by tribes in central Asia and are usually a rich, dark red with a short, cut pile. The design motifs are simple geometric forms without symbolism. The outstanding rug is the Bokhara. Others are from Afghanistan, Beluchistan, Samarkand and Turdestan.

CHINESE RUGS

Chinese rugs of the seventeenth and eighteenth centuries were beautiful, but the modern rugs are often without merit in design, for in them naturalistic dragons, clouds, waves, trellises, flowers and birds usually blue, are scattered around over light backgrounds in confusion. A promising new development has been the production of one-tone, textured, patternless rugs. The borders are usually narrow, the central medallions or small spaced motifs are almost always symbolic. Soft yellow, gold, cream and apricot are the favorite colours.

INDIAN RUGS

These rugs vary considerably in different parts of India. their designs have been influenced by beautiful fabrics, sculptures and architecture of the country. The least realistic designs are usually the best. Indian rugs display great variety in colour and texture. Numdahs which are fetted rugs, are frequently made of animal hair. They are often embroidered with view and floral designs in bright colours.

ENGLISH ANTIQUE RUGS

They are more sedative and consetvative. Floral emblems placed almost primly in separate squares of green, crimson, rose and gold identify the Georgian era; bouquets of cabbage roses on light or dark grounds typify the Victorian period; classicism is revived in the arabesques, swags and urns of the Adam and Regency period.

EARLY AMERICAN RUGS

They are unpretentious but are colourful. Hooked and braided rag rugs and embroidered rugs coloured with homemade dyes are a proud part of the craft as well as decorative heritage.

MEDITERRANEAN AND SPANISH RUGS

The rugs alongwith the furniture from this part of the world, have a Moorish quality. Most famous are the Hispano – Moresque, a medley of geometric from the Moors; animal, human and floral forms from the Gothic motifs from the Italian Renaissance; and Spain's own heraldic emblems and coats of arms.

CARPET INSTALLATION

For wall-to-wall installation of carpets, the base should be smooth, level led and without gaps. A carpet cushion or underlay should always be added as it adds to the life of the carpet, absorbs traffic noise and improves thermal insulation. The various carpet cushions used includes felt padding, sponge rubber, urethane foam and foam rubber; urethane foam being most durable, though expensive. The carpet is then either stretched – in over the cushion or glued down using adhesives. Aluminum/brass extrusions are used to protect the edges of the carpets at junctions with other floorings.

USAGE

Carpets and rugs are widely aspired for any residential interiors for their aesthetic advantage, though kitchen and bathrooms where there is likely to be certain amount of splashing or spilling, they are not an ideal choice. They are equally preferred for high traffic areas too, such as commercial places, offices and public institutions.

Maintenance and Care: Stains should be removed immediately. Vacuum cleaning is necessary at intervals varying widely with uses, and less frequently, dry cleaning or shampooing which are best done in-situ, are required.

For average carpet wear, the following care is suggested:

1. *Daily:* Once brush lightly with a hand-push sweeper.
2. *Weekly:* Thorough cleaning with a vacuum cleaner to remove deeply embedded soil and grit.
3. *Periodically:* Depending upon traffic, most carpets need to be cleaned by professionals. You can clean your carpets by yourself with wet/dry treatments; but home cleaning cannot equal professional cleaning.

Good care extends the life of soft floor coverings. Where and how you live determines the amount of cleaning needed. Such things as the cleanliness of the air, the condition of the grounds around the home, and the number of family members as well as the family activities will affect the amount of soiling.

Chapter 10

SOFT FURNISHINGS

Fabrics are an integral part of every room in the home. They contribute so much to the general atmosphere that they should be chosen carefully and used aesthetically. Every note of texture and colour, from the functional dish towel hanging informally in the kitchen to the most elegant draperies in a formal living rooms, adds to the personality of a home, and expresses the taste and culture of the family living in that home.

The basic requirements for the various fabrics that are selected will differ, because each one will be used under a different set of conditions. Few of these fabrics alone may have every desirable characteristics. Therefore it is important to know how a fabric will be used in order to select one that has the most desirable and appropriate properties.

The textile industry and most department stores tend to separate decorative fabrics from regular yarn goods, but there is also some overlapping in these arrangements. For example, satin and velvet are considered dress materials but they are useful in several areas of home furnishings. A few decorative fabrics may be used for certain articles of clothing. Sheeting, muslin, casements and towelling materials are often sold in meters in some divisions of the textiles or furnishing stores. It may be necessary to look into several retail outlets or furnishing stores to find exactly the right kind of fabric for a specific purpose. Once identified, the required amount of fabric can be bought in desired length.

The other factor that influences the purchase of furnishing fabrics is their width. Widths of these fabrics vary considerably in different kinds of materials. In general, decorative fabrics tend to be wider than the dress materials. Cottons and linens as dress materials are usually forty five inches wide, but in decorative and furnishing fabrics, forty eight to seventy two inches is a more standard width. It is, therefore, important to know the exact width of a fabric when finalising the required length of fabrics for anything one plans to make.

HISTORICAL BACKGROUND

Any study of fabrics is incomplete without tracing its historical background. The earliest form of weaving was probably the interlacing of reeds for meeting the shelter needs. This was followed by basket weaving. Later fibres were woven into cloth which could be substituted for the pelts worn by the early man. The discoveries by the Archaeologists have revealed evidences of spinning and weaving in the very oldest homes of pre-historic man. Ruins of the Stone Age show fragments of fabrics of linen and wool, some of them decorated with designs of human and animal figures. Spindles, looms, and needles have been discovered from the period of Bronze Age. These tools were similar to the ones used by some tribals even today.

In the East, the great skill of spinning, weaving, dyeing and ornamenting fabrics of wool, flax, cotton and silk gets unfolded from the earliest times of ancient history. The whole process of textile making is depicted on the walls of the ruins of Thebes, Babylon and Nineneh. Ancient Egyptian mummy clothes and coptic textiles are among the finest in existence even today. The knowledge of spinning and weaving spread westward from the east. Greece and Italy taught Spain, France and Finlanders, from them Germany learnt, and in turn taught England and Scandinavia.

In America, ancient Peruvians wove cloth of fine conventional designs and of exquisite colours, which has lasted atleast a thousand years. In the homes of the cliff dwellers of the south western United States, fragments of different types of textiles have been discovered. Some American Indian and some Mexican Indian tribes continue to work even today as primitive weavers to create similar designs of the bygone days. The powerloom which was invented early in the nineteenth century works similar to handlooms. Therefore, it is not basically different from handlooms, although improvements now equip it for weaving much more intricate designs and patterns.

Textile Terminologies

To choose wisely from the available textile products in the market, it is important to know some textile terminologies. With the vast number of textile products that are available, customers are often bewildered by different trade names, and other terms used by the manufacturers and traders. To overcome such situations, it will do good to learn certain terminologies used in textiles. The study of textiles can aid a customer not only in the selection of textile products, but also in their use to the best advantage. By doing so, the customers are in a position to get maximum satisfaction in all their purchases. Keeping this mind, some of the common textile terms are mentioned here as an aid to customers.

Fibres

The smallest structural units in a fabric are called the fibres. Fibres can be natural or man-made/ synthetics. The natural fibres are cotton, silk, wool, flax, hemp, jute, kapok, asbestos, rubber etc. The man-made fibres are rayon, acetate, polyester, cashmilon etc. No single fibre has all the desirable qualities that would make it useful for every type of fabric used in the home. Raw materials may be selected on the basis of their cost, availability, texture, strength, resistance to soil, ease of care etc. In many cases, the manufacturers try to incorporate several desirable qualities into the fabric by blending two or more major fibres.

Yarns

Fibres are twisted or spun into yarns in preparation for the weaving or knitting of fabrics. The construction of the yarn influences the appearance, durability and serviceability of the fabric. Shorter fibres need more twisting to form the yarn than do long filaments and the way the fibres are twisted as well as the amount of twist will influence their performance. In some cases a high twist is desirable for strength and durability, but this exercise decreases the lustre of the yarn. To retain the lustrous quality of some fibres, the long smooth filaments are sometimes given very little twist. Unless a strong, durable fibre is used, such low twist yarns will quickly show signs of wear and tear.

Construction of cloth

There are various ways of constructing a fabric and the methods used will affect not only the appearance but also the purposes for which the cloth can be used. The major methods of construction are weaving and knitting.

Weaving is the principal method of constructing cloth, although a few fabrics are made by other means. The various basic weaves include plain, basket weave, twill weave, herring bone, sateen, dobby, Jacquard, Leno weave. Novelty weaves with beautiful textures are made by combing coarse and fine yarns or twisted and slub yarns.

Other constructions include braids, bonding, felt, nets and laces, non-wovens, tufting. Felt is made of pressed matted fibres, and jersey is knitted. Knotted threads are used to make nets. Sometimes fibreless fabrics are made from fluids which become sheets by solidifying as in the cases of rubber and plastic sheets.

Dyeing and printing

Colour can be applied to textiles at any stage of its manufacturing – before the fibres are formed or before spun in to yarns or before used on the loom or piece-dyed after weaving or knitting. Colour fastness is a difficult quality for the manufacturer to guarantee. A drapery fabric might be perfectly satisfactory under most conditions but if it is exposed to very strong sunlight for many hours on each day, or to excessive soil and fumes, it needs frequent laundering. Such harsh conditions could cause a colour change faster than in more normal conditions.

Various methods can be used to print designs on fabrics, such as block printing, photo engraving, resist printing, discharge printing, roller printing, screen printing and warp printing. In some fabrics, the pattern is printed on the warp thread only. In machine printing patterns are generally printed on fabrics by a series of copper rollers each of which print one colour. Thus beautiful and variety of colours and prints are produced on fabrics.

Elements of Art in Fabrics

It was mentioned earlier that a variety of fabrics are made available in the market and these are produced in different ways. The main purpose of all these is to create beautiful fabrics, and beauty of fabrics, depends largely on their expressiveness, feature, colour, and pattern.

Expressiveness

All fabrics are chosen mainly because they express a certain purpose or theme or idea. A single outstanding idea can be portrayed on the texture, colour and pattern of a fabric. When these three elements fuse to give one impression, the effect is enhanced, and dramatized and a distinctive fabric is created. Dignity, delicacy, sophistication or almost any theme can be found in fabrics. The markets of contemporary times provide a variety of fabrics, which a customer can choose to suit his requirements and ideas.

Texture

The Texture of a fabric is the most significant quality which determines its character more than any other element. Textures can be felt as well as can be seen. The modern trend is to emphasise textural variety on fabrics in order to obtain interest and beauty without employing pattern. Textural variety in fabrics is created by combining yarns of different thickness or of different sources. Embossed fabrics may also exhibit textural variations.

Colour

Fabrics are pieces which provide most of the colour interest in a room. Therefore subtle and beautiful colours should be sought in them. Overdoing with the colours in the fabrics, however, might result in confusion/irritation. Finely related colours in drapery, upholstery, and rugs are essential for harmony and beauty. A knowledge of colours, their qualities, and the various colour schemes, can help a customer to choose suitable fabrics for the home.

A detailed study on colour is dealt with in a separate chapter and can be used to learn more about colours, its usage and the schemes for furnishing various rooms.

Pattern

Weaving, printing/embroidery or putting patchwork are the ways of producing decorative patterns on fabrics. Printed patterns should suit the textiles on which they are stamped. Woven patterns that grow out of the weaving process itself are more distinctive than those that are printed on finished fabrics. Thus different types of patterns on fabrics can be created through various means.

Stripes, checks, floral or dots, are the safest patterns seen on fabrics. Stylized patterns and modern geometric patterns are often desirable in household furnishings. Ferns and other large foliage patterns are among the best and popular designs. Floral patterns are difficult to select as they are more often poor than good since the naturalism and sentimentalism of Victorian times still persist in them. Sometimes birds and animal patterns are also made on fabrics which can be used in children's rooms. They can also be combined with patterns like baloons, Gift packs etc. to decorate a nursery room. However floral patterns look the best in laces or on net fabrics. In any case, the choice of pattern depends mainly on one's taste and the theme of the room in which these are going to be used.

Windows and window dressings

Strictly speaking, windows along with walls, floor and ceilings form the shell of a room and may be treated to blend unobtrusively into the background. On the other hand, windows are one of the first things to catch the eye when you enter a room. Windows are really a decorating opportunity for setting a room's mood and style. Window treatments can cover faults, create illusions, or become a focal point. The mood set by window can be quiet reserved or wild as a night club, luxuriously formal, contemporary, classically charming and nostalgic. Window treatments offer large units of colour, texture and styling in a room. Whether to provide a inconspicuous background or to make a decorating asset, it should be remembered that the basic purpose of a window is to provide light and air. At times it may be desirable for sunshine to flood the room; at other times, it may be appropriate to block out strong sunlight. It may also frame a lovely view or exclude an unsightly one and it may be the means of ensuring privacy where privacy is needed (Figure 10.1 – see colour plate 23).

The simplest window treatment is sheer curtains to soften the stark outline of window frames, filter the sunlight and cover and block harshness of glass reflection at night. More elaborate treatment with knowledgeable choice of fabric and style can add immeasurably to the beauty of the rooms and the house.

There are a number of window dressings one can do to enhance the beauty of the windows and thus of the whole room. These range from starkly simple to strikingly elaborate. One can hang under curtains (glass curtains) or over draperies or both: lowered shutters or filigree panels, window shades in a choice of roller, traditional or modern styles to roll up and down. The treatment can be short or long; the material sheer or opaque, lined or unlined, plain or patterned; the decorative effect enhanced by ornamental nods, valences, sways. Appearance of course has a place, but what a person wants a window to do is the primary deciding factor. Windows can be dressed to disguise when not desired or to open up views when desired.

Properly dressed, window can help to do a disappearing act with architectural faults, balance out-of-scale room to correct proportions, conceal unsightly radiators or air-conditioners, and screen out the interior view from outside. Window dressings can also give a bright look by not concealing the day-light or provide a dark room by total concealment or a subdued light effect by partial concealment. Thus window dressings can do wonders in a number of ways.

Window as part of total Environment

Before deciding on a window treatment one should consider the room's size. It is important that the treatment should be in scale with the size of the whole room. In a large room with small windows, the treatment should extend beyond the window perhaps even cover the entire hall to make it proportionate to the room. In a small room, the style shall be simple and light in effect, so that the room seem larger and so the window treatment should not make the room to appear smaller.

If the ceilings are high, a deep valence can be installed to lower them visually. In a low-ceilinged room, a ceiling-to-floor treatment will act as a counter balance. When windows are very narrow, draperies hung beyond the frame will have a widening effect. A pair of windows that are too close together can be made to look like one by extending the side draperies beyond the windows and hanging the central panels conventionally. If the intervening space is very small, a single panel may suffice.

For the too short windows, it will look good to hang curtains from the ceiling to the floor, perhaps adding a valence from the ceiling to the window top. A concealing valence and small size curtains are the practical solutions to the problem of two windows of different sizes on the same wall. The shape of the window affects the curtain design, especially when it is unusual. A window with a curved top line may have the draperies curved to fit, but curved top windows are often concealed behind straight top draperies.

Exterior Window Treatments

Exterior window treatments are usually not emphasized or decorative in nature. Some of these treatments are awning or shades which protect windows, from rain, sun and wind. Depending on the climate, shutters are seldom used or added to temper the light, heat and cold. Griller made of wood, plastic, aluminium and steel are also used. They also control privacy, sun and wind. Overhanging roofs are exterior shading devices that do not control privacy but do offer some protection from the sun or cold.

Hard Indoor Window Treatments

Hard indoor window treatments combine function with beauty and one available in a wide variety of styles and colours. Any one of them can be used to provide a complete window treatment all by itself. They can also be used with draperies, curtains, valences and cornices to create variety and interest to the room settings. These include –

- blinds
- shutters
- screens
- panels
- shades
- curtains etc.

Blinds

Venetian Blinds: They actually originated in China. At present they are available everywhere in a variety of styles and materials. They adopt well to new and glamorous decorative treatments. Originally there were split bamboo and match-stick blinds. Today they have undergone vast improvements and are available in many colours, textures and thicknesses. These blinds made of adjustable horizontal or vertical slats of wood composition, metal or plastics admit light and air while giving privacy. They have a trim tailored look. There are various other ways of window treatment, but there is still nothing that beats the blinds for adjusting light without hindering ventilation. Some blinds even exhibit beautiful patterns of a painting when the slats are closed. Blinds can be used with or without draperies, specially in summers (Figure 10.2 – see colour plate 23).

Venetian blinds have been supplemented by newer variations called shade-cloth vertical blinds, which consists of vanes of shade cloth. Approximately one and one-half to six inches wide, they can be cut to fit regular silted windows or the floor-to-ceiling type.

These blinds are, however, difficult to clean, and are liable to become unworkable at times as parts wear out. But, when well maintained and regularly wiped to clean, they last for a long time.

Shutters

An effective foreign custom is to have two narrow hinged folding wooden shutters inside each window instead of shades and curtains. Although shutters are expensive, they have a long life and require little maintenance. As well as being decorative, they can disguise problem windows, conceal air conditioners and radiators and enhance the architectural elements of a window. They allow maximum flexibility in light control and ventilation. They ensure privacy.

Shutters can be purchased as ready-to-use, painted with natural wood-like finishes. They are also available as unfinished-ready-to-fit attachments which can be treated later to suit or match the exact requirements of a room. Shutter panels are designed with movable or fixed louvers (sloping slats) with cane, mesh, solid panel, stained glass, perforated hard board or fabric inserts. They can be adjusted according to the light requirements. Shutters may be used with draperies or in place of draperies. They come in a wide variety of stock sizes, or can be made-to-size orders for windows of unique dimensions. These shutters provide ideal treatments for windows during cold seasons and also as security panels.

Shoji Screens

Decorative screens are painted or covered with wall paper that are of full window height. Shoji screens originated in the orient. They are paper screens serving as wall partitions or sliding windows. They are used in many oriental homes and the westerns use the Shoji Screen to establish an oriental theme in a room.

PLATE 23

Figure 10.1. Window Dressing

Figure 10.2. Venetian blinds
for windows

PLATE 24

Figure 10.3. Curtains for window dressings

Figure 10.6. Idea for Household Linen

Panels

Panels slide across or away from windows. Panettrac is a sliding fabric panel that looks like a Shoji Screen. Flat fabric panels are attached to slides in a multi-channel tracks with velcro fastening tape. Panettrac is ideal for use with contemporary and oriental furnishings. Variety in panels can be achieved through the fabrics with different colours, patterns, stripes, checks etc.

Shades

No one who notices decorating trends would have missed what has happened to shades. Gone are the dowdy old reliables, plain, green and white. And in their place has come the most extraordinary variety imaginable. Shades fit windows of all sizes and can be made to suit any decorative finish. There are four basic types – the pull-down shade, the Austrian shade, the Roman shade and the bottom up shade. All of them come in plain and textured material that is either opaque or translucent. The opaque material ensures complete darkening of the room, and the translucent material allows light to filter through.

Shades come in solid colours as well as in striped and in patterns. The colours range all the way from the most unobtrusive, subtle, neutral tints to vivid decorative shades. They can be simply tailored, accented with a decorative trim, or have a custom look with shaped hems, Scalloped borders and matching cornices.

Austrian and Roman shades are not attached to shade rollers, but are regulated with a cord and a pulley. The Austrian shade is shirred and draped when lowered or raised. A sturdier variation on the same theme is the Roman shade that falls into pleats when it is raised.

Bottom-up shades are mounted at the base of the window and drawn up from the sill on a smooth pulley mechanism, which provides flexibility in light control, privacy and ventilation. These shades are recommended for cathedral windows or for covering an air conditioner.

Roller shades are practical, as they are easily adjustable to give privacy and to exclude light. Although there is structural unity in having all the shades at the same level, a sheet - in feeling results from keeping them all half way down the windows. It is often advisable to keep them rolled up out of sight unless utility demands otherwise. As a rule it is better to have roller shades in light colour, preferably like the walls or the wood work. Roller shades are sometimes made by tacking oil cloth, chintz or other fabric to rollers.

Shades can be made with fabric laminated to shade cloth. Readymades in a host of exciting colours and designs are also available in the markets. Some are trimmed and adorned with a painted scene for the fool-the-eye-effect. Bamboo or reed roller shades in natural colours can be used for indoors to induce a tropical, studio or garden effect. Now-a-days chicks are also used as shades for windows or balconies to curtail the extra heat and light coming from outside.

Curtains and draperies

Curtains and draperies are the conventional ways to decorate the doors and windows. Within this convention, however, lies an enormous variety of decorative window treatments, including some quite unconventional ones. Even ordinary frilly curtains and tailored draperies can assume some surprising and imaginative forms. Conventional window treatments can also be effectively adapted for other decorative purposes. In all, curtains and draperies are categorized under "Soft window treatments". There are so many types of curtains and draperies currently popular that these two terms have rather broader meanings.

In general, curtains are made of sheer, semi-sheer, and light weight fabrics. Draperies may be used alone or over some type of curtain. The term 'curtain' refers to various types of fabric coverings that may extend over only a part of window, all of it, or from the top of the window to the floor; usually they are thought of being hung next to the glass. On the other hand, draperies as a rule, extend atleast from the top of the window to the sill. More usually they extend below the sill to cover an apron if there is one to reach the floor. Both curtains and draperies can be designed to draw back and leave the window completely uncovered (Figure 10.3 – see colour plate 24).

Curtains and draperies are used for a number of reasons. In general, they serve the following purposes:

- Control light, air and view
- Provide privacy
- Add a feeling of warmth to the room
- Lend softness to the atmosphere
- Reduce noise in some areas
- Create illusions in terms of length, breadth and height
- Give a theme to a room
- Add to the aesthetic appearance of a room.

FACTORS IN THE SELECTION OF CURTAINS AND DRAPERIES

As mentioned earlier, curtains and draperies are used for a number of reasons. Whatever may be the purpose, curtains and draperies should be used in such a way that it allows a window to fulfil its major functions of admitting air, light and view.

Since curtains and draperies are used for varied reasons, it is important that they are chosen carefully. By considering the factors that influence their selection, a person is able to make wise purchases, without wasting the valuable resources like time, effort, money etc. The factors that influence the selection of curtains and draperies are, therefore, elaborated in the following paragraphs.

Style

Initially window hangings were regarded as utilitarian furnishings only. In the last 300 years or so the thinking has changed and gradually they are thought of specifically as decorative items, which has emerged to give more meaning to this kind of window treatments.

During the 10th century, interior decorators started focussing on window as part of their professional input. Although decorators of this period preferred pale fabrics, they were usually of rather light weight ones such as satin or tafetta. The drapery itself tended to be extremely ornate replete with overlapping swag, extravagant cascades and elaborate trimmings. In victorian times, the window treatments were considerably simpler. Today's designers still draw heavily on 19th century arrangements of curtains and draperies.

In the 20th century, decorators were as a rule, content with a single layer of curtaining, but it was still not uncommon to see blinds, sheer curtains and draw curtains, or ornamental drapery, all at the same window. While most modern draperies are crisply tailored, swags and carcades have by no means disappeared, and though we seldom fancy lace on our windows. Modern casement clothes with their open and sometimes intrinsic weave amply demonstrate that we appreciate the charming effect of light filtered through a flimsy textural fabric. Today, further more, we can find bright prints and stripes and checks woven in bold patterns and exotic of colours that were unknown to early decorators.

Styles for formal settings: For a formal living or drawing room of whatever style, it is preferable to have full length draperies, probably made of opaque material. In a traditional room, it would look good to use luxurious fabrics, perhaps figured or damask brocade or cut velvet. In modern setting, draperies tend to be somewhat self-effacing. They are generally made of plain coloured, even white fabric, the visual effect being supplied by the fabric texture and the heavy linings. Heavy silk or simple patterns on white jacquard weaves are typical.

Styles for informal setting: For cottage type living rooms, and for family rooms and ornamental draperies and opulent fabrics are inappropriately luxurious. Where a real casual requirement is wanted as in kitchens and bathrooms, curtains should be brief, reaching no further than the top of the window sill. This length besides being unpretentious facilitates dusting and protects the curtain from unnecessary soiling. Bright cotton prints are a favourite curtain material in kitchens, but open weaves and printed sheer fabrics are also appropriate especially when good natural sunlight is important.

Pattern

Though plain curtains are common, the patterned ones are more popular as they provide variety and a higher aesthetic value. The patterns can also be used to signify certain themes and to express various meanings to a room. Decorative pattern is usually produced by weaving, embroidery, printing or patch work. Printed pattern should suit the textiles on which they are stamped. Woven patterns are more distinctive than the printed ones that are done on woven fabrics. The patterns seen on furnishing fabrics that are created in either of these processes can be grouped under the following categories.

- The overall pattern
- Small neat pattern
- Bold sharp pattern
- Floral patterns
- Geometrical patterns.

The *overall pattern* is created when the pattern covers the entire area of the fabric. Here the design and its components are so small and so close that one does not even realise it is there until he gets right upto it. The effect is so much the same as one would get with a plain fabric of crimpy finish. Practically no background space is made visible. The closely woven or printed pattern seem to merge with the background space (Figures 10.4).

Small neat pattern is the next type of pattern which is also spread all over the fabric. It is different from the previous one in the sense that they are evenly spaced and a certain amount of background space is made visible. The design is, therefore, appears more tidy, which are normally woven or printed in a pale colour against a dark background or vice versa. Thus they make a striking contrast when looked from a distance.

Bold sharp pattern is another type of pattern which can be done as checks, stripes, animals or broken up patches of colour or sheer abstract patterns. Most of these are strong on colour. These pattern can be done either in two or more colours. The pattern effects are created either at the time of weaving, or printed after the fabric is woven. The most popular of all patterns are the natural *floral patterns*. Many of them are also projected as stylized patterns of the natural ones such as roses, chrysanthymums, dahlias etc.

Some of the furnishing fabrics also have *geometrical patterns* such as squares, triangles, circles etc. They are evenly spaced and so appear attractive. Such patterns give weight to the fabric. They become more conspicuous and contrasting when the patterns are made in black and white. Sometimes a combination of two or more of these patterns is also produced to give variety, and to provide striking results.

The use of pattern or lack of it in the drapery material should depend on the amount of pattern already in the room. A patterned rug or a patterned wall paper calls for plain draperies. If a room has uninteresting furniture or is dull for other reasons, it should be enlivened by an attractive contrasting pattern in the drapery fabric. Since draperies hang in folds, patterns should be viewed in folds. Whatever may be the reason, it is necessary to keep a balance between plain and patterned areas in a room. As a rule, pattern should cover atleast one-third of a room, and the remaining two-thirds as plain.

Though a person can use one weak pattern (all over or small-neat) and one strong pattern (bold or floral or geometrical) quite successfully in large rooms, but it is not advisable to have two weak or two strong fabrics side by side. The only real exception to this rule is – if one of the strong pattern is a very plain one like a simple stripe or a check in not more than two colours.

Patterns also give some effects in a room. A pattern with a wide upright stripes make the wall to look higher. A pattern with stripes going across make the room look wider. In a really small room, draperies with large pattern can be overpowering. In a large room, on the other-hand, a small pattern can almost disappear, or look dull and feeble. Thus it is important to keep the size of the room in mind while deciding the type of pattern in the room furnishings.

Texture

The next factor in the selection of furnishings is texture. One or two different textures do make a room more interesting, and can give a different feeling in the room. When everything in a room has a satin smooth finish, one would feel being in a vacuum. Otherwise if everything has rough hard texture, it would make a person to feel being surrounded by pebbles. So addition of a little textural variation between curtain or upholstery material makes it to look subtle, and at the same time, attractive. Certain textural qualities are generally associated with particular colours, such as dark red, emerald green etc. Bright gold and purple suggest luxury and elegant of textures, when used in correct proportions. Texture has assumed greater importance these days through the use of thicker novelty yarns of various materials.

Delicate rooms require such fabrics like silk, satin, chiffon, taffeta, velvet etc. Sturdy rooms need natural cottons, fish net, casement cloth, voile, gauze, organdy, coloured sheeting, handlooms fabrics etc. Elegant stately rooms require raw silk, satin, damask, brocade, metallic cloth, linen or other high-grade textiles. Modern rooms can use either the new sleek shiny composition fabrics or the boldly textured novelty weaves or polyester blends which can provide constrast for modern smoothness. Other fabrics and weaves are terry, pebble weave, jersey, cheese cloth etc., which are also found in the modern market. Rubber, plastic or cellophone sheets are meant for shower curtains and may also be used for bath and kitchen windows.

It is important to remember that the texture of the drapery must agree with the mood and style of the room. Variety in texture is necessary, but extremes of contrasting textures must be avoided, because extremes of contrast are incompatible. Texture patterns as well as coloured patterns should be considered in selecting fabrics.

Proportion

While deciding window treatments, it is advisable to take into account the line and proportions of the room. For instance, short curtains are more suitable for the lines of a cottage rooms. When additional height is desired in a room, draperies and curtains are hung straight upto floor level. Such curtains create a vertical line effect which gives an illusion of extra height to a room. In this case, the line is formed by the silhouette and the folds of the draperies. Rooms with extra high ceilings are improved by the use of valances or horizontally placed pelmets, which also appear to reduce height. It is advisable to avoid a square line on window treatment. The preferable line and the most pleasing to eye is seen on vertical rectangle, an area that is twice as high as the width of a window.

Another factor for consideration is the number of windows. Whatever treatment is chosen, must be repeated as many times as the number of windows. To avoid the repetition of a striking, bold treatment for too many windows in a room, the drapery may be treated as part of the background and repeat the exact colour of the wall. Another way of handling too many windows is to treat them as a single unit in their curtaining. A single wide venetian blind or a reed roller shade can be made to cover two or more windows. In such a treatment, all the intervening wall space is hung with drapery fabric.

Fashion

Like in other areas, fashion affects window treatments too. To meet the challenge of changing fashions, one should be alert to changes by visiting shops, exhibitions, fairs etc. Acrylic yarns and plain coloured fabrics tend to be fashionable at one time, and patterns are in fashion at other times. When patterns are used, it should be soft and not overdone as in victorian styles. Heavy linen fabrics, and jacquard are very much in use these days. A variety of literature in the form of pamphlets, catalogues, advertisements and articles in magazine and leading news papers are also good sources of information for everyone of us. Thus it becomes easier to get familiar with the changes and fashion trends. What is important is to understand these trends and make alterations in the existing stock without incurring much expenditure.

Space

Space, the art element, can be used to alter the size of a room visually. It can either increase or decrease the dimensions of a room. Space can also be altered to give a room a quite feeling of rest and repose. When the space in a room is adequately used and furnished, it adds to its beauty. When the room is over-furnished, it looks overcrowded. Similarly when the space in a room is not aptly used, it looks empty. Large openings through doors and windows in a room encourage the eye to explore the distance beyond – especially when the same material or colour is carried throughout. Cool and light value colour will also add to the spacious feeling. Use of floor to ceiling glass curtains and illusionistic devices such as mirrors, paintings with deep perspective and scenic wall paper can also suggest extra space in a room.

Durability

Ordinary cotton draperies have a tendency to get bleached out by the effect of sunlight, and this in turn may affect the durability of the fabrics. Thicker draperies are preferred as they are more durable than silk or silk cotton fabrics. Acrylic yarns are often introduced for durability in combination with cotton or rayon or silk fibres. Expensive draperies like velvet, jacquard etc. are long lasting and therefore, should be bought with consideration to other upholstery fabrics used in the room. Keeping in mind their respective colour and texture, children rooms should have cheaper fabric that need to be washed very often. Imitation of costly fabrics such as cheap satin, artificial silk and damasks should be avoided, since they prove to be not only uneconomical in the long run, but- also spoil the entire look of the furnishings in a room.

Personality and personal preferences

The owner's taste, personality and needs are such important factors in the selection of window treatments that professionals, at times, may not be able to provide such services. Gaiety is expressed through crisp, airy floral prints, Elegance and restraint can be brought through glossy smooth heavy draperies. Young ones are pink, aquamarine, copper orange, mauve and sapphire. Blue and some types of checks lend informal feelings, besides being masculine. Sober fabrics have tints of pale blue, ocean blue, sky blue, earth brown and metallic green. Purity and formal feelings are expressed through the use of white and cream coloured fabrics. Whatever may be the effects of colours, it is important to keep in mind the individual likes and dislikes while choosing the colours for personal rooms.

Colour

Most of the colour interests are generally provided by the furnishing fabrics used in a room. One or two colours are more effective than a larger number of colours. Whatever is the number of colours chosen, subtle and beautiful colours should be sought in them. Finely related colours in drapery, upholstery and rugs are essential for creating harmony and beauty. When different colours are used in a room, some draperies containing all these colours, can help to unify them. More information on the use of colour is discussed in details in the two chapters on colour.

Mood and theme

The rooms can be made to appear sturdy, stately, dramatic, informal, elite, severe or whatever is wanted, by the use of appropriate window treatments. Thus the mood of the room suggests the type of treatment suitable for its windows. A reed roller shade used alone creates a masculine or studio effect. Sheer or ruffled or delicate net curtains suggest femininity.Long rich draperies finished at the top with regular pleats and looped back showing long glass curtains connote stateliness. Long silk draperies with an extra foot of material lying on the floor express luxury. Short full curtains with gathered valances are informal. A room with ruffled valences and curtains appears artistic.

Cost

For most families, the important factor for consideration is the budget allocation for the purchase of household furnishings. Thus the cost of fabrics also influence the choice of furnishings. The cost of fabrics for curtains might be lower as compared to the expenditures for other furnishings, because draperies need to be changed every five years or so. In almost any room that needs reviving, the most effect for the money spent is obtained by the acquisition of new curtains. When shopping for new curtains, a sample of the existing curtain material should be carried. By doing so, one can select suitable and matching material to the old ones, some of which may still be in good condition.

It pays well to conduct a market survey before the final purchase. All the nearby markets, as the wholesale markets, or the markets known for selling a good variety of furnishing fabrics should be visited and their prices and the variety available should be compared. Only after analysing the information gathered from these sources, the final decision on the choice of fabrics should be made. Some shops also provide the services of stitching the draperies and curtains, when the purchases are made from them. They may also give discounts on the final bills. A customer must take into account all these factors while purchasing the furnishing fabrics for the home.

Selecting furnishing fabrics is not as simple as it appears to be. A large number of factors, as discussed above, influence the final decision on the choice of fabrics. A good knowledge of the fabrics and familiarising oneself with the markets providing these fabrics can help a homemaker to simplify the process of selecting fabrics for furnishing the home.

TYPES OF CURTAINS AND DRAPERIES

Beauty of a room depends largely on the amount of curtains and draperies used and the way they are placed. The use of a room is the chief factor in determining the type of curtains and draperies to choose. Just as modern houses are planned for modern living conditions and new materials, so are the designs for the interiors and furnishings planned. Curtains and Draperies can be made in a number of ways to suit one's personality and need and the requirements of a room. These types of curtains and draperies are discussed in the following paragraphs.

Sheer Curtains

Many interiors are most successfully flattered with a minimum of window dressing. This does not mean minimum fabric, however; soft sheers look best hung in deep, luxuriant folds. Sheer ruffled curtains, crisp and smocking pay a lovely compliment to any room. A contemporary interior can benefit just as surely from any open work curtains. Some are so richly textured that they appear to be hand woven. For a high ranch window and door in a child's room, sheer cafes may be all the curtaining you need.

Draw draperies

These are mounted on traverse rod and so they can be drawn, open or closed. Different types of traverse rods permit wide variety of treatments. Overdraperies may be used alone or with glass curtains roller shades or venetian blinds. They should reach the floor and ordinarily they should be made so that they can be drawn over the windows to exclude the blank night view. Some draw curtain panels have such provisions that they can meet at center of the window (two way draw); others draw one panel across an entire window area (one way draw). Curved traverse rods are also available.

This treatment graces more windows in more homes than any other and harmonises with any mood or period. They are easy-to-find in readymades and easy to make in custom draperies. The style possibilities lie in the fabric choice and here, anything goes, from the plainest cottons to the most, sumptuous damasks, antique satin and velvets with only an occasional exception, draw draperies are appropriate for all types and styles of windows and rooms of many periods. In a formal setting, these are designed to ensure privacy in the evenings and are thus more or less opaque. Generally they are left open during the day and can be drawn by hand or by chord after dark.

Curtain/drapery combination

The combination of sheer as under curtains and heavier fabrics as draperies more than doubles the beauty of either one, blending as it does light-filtering softness with dignity. Such combinations with plain fabrics become more striking when the window opens to view a beautiful landscape or a garden or a lake.

Tie-backs: Draperies which are tied back in graceful folds lend dignity and formality to a window and thus to a room. Panels are draped aside and held to the frame or the wall with extra pieces of fabric or special fixtures. Tie-back curtains are often made of sheer fabric – self fabric decorated with ruffles and are extremely ornamental (carved wood, embossed metal or beading). Placement of the tie-back may be high, low or centered, as it appears most pleasant to the window and room proportions.

Cafe Curtains

This many-talented window treatment can be dignified; Quaint or casual cafe curtains are short curtains that cover a portion of a window. They hang in layers of different colour to pick up several shades in a room. They act as cover ups for radiators and air conditioners; they also give privacy with no sacrifice of light. They are often hung on decorative rods by means of rings, clips, hooks or loops.

Ruffles

For softness, and feminity, the ruffle has no rival. Their charm can be expressed in tiers, in a single ruffle at the bottom of a plain curtain or as a valance. They are the delight of a young girl's domain. To ruffle effectively fabrics should be light weight.

Valances

As the top part of the window treatments are often unattractive in appearance, they are sometimes concealed by means of valances. Just the right touch of top interest can make a window treatment distinctive of one's own taste. The decorative finish at the top of a window may consist of a gathered or pleated ruffle or a shaped piece stiffened with buckrum. A valance board is a narrow shelf to which the fabric curtain is attached. Valances are the coverings glued to plywood. Instead of a covering, the valance can either be painted or stained. They can assume sinuous shapes or to absolutely straight lines. Curves seem best suited to period rooms. Straight lines are most in keeping with the contemporary settings. To the basic framework, tassels, braid and ball-fringe can be added.

Swags

A simple type of valance, consisting of a length of pliable material carried across the top of one or more windows is passed through festoon rings or holders made of plastic, wood or metal, possibly tied in a knot and allowed to hang down to whatever length that is desired. It is usually used with short pieces at sides to form cascades. A sway is similar in decorative effect, to a valance, except that it is done with a free-flowing folds of fabric and so is much looser in line. Swags may be used alone or with blinds, shades, curtains and draperies. It is suitable more for formal window treatments than for informal settings.

Cornice

A window treatment may be finished at the top with a cornice which is a stiff box-like treatment. It is a wood or composition moulding or frame, from 4 to 8 inches wide painted or covered with some material. A decorative tin or wire arrangement may substitute for the usual wood cornice. Fabric covered cornices are sometimes decorated with fringes and braids.

Glass Curtains

They are made of sheer fabric in simple straight lines and hung next to the window. They may be used with or without draperies. Usually they cover the whole area of a window. Net, organdy, dotted swiss, voile, celanese, scrim, marquisette, rayon gauze, madras or 1 aces are used for this purpose. Glass curtains give daytime privacy, soften the light and modify the harshness of the frame and the shade. Plain white translucent material is to be used. They provide the most desirable effect.

Ropes/cords

Curtains and draperies are omitted where ropes or cords are hung as a substitute, particularly in summer. They are effective over venetian blinds and to develop some period effects, western or masculine themes. When used in traditional colours, they provide an ethnic look to the room. Such effects can be observed in State Emporia and handloom retail outlets.

Jabots

They are the pleated or draped lengths of fabric that hang down the side of the window. Jacobs can be up to floor length or only a foot above the floor level. They are used more for decorative or other effects than for concealing light or view. Thus some other window dressing may have to be provided if the view has to be concealed.

Pinch-pleated/French draperies

They are the draperies that are pleated at the heading. They are custom made or made with pleated tapes. They are used for doors and windows which run for the entire length of a wall in a room.

Shirred Curtain

A shirred Reading is used when there is no valence. A five-inch crinoline in-heading is generally used for this purpose. This consists of four rows of shirring and gives the appearance of smocking. Thus they provide a decorative look to the window dressing.

Sash curtains

Sash curtains are similar to glass curtains but mounted on a rod attached to the sash or the window frame. They may cover only a part of the window. Sometimes it is held taut with rods at the top and the bottom.

Tier Curtain

For producing tier curtains, two or more horizontal rows of short curtains are mounted so that they overlap and give the desired effect. The name tier curtain is given on this account.

Casement Curtains

Casement Curtains are the ones which are used as sill or apron length draw curtains, often made of standard casement cloth or a plain medium-weight opaque material, usually function in place of shades and blinds to give privacy. They may be used alone or along with draperies.

Fixtures

There are so many different kinds of curtain and drapery rods (Figure 10.5) in the market that it is important to plan the window treatment carefully and to choose the fixtures that are most suitable for the desired effect. Traverse rods allow easy flow while drawing curtains and draperies across the window. On a two way draw rod, the curtains will meet at the centre. However, for some openings, such as a corner window, one way rods are more useful. It is desirable to clear the window glass when the draw drapery is open, the drapery rod should be one-third more than the total width of the window. Ornamental rods are used for full length draperies and looks good. Therefore, they serve as a substitute for a valance. At present, a large variety of rods made out of materials such as wood, metal and plastics are available in the market at varied prices. One can choose these according to their taste as well as their budget.

Trims

Similar to the buttons and belts a person chooses for a dress, trims can make the window treatments unmistakably one of its kind. They can supply colour-contrast, interesting shape, and the neatness of line that is one of the meanings of "trim". Rickrack is an all-time favourite braid that does all the three functions exceptionally well. It comes in big and small scallops, and in a multitude of colours to mix-and-match. Contrasting fabric or a solid colour picked from a print, can be used on a plain pleated drapery as edging and tiebacks with remarkable effects. Fringe, braid, ball fringe, metallic materials in many forms and colours, are in plentiful supply and so the trims are among the highest of fashions at present.

FABRICS FOR WINDOW TREATMENTS

The fabrics used for windows contribute to the character of the room. The new era of textile technology has produced creative fabrics for window treatments. Of all the new fabrics, man-mades such as nylon, acrylic, rayon, polyester, plastic and fibre glass are the most popular for their qualities such as availability, variety, colours fastness, durability and easy to wash and maintain. Fabrics made from manmade fibres are being improved all the time.

While purchasing furnishing fabrics, it is important to consider fabric durability, resistance to sunlight, and soil, possible insect damage, suitability, total cost, problems of construction, shape, stability, maintenance and ease of laundering and cleaning. Finishes affect appearance and performance. Excess starch and sizing are undesirable. They are removed after the first wash or cleaning and the fabric loses its shine and strength. Chemical treatments are available for the fabrics that make them permanently, crisp, crease resistant, moisture and stain repellant, resistant to yarn slippage, washable, shrinkage controlled, non-in-flammable, fade resistant, and wash and wear. A customer can look for these qualities as these information are found on the label.

Figure 10.4. Samples of patterned furnishing fabrics available in the market

The stiff and cumbersome draperies of the past are out, whereas the trend is for fabrics that are soft and supple, so that they fall in graceful folds. Even in the formal treatments for the traditional rooms, lightness and grace are required. This is available in the fabrics of today that also resist sun and soiling and do away with the need for lining though in some cases it is still desirable.

In drapery fabrics, anything goes so long as it is congenial to the room setting (Figure 10.4). Sheer and open-weave and light weight opaque curtains, are equally popular and are used with or without accompanying draperies. Non-Iron sheets come in beautiful designs and colours and make superb and coordinated curtains and draperies when matched to those on the bed. Time-honoured chintz, for colonial and English or period rooms, has lost none of its appeal.

Damasks and antique satins, as well as the recently-introduced practical velvets, are elegantly impressive as windows. Patterns, whether vividly modern or quietly traditional, are most effective, particularly with matching wall papers. Modern fabrics and dyeing methods make all of them sensible and practical choices. It is good to hold the fabrics being considered for window treatments upto the light and check translucency or opaqueness. A sheer fabric that will transmit daylight is a good choice for a north-facing kitchen window.

To create a private living and dining room in the evenings when the lights are bright, it is advisable to consider a tightly woven opaque fabric. Tightly woven draw fabrics are also a good choice for a bed room to keep out the morning sun.

MAKING DRAPERIES

It is not difficult to make one's own draperies. They should be cut extra long so that they can be altered and also to give an extra allowance, because they shrink from cleaning and from exposure to the air. An extra half a meter of material can be concealed in a curtain making double hems at the top and bottom to meet such situations. Draperies and curtains of all types should be made to a full length, a width or more of the material being allowed for each side of the window.

If the window is narrow, it is enough to have drapery on only one side. An ample quantity of a cheaper fabric is far more effective than a narrow curtain of more costly material.

A few miscellaneous observations are worth mentioning here. Selvages must be cut off some materials before hemming to allow curtains to hang well. Hems intended for the insertion of curtain rods should be sufficiently wide to allow easy movement. Lace or Net curtains should be hemmed by hand, as it is almost impossible to rip machine stitching on net to make alterations when necessary. Small weights are sometimes sewed along the hems of draperies to make them hang well. The lower lines of lining and drapery fabrics should not be sewed together for each one should be free to shrink or stretch independently.

Lining

It is not necessary to line draperies, particularly if they are very full and of thick material but lined draperies usually hang better. They also last longer as the interior fabric is protected from the sun and soil, which rots and fades it. Costly and delicate fabrics are usually lined. Linings need not be made of the usual cream or beige sateen; a turquoise fabric might well be lined with a contrasting yellow material. Sometimes when the lining is done neatly, and with another type of material, the draperies can be reversed to provide a different combination.

Trimming

As a rule draperies do not need any trimming. Some glass curtains are improved by ruffles, however, and curtains like chintz sometimes look well with white cotton balls or thick fringe. In coloured fabrics, trimming should be either a perfect match or a desirable contrast. Too much trimming can be as cheapening to draperies as to dresses. Naturally never more than one kind of trimming should be used.

Tie backs, if used, should be placed so that they divide the drapery about in thirds, never in halves. A tie back is generally made from the drapery material, but a cord may be substituted. Sometimes a tie back is looped over a nail with a rosette of crystal or metal as its head. Wood, metal, or plastic arms are also used to hold drapery back. The modern sale outlets provide a large variety of materials for this purpose.

THE HANGING OF CURTAINS

It is a better design to get the curtains hang straight than to loop them back tightly. They should extend to a structural line in the room. A good line is created when the curtains hang to the floor although in informal rooms it is consistent to have them come up to the window sill. Curtaining the windows becomes more puzzling when the openings in the room are not uniform. Narrow high windows, French doors and air-conditioner underneath windows complicate the problem. The usual solution is to hang the curtains to reach the floors on all windows. Curtains long enough to remain a foot above the floor are used only in very formal or social rooms. Curtains may be hung inside the wood trim or outside depending upon the effect desired. They should be hung to cover the trim when it is desired to secure the maximum amount of life, when the wood work is unattractive, or when the proportions of the window or group of windows need to be improved by the effect of additional width; the curtains should be hung to cover whole length of windows.

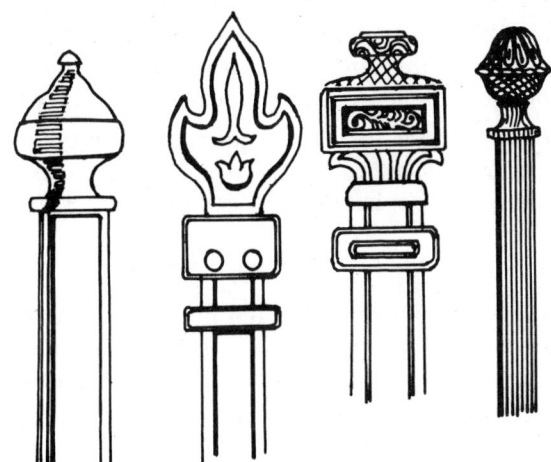

Figure 10.5. Ornamental rods used for hanging curtains

Whether curtains are made with or without a valence depends upon the shape of the windows. Valences tend to make the window appear shorter and wider and they will make a group of windows as one unit. If it is desired to make windows appear longer, the curtains should be long and the valence should be omitted or else used in the form of a valence frame hung above the casing of the window covering it completely. In that way, the actual length of the curtains will be increased as well as the apparent height of the window. In most homes the straight valence is the best type. Draped valences are shifted to formal rooms but when they are used in modest homes they appear pretentious.

UPHOLSTERY FABRICS

It is wiser to shop for upholstery, slip-cover, and drapery fabric samples all at the same time. A plain rug and a plain wall allow either the upholstery or drapery fabric to dominate. When the upholstery is featured, usually two chairs are covered with interesting patterned material, other pieces with stripes or similar motifs, and some with plain or mixed fabrics. A room with a patterned-rug or patterned-wall paper usually necessitates plain upholstery fabric. Sometimes it is best to upholster all chairs in similar colours but in different textures. A number of factors affect the choice of fabrics.

Naturally the mood of the room and the lines and size and material of the chairs affect the choice of upholstery fabrics too. A Neo-Classic room might employ striped light blue and lavender satin on a Sheraton sofa. Oak furniture suggests heavy fabrics, whereas metal chairs are in accord with sleek composition fabrics. Thus the material of the furniture and the theme of the room influence the choice of furnishing fabrics.

The common upholstery fabrics are the printed textiles like chintz, cretonne, linens, Jouy prints, and wrap prints; the decoratively woven materials like damasks, brocades, brocatels, armures, reps, and denims; the pile fabrics like velvets, velveteens, corduroy, plush, and friezes; the smooth silks and satins; crewel and needle-point embroidery; mohair and other wools; and the durable quilted cottons. Plastic materials are used, especially where there are children and dogs. Leather is also used on upholstery for men. The purchaser should realize, that smooth flat-surfaced materials such as satin, damask, rep, brocatel, brocade, and tapestry show wear sooner than a material such as frieze, which is a pile weave with loops uncut. All firmly woven fabrics wear better than loosely woven ones. Fabrics that are easily cleaned are preferable. A buyer should have dyefast, mothproof, and possibly flameproof guarantees for all their purchases.

Various techniques of upholstery are tufting, quilting, cording, the use of buttons or nails, or trimming with welting, tape, fringes, or braid. Any of these must be employed with restraint. Only one fabric should appear on one chair; having the back different from the seat destroys unity.

BED SPREADS

These are also a part of the household furnishings. It is important to make sure that the bed-spreads suit the character of the room where they are used. In earlier times, there were hand-made counterpanes such as patchwork quilts, candlewick spreads, woven coverlets, or peasant spreads from other countries. Indian prints, chintz, calico, plaids, or checked materials are also suitable for bed covers in such rooms. A bed room of feminine type may have a satin, taffeta, or similarly fine bedspread. For a room shared by a man and a woman the bedspreads and other fabrics should not be too feminine in feeling. The bedspread in a man's room could be heavy and rather dark, such as brown corduroy. Some materials for bedspreads are cotton taffeta, chenille, up-holsterer's sateen, plain English broadcloth and arras cloth. Slip-cover and quilted materials are effective and cover the beds evenly.

A couch-bed, at present called a Hollywood bed which usually consists of a regular twin-size bed without a visible frame, is easy to cover. A loosely fitted boxlike slip cover with a big box pleat at each corner, or a large sheetlike spread that reaches to the floor at the ends and sides of the bed, is desirable. For a couch-bed in a combination study and bedroom, the spread should be a darker colour and more restrained than for one located in a bedroom. Plain colours are usually the best; however, mixtures, stripes, plaids, or checks are satisfactory patterns because they produce a desirable tailored effect. It is often well to have plain cushions with a figured cover, or vice versa, so that the amount of material of one kind will be less (Figure 10.6 – see colour plate 24).

The fabric for a couch-bed spread should be pleasant to touch and heavy enough so that it will not wrinkle easily. Heavy velour, corduroy, monk's cloth, crash, denim, and printed mohair make good couch-bed covers in rooms where they are appropriate. A bedspread must be *functional*. In a small home, bedspreads or couch covers should not be too good to be used freely. If a busy woman wants to lie down for a few minutes' rest without removing her shoes, the spreads should be dark enough to permit her to do so. There should be a folded shawl or blanket of harmonious colour and pattern on the bed in cool seasons for the comfort of anyone who lies down during the day.

Making a bedspread is not a difficult job. A spread should be made very long to cover the pillows adequately. It is often advisable to make the main bedspread separate from the mattress flounce, dust ruffle, or valance which conceals the box spring and the space under the bed. The top counterpane and the side ruffle need not always be of the same material. When making a spread for a bed with an ugly frame it is well to consider making slip covers for the head and foot boards, too, from the same material. Quilted fabrics are especially effective for this purpose. Separate shaped, padded headboards and sideboards, to be covered with the same material as the bed-spread, are usually procurable, or can be made to order.

Bed Linens

It pays to invest in good-quality bed linens because they wear longer and present a more attractive appearance. Sheets, pillowcases, mattress covers, and mattress pads must be laundered frequently. Poor quality products become limp and sleazy after a few washings. Therefore, it is important to select good quality and durable fabrics for this purpose. Before buying bed linens it is most important to know the exact size of the mattresses on which they are to be used. There is considerable variation in this respect.

Sheets may be either fitted or of the traditional flat type. The fitted or so-called contour types are shaped to fit the mattress. Bottom sheets are shaped only at the two bottom corners. Innerspring mattresses are usually six to seven inches deep; foam rubber mattresses are generally four and one-half to six inches deep. Fitted sheets are made for both types. Sizes for flat sheets are usually expressed as the torn size before hemming. The actual length of the finished sheet will be several inches less than the measurement on the label. It is important to use a flat sheet large enough for the bed.

For a bottom sheet there should be sufficient amount of material to tuck under the mattress, at least five inches at the foot and at the head of the bed. For a top sheet there should be sufficient length for a five-inch tuck-in at the foot and an eighteen-inch turn-back over the top of the blanket. A twelve-to-fifteen-inch overhang at each side is necessary for comfort.

Types of Sheets

Fitted sheets provide a smooth, wrinkle-free surface, which many people find most comfortable. Because they stay in place, less time is required for making the bed. They may stretch taut over the mattress and make wrinkles disappear. However, they are more expensive and may not be adapted to beds that vary slightly from standard sizes.

Sheets are available in a variety of fabrics – some woven, some knitted. Cotton is the most widely used fiber because it is very comfortable for sleeping, but the man-made fibers are becoming more popular. Nylon tricot sheets have a silky texture and they are easy to launder. The blends of cotton and polyester are becoming very popular because they require little or no ironing. Combinations of 50/50 per cent cotton/polyester and 65/35 per cent cotton/polyester have produced fabrics that seem to meet the demands of durability and easy care.

Cotton sheets may be muslin or percale. The basic differences between the two types are weight, type of yarn, and texture. Muslin sheets are heavier and woven with thicker carded yarns. The thread count, or the number of threads in one square inch of fabric, is one way of designating different types of sheeting. A heavy muslin would have 140 threads to a square inch; a medium weight would have a thread count of 120. Lightweight muslin with a thread count of 110 is available but not generally recommended. Percale sheets are lighter in weight and generally made of smooth combed yarns. Percales vary in the quality of yarn and the closeness of the weave. Fine-quality percale has a thread count of 180; an even finer quality has a thread count of 200 or more. Top-grade percale is expensive, but it has a smooth, luxurious texture.

The choice between muslin and percale may depend on several factors. A medium-weight muslin sheet is strong, long wearing, and economical. Good-quality muslin stands up well in commercial laundering, but when charges are based on weight, the upkeep of a heavy muslin sheet will be comparatively high. If sheets are laundered at home, percale may be easier to handle because of its lighter weight.

Well-made sheets will have strong taped selvages at the side edges. Hems will be smooth, with small, firm, even stitches. It is customary to have a one-inch hem at the bottom and a three or four-inch hem at the top. However, some sheets are made with equal-size hems at the ends, which makes it possible to reverse the position of the sheet. If these hems are narrow, however, they are not usually as attractive as the one with a deep hem at the top. If one has sheets of different sizes and different types in the linen closet it is helpful to have some identifying label stitched at a bottom corner. Many manufacturers provide some sort of a marking, or sheets can be easily marked with coloured tape, thread or a laundry marking pencil.

Design

Colour and pattern have become important style features of bed linens. Fashion has moved into bed linens. In fact, many consumers buy extra sets of sheets that they really do not need because they like the printed design or the colour of the sheet. In most cases reliable manufacturers use dyes that stand up well even with commercial laundering. However, in poor-quality sheets the colours may fade. Bright, fresh designs on bed linens often contribute to the attractiveness of a bedroom, but this is another matter of personal preference. Of course the design and the colour scheme should blend with the decor of the room for the most charming effects.

Requirements

Specific needs for bed linens will vary for each household, depending on the number of beds, the number of different sized beds, the types of sheets (flat or fitted, white or coloured, and so on.), and how the laundry is done. A good basic rule to follow, however, is to allow six sheets for each bed and three cases for each pillow. Thus there will be two sheets in use, two in the laundry, and two in reserve. Linens will wear longer if they are rotated in use, so the reserve supply should be used as frequently as the others.

Many homemakers like to protect mattresses with heavy muslin covers and pads. These are especially desirable in times of illness, when food or medicine may spill and spot a mattress.

TABLECLOTHS AND PLACE MATS

A good-looking table top, which has been made heat resistant, needs no cover of any kind. However, tablecloths and placemats which meet the following requirements, are a good buy.

Expressiveness: Informality is the theme of most meals of today; therefore tablecloths have been replaced by small mats. Fabrics, fiber, tin, wood, cork, plastics, and string are employed for mats, consequently they vary greatly in character. Large damask linen cloths are still used for *formal* occasions and sometimes for buffet service. They are, however, difficult to procure and arduous to launder. Other formal covers are made of embroidered linen, appliqué, lace, or cut work, in white, cream, or very pale tints. It is obviously incompatible however, with the textures of copper, pottery, autumn leaves, zinnias, or fruit.

Colour: In the average home it is well to use brightly coloured napkins and place mats or small tablecloths, especially if the glass-ware and dishes are colourless. Some designers feel that snowy white tablecloths and mats are unresponsive backgrounds for their appointments and food and, decoratively speaking, consider them as white elephants in most rooms. Some white linen should be dyed pale yellow or other tints to harmonize with the surroundings. The tablecloth sometimes acts as a link in colour between the table and the rest of the room.

Pattern: Decorative design is, of course, not necessary in tablecloths or mats, but if patterns are used they should preferably be stylized or geometric. Naturalistic motifs are structural if they are concentrated in definite areas such as borders. Stripes, plaids, polka dots, or simple borders are the most attractive departures from plain tablecloths. Period designs of merit, for traditional homes, are sometimes procurable. Plastic mats on which pictures are reproduced are absurd.

The best tablecloth designs in lace work, drawn work, cut work, or in other types of patterned cloths are those that follow the lines of the table and are controlled and confined to small definite areas. Some of the most structural lace designs consist of small units containing solid areas. Most of the lace cloths cannot be considered as background for they attract too much attention.

Table Linens

In the selection of table linens there is far more freedom of choice than for other house-hold linens. With a trend toward informal service and many easy-care materials table settings should be both attractive and functional. With laminated table tops that require only wiping with a damp cloth, and with paper napkins, there may be no need for any kind of table covering. This kind of table service, however, lacks a certain warm and gracious quality. Various place mats and tablecloths require little or no care, yet they lend a charm and an aura, of hospitality to meal service. This is an area where individuality may certainly be exercised, but today it requires little effort to set tables that are gracious and inviting.

Placemats and even table cloths in plastic or plastic-coated fabrics have become more glamorous and more sophisticated in styling. Lovely textural effects and interesting colour combinations are rapidly removing the stigma that has for so long been associated with plastic. If there should be no objection to their use even on special occasions for everyday use, especially for breakfast and lunch, they are attractive and require minimum care.

The more traditional mats and cloths are made of linen, cotton, and rayon, along with synthetic fibers. Linen damask cloths have, for many years, been the most desirable cloths for rather formal, elegant settings. The lovely texture of linen provides a suitable background for beautiful china, glassware, and silver. Linen cloths do not become fuzzy with use, they do not absorb stains quickly, and they launder easily. Cotton and rayon have been widely used because they are less expensive, but neither of these fibers offers the beauty and the lasting quality of linen.

Lace is often used for tablecloths that may be either formal or informal. A good lace cloth has many practical advantages. It can be laundered easily, it requires little or no ironing, and it is elegant enough for very special occasions. Spots do not become immediately prominent. A lovely handmade lace cloth is indeed a family treasure. Machine-made lace cloths, although not usually as beautiful, do offer similar advantages.

The synthetics in modern table coverings present beauty and elegance with appealing minimum-care features. Most of the synthetics do not spot easily and require little or no ironing. All of these characteristics are making them more important in the field of table "linens." Be sure to choose these "linens" with a soil release finish.

Requirements

Naturally the requirements for table linens will depend on how family meals will be served, the kind of entertaining that will be done, the size of the tables, and the general decor of the home. However, table linens might be considered as accessories in that they complement the decorative scheme of the room. From this standpoint, they should be chosen with care so that they will provide a suitable background for the other table appointments.

Slip covers

These are an important item in home furnishings. Furniture can be purchased in the muslin or satin stage before final upholstering and covered only with slip covers, for economy and convenience. Slip covers are also good for covering old furniture which may not be appealing, or worn out or inharmonious with the decorative scheme of the room.

Thus it can be seen that soft furnishings play a big role in household furnishings. They are of various types and are available in a large variety of materials, colours, textures and patterns at high, medium and low prices. Any one who is in the job of furnishing the home interiors, can choose these according to his choice, budget and requirements, since modern sales outlets are there to cater to these needs.

GOOD TASTE AND PERSONALITY EXPRESSED THROUGH THE CHOICE OF FURNISHINGS

When an individual chooses an object for her home, it expresses two things – one, she is satisfying some need or desire, and second, she, through the qualities possessed by this object, is unconsciously stating her personality to everyone, with the power and insight to interpret the meaning behind the choice. One's clothes, home, pictures, books, furniture and furnishings, all silently exhibit to the others just what sort of a person, she really is. The objects speak of her interests. They prove or disprove her sincerity. They display one's imagination or lack of it. It is, therefore, necessary to make a definite effort through knowledge and appreciation gained from it, to express in her choices, her best personal qualities.

Things like wall paper or metal carvings that look like wood, or too shiny fabrics imitating costly silks and satins, all these are avoidable, if their significance is understood. For, it would take an unusually strong character to remain true to high ideals of truth and sincerity if dishonesty were the keynote of their surroundings, since mere belongings influence in forming character.

Objects like furniture, wall hangings, pictures and decorative objects may suggest either a masculine quality, feminine quality or they may be intermediate. Normally the same things may not be chosen to furnish a bed room for a girl, a man's room or a guest room. The selection of a little lighter type of furnishings, a slightly smaller, finer pattern in the drapery material, and a little more grace in the lines of the furniture and other objects, would provide a feminine quality. The colours in a feminine room should be somewhat different from those in a man's room. They should not be lacking in character, or should appear weak, but the colours may well be lighter and the textures, finer. A man's room need not be dark or heavy to to be masculine in quality, but it should have no appearance of "daintiness"; it ought to be more solid than a woman's room, and somewhere a forceful note of contrasting colour would provide the desired effect.

In the case of a master bed room, or a guest room, it should be intermediate i.e. just between the masculine and feminine, so that both or either a man and/or a woman would feel at home in it. A transition quality should be present, which may be achieved by selecting furnishings neither distinctly light nor heavy, patterns neither very small nor very large, and colours, neither dainty nor heavy. It is through striking a middle ground in form, pattern and colour that one secures a room in which either a woman or a man, or both, would feel at home.

Besides expressing a masculine or a feminine quality, objects and furnishings would give a social or a domestic feeling, or be impersonal. Domestic quality is the outward expression of the love of the home and family, and is usually informal. The social idea is usually expressed more formally than the domestic. When the term 'social' is used to define one of these group expressions, it must be understood that the more limited sense of the word is intended, as referring to characteristics resulting from an interest in the conventions of a formal society. This expression will vary according to the social standards of the individual. If he is a person of taste, if he is sincere, and his standards are high, grace and charm and fine quality will be reflected in the choice and arrangement of his furnishings. On the other hand, if he is insincere or a social climber, that will become apparent in the things he selects, for they will be ostentatious.

It is important to understand here that the questions of expense, of good or bad taste, of richer or poorer materials, never enter into these attributes of objects such as social or domestic, masculine or feminine, personal or intermediate. It is simply the individuality of an object, just as it is individuality which gives a distinct character to each of four different types of persons. One person is devoted to the home and the family, the second, interested in social life, the third having some of the traits of both, and finally the fourth, a colourless individual with no imagination. Though the actual furnishings of the houses of these four persons would change with their changes of fortune and their acquisition of taste, the essential quality would always be the same. The possessions of each individual would reflect his personality because he could not help surrounding himself with the things that reflect him.

While observing the quality of a man who has keen interest in the home and has good taste and average means, his house and the objects including the furnishings he has selected, would reflect domesticity. The home-like quality is expressed through the low lines of the structure, which tie the house to the ground, making it seem secure and informal. The plantations around the house, heightens this effect. There is an over-whelming air of friendly hospitality and complete absence of formality or indifference. In such a home, the emphasis is on informality, and the casual friendly air of the exterior can be observed in the furnishings and their arrangement. Along with the furnishings, a feeling of companionship has been suggested by the presence of comfortable furniture as well as by the other objects like books, accessories etc., which tend to stimulate interesting thoughts. The comfortable furniture in such a room, is arranged to facilitate easy conversation. A friendly fireplace, and the appropriate use of patterns, accessories and textures suggest sturdiness and informality. It is not difficult to achieve a home-like atmosphere in homes contemporary as well as in traditional styles. This makes it clear that the characteristic quality of a house, does not depend upon the amount of money spent, but upon the choice of the textures and designs and their arrangements. It is not difficult even for a person with poor taste to reflect a comfortable and a home-like appearance, though not very beautifully.

Some individuals, who can be called the "social" type, are influenced by the conventions of more formal society. As compared to the domestic individuals, they are more formal and conventional. Some of the characteristics that help to mark the house as a social expression in architectural features such as the symmetrical plan, the sharp contrast of dark shutters against white walls accentuating the formal balancing of the openings, the semi-circled arch above the windows and doors, the studied grace in the design of the entrance, and the formal placement of plants around the house, etc.

It is difficult to say that the aesthetically good expression of architectural features in one house is better than the others. These types such as social, domestic and impersonal are so different that they cannot be compared with one another. While one type of expression satisfies some people, another type may be more appealing to some others. One's choice between these is a matter of temperament. Whatever may be the choice, one should attempt to show his personality in the most consistent and the most beautiful way possible.

The third type of personality is the combination of the first two, i.e. the domestic and the social inclinations balanced. There will be more emphasis on comfort along with some amount of formality. The expression may be through simplicity and good taste, or costly and formal. In both the cases, however, there is something of the intimacy and informality of the domestic rooms.

The last, and the fourth type individual represents the type most unlike the other three mentioned earlier. He is the unimaginative personality and is colourless and uninteresting. Since there is no scope for imagination, he just follows the crowd. The rooms would be stereo-typed and lacking in personality. There would be nothing to induce interesting thoughts or encourage conversation, no charms, no books or no magazines, and there is nothing interesting to leave an impression that people enjoyed living there. The house is impersonal, and at times, cold.

The modern family with its life of many outside activities and interests, tend to make a house with a corresponding simplification of the furnishings within the home. It is interesting to see that masculine, feminine and intermediate qualities, or domestic, social and impersonal characteristic are definitely expressed in modern houses too. In addition to these, a room also shows good or poor taste. It is, therefore, necessary to aim at developing and acquiring a good taste. Appreciation based upon knowledge will aid in discriminating between honesty and impersonal, between simplicity and gaudy. It will also lead to an understanding of fine quality so that one may express his individuality in the most beautiful way.

Chapter 11

FURNITURE

I̶f architecture is the structure or bare bones of space of a house, then furniture is the muscle. It fleshes out the room and makes the empty space habitable and functional. Thus, furniture lends a certain character to a room which determines whether we call it "modern" or "traditional"or habitable.

Furniture is the nucleus around which the entire decorating scheme resolves. It is generally seen that features such as accessories, soft furnishings, wall hangings, foot mats and floor coverings are planned and selected only after the furniture items for a room are finalized. A scantily furnished room will appear empty and barren. When a person enters a room, it is the furniture arrangement that makes an impression and extends a warm and friendly greeting. It can be said that it is the furniture which gives character to a room and other features add to its ornamental value.

The foremost relevance of furniture is its function or utility. Furniture in a room reflects a great deal about the occupants. What we choose to live with, will reflect our educational, ethnic, social and economic background, besides indicating our age and cultural development. The furniture we select, its quality and style, the harmony or discord and its overall effect transmits this subtle information. It gives clues to where we stand at a given moment of our lives. It speaks silently about how we think, what we value and who we are.

Careful preliminary plans of orderly arrangement are necessary for the creation of beautiful, comfortable rooms. Just as the artist works with paints and brushes to compose a picture and the land-scape designer uses plants, so is the interior designer who constructs the space with furniture and other accessory items.

BRIEF HISTORY OF FURNITURE DESIGN

The development of civilization presents a fascinating approach to the history of furniture. Throughout the ages, homes have represented a way of life that determines the decorative modes. A brief history of the furniture design briefly discussed in the following paragraphs, will give an idea about its gradual evolution from the scratch to the present form.

In medieval times, furniture hardly existed in the western world. The precious little that did exist was more-or-less identical in all countries. Medieval castles were stone fortresses which were mainly for protecting from the enemy rather than for occupant's comfort. Initially the flat surface of the bark of a cut tree was used by people either as a seat or a table. A typical household inventory included a few simple chests (with no drawers), stools and settle benches. Tables consisted of huge boards temporarily laid across trestle bases and dismantled after meals. Chairs were stiff-backed, not upholstered, and consequently uncomfortable, reserved expressly for the Lord of the Manor.

Furniture forms evolved only gradually. The furniture started standing on legs, accommodated hinged doors and finally arrived the drawer. However, the furniture was still massive, heavy and immobile. Then came the *Renaissance* which marked the beginning of an architectural and decorative style called *Baroque*. It added little to furniture comfort and changed the face of furniture design. Court style gradually filtered down to the provinces, first to the homes of the country nobility and finally to the public at large.

The eighteenth century is considered the great age of furniture making both in England and in France. Design reached a degree of excellence. This century dawned with the graceful *Queen Anne* style (1702-1714) and reached its maturity and waned with the elegant Georgian styles that had become better known by the names of its cabinet makers – Chippendale (1718-79), Thomas Sheraton (1751-1806) and George Hepplewhite (1770-1786) (Figure 11.1).

During this time trade was flourishing and hence a greater number of people got exposed to better furniture with bigger homes. The furniture then was evolving from heavy, rigid, blocklike forms into objects of great refinement and beauty.

Table 11.1: Use of Materials and furniture style at various periods

S. No.	Period	Materials	Style
History of Furniture Design			
1.	1100-1660	Oak	Romanesque Gothic Renaissance
2.	1660-1714	Walnut	Late Renaissance Baroque (Louis IV William & Mary, Queen Anne)
3.	1714-1760	Mahogany, Walnut, Laquered Gilded & Painted wood.	Boroque Rococo (Chippendale, Louix XV
4.	1760-1830	Satinwood, Mahogany, Painted & Laquered. Hepplewhite, Sheraton	Classical Revival Louis XVI
5.	1830-1900	Rosewood, Oak, Mahogony, Satinwood, Plywood.	Victorian, Art Nouveau
6.	1900-1970	Rose Wood, Teak, Oak, Mahogany, Ply wood and Plastics.	Victorian, Classic and contemporary
7.	1971-2000	Pure wood, Ply wood, moulded plastics recycled wood of agricultural waste.	Conventional/Traditional, Period, contemporary, Eclectic

After the turn of the century, furniture design declined. Novelty and oddity held sway; consequently, much of the furniture is less refined and more vulgar than the designs of the previous

Louis XV

Louis XVI

Sheraton

American Colonial

French Provincial

Federal / Impire

Chippendale

Hepplewhite

Victorian

Spanish

Adam

Queen Anne

Early American

Figure 11.1. Major classical furniture styles

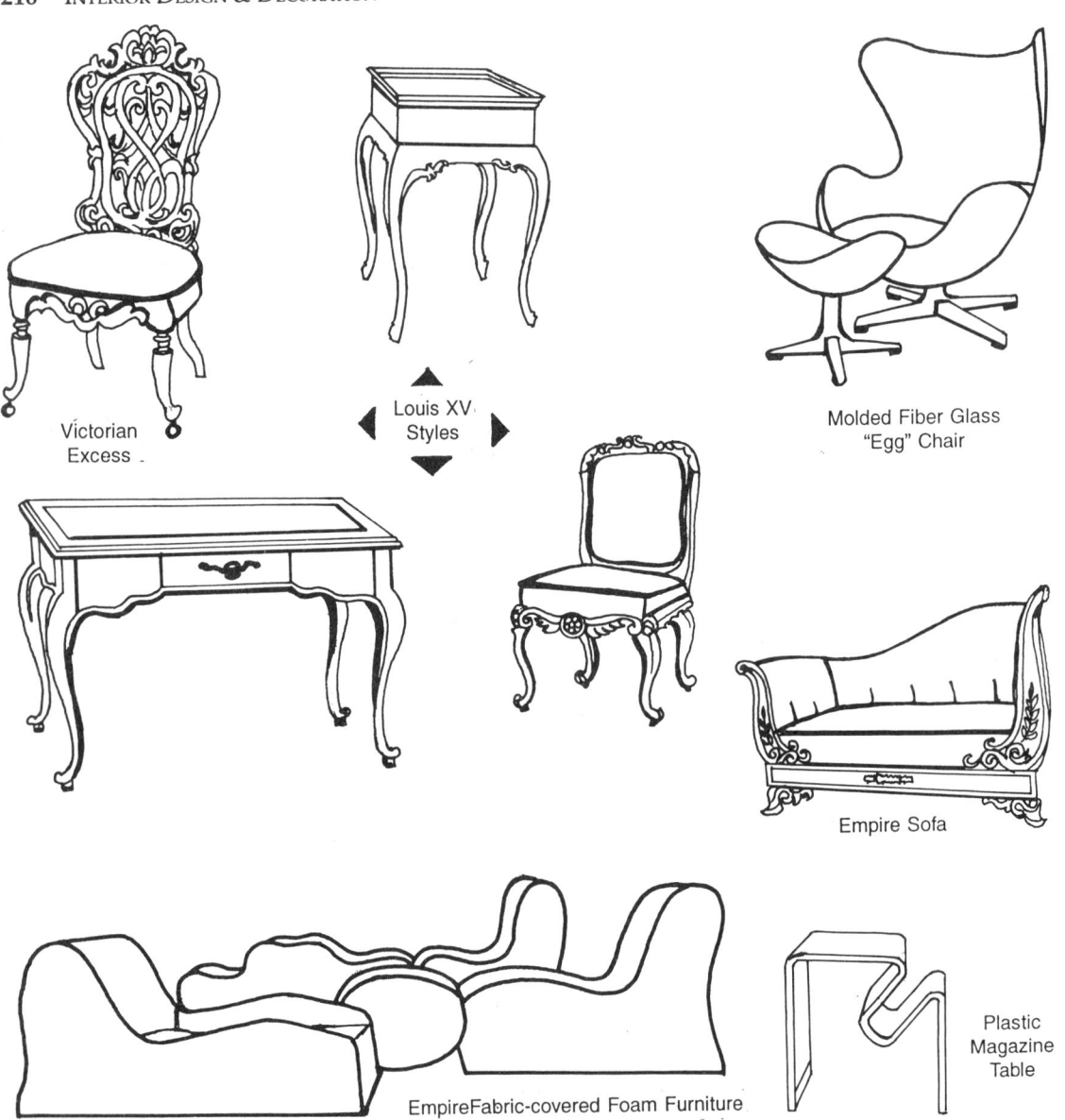

Victorian Excess

Louis XV Styles

Molded Fiber Glass "Egg" Chair

Empire Sofa

EmpireFabric-covered Foam Furniture

Plastic Magazine Table

Figure 11.2. Free form curves in furniture designs

century. The style of our times is truly international. It reflects the advances achieved since the eighteenth century, when furniture was painstakingly made by hand, and the nineteenth century, when machines took over the task, to the twentieth century where new materials and technology have altered our production methods and our whole way of life.

New principles and new materials have led to the creation of new stylistic forms. A design movement is an outgrowth of the changing social, economic and technological climate of a given period of time; if it is to have a lasting effect, it must epitomize the moment. Otherwise, a fad will result, like a plant placed in shallow soil which withers at the first hint of harsh conditions.

FURNITURE REQUIREMENTS AND THEIR ARRANGEMENTS IN THE HOME

Every house and every room in the house has certain furniture requirements without which the house and the rooms give a sense of being incomplete. Therefore, one should exercise caution while selecting furniture. Furniture is an expensive item and bought after careful analysis of one's budget and requirement, and so, it is not possible to change it frequently. Besides selection, the placement of furniture should also be given careful thought. Even less expensive furniture items if arranged in an artistic manner, may look attractive or vice versa. Arrangement of furniture should be such that it does not interfere with the family's day-today household activities. While arranging the furniture, the aesthetic consideration as well as the functional activity centers should be kept in mind. The furniture arrangements for different rooms are different, as their functions differ in many ways.

Living Room

The living room should be the kind of room that is implied from the name, for every member of the family who are to "live" in it. The living room should express the spirit of home to the family and of welcome to the friends of the family. Therefore, the environment should be pleasing and welcoming to the outsiders too. It should be planned in such a way that other collective activities can also be performed there without affecting this function. The living room is of paramount importance since it should provide for all the *quiet recreational activities of the family* such as *reading, conversation, indoor games* and low music. If a family consists of more than two persons some additional area is usually needed, such as a family room, book room, study, den or modified dining room. Thus the living room furniture need careful consideration (Figure 11.3 – see colour plate 25).

The furniture should be so arranged that the living room has a center of interest. The center of interest may be a fireplace, a sofa or any other distinctive piece of furniture, a hobby collection or a group of pictures. A secondary center of interest, is also desirable, particularly in a large room.

How a living area is furnished will depend upon how the room is to function and whether or not there will be other areas in the home that might be used for certain activities. A living room comprises of several group areas. Two points of emphasis in the living room, the center of interest and the principal group of seating furniture, can sometimes be combined, thus securing an unified effect.

Major conversation group

In a modern living room, a conservation corner is arranged with an L-shaped sofa, with built in seats, or with sectional seating units that turn a corner satisfactorily, accommodating several persons seated on each side of the corner. Where a window is the nucleus of the main conversation group, a sofa or love seat facing the window with two comfortable chairs of different types at right angles can be placed at the ends. The sofa may also be at right angles to the window wall with two identical chairs opposite to each other. In any case, a table or some other surface can conveniently be placed for the refreshment to be provided.

A reading or study group consists of a table and a chair or two which is convenient. Good lighting for both day and night is essential. Therefore, this group should be located near a window. A side table adjacent to the chairs should be placed to keep a lamp. A special writing group needs a large desk and a chair that is comfortable for writing. It should have a good lamp and a nearby window along with bookshelves. Reading and writing groups should be as far away from the conversation group as possible to provide privacy and convenience. *A music enjoyment group* may consist of a music system, a radio, flanked by two easy chairs, and a low stool. Audio tapes/CD's, phonograph and records are often placed close to a sofa. The person who likes to lie on a sofa and listen to the music should have a low end table that can be reached comfortably. A piano if kept in the living room should stand with its keyboard parallel to a wall. It is permissible to place a chair or a table in the curve of a grand piano. A sofa and a piano should not usually be placed at the same end of a room otherwise it gives a very heavy look on this side of the room. A television enjoyment group may consist of Television, VCP/VCR/Movie tapes cabinet and comfortable sofa/chairs or rugs on the floor with lots of cushions.

A tall, important article like a bookshelf, a showcase or a cabinet is sometimes needed across the room to balance the fireplace or the place meant as the center of interest. Several specially designed stickball units, such as a television cabinet, record player, picture projector and book-shelves may be combined to save space and give height to their arrangements thereby occupying very little floor space.

Dining Room

Dining is an everyday occurrence, but an attractive and convenient dining area can make every meal special. It should be a pleasant place in which to eat, talk and relax. Suitable furniture, flattering lighting, gracious and efficient service and proper ventilation contribute to an attractive and comfortable dining area. Besides, a dining area should have a distinctive atmosphere, express-ing something of a family's personal taste in decorating as well as dining.

Dining furniture varies with place and time. In the East, diners sit on cushions, in rural India they sit on low wooden stools or floor mats and in ancient Rome they reclined on couches and the modern western world likes to dine on a table and chair. Seats, an eating surface, serving area and storage space are the essentials usually consisting of a table, chairs, cabinet, buffet and trolley. From this, a wide range of types, styles and designs for all kinds of dining areas, traditional or modern, formal or casual, simple or lavish have all evolved.

The dining table is the focal point of a dining area. Standard dining tables are rectangular, square, oval or circular in shape, extensions or drop leaves may enlarge one dimension, changing a square to an oblong or, a circle to an oval. Traditionally, the rectangle is the favoured shape, but an oblong or oval table is considered most pleasing in a rectangular dining room. A narrow rectangular table may be placed flat against a wall, with a chair on either side to save space. Optimal table dimensions are essential for dining comfort and depend on the size of the room and the number of diners. 36 to 45 inches is the maximum width for a small or a medium-size area to allow passage around the table and the height varies from 28 to 30 inches for a standard table.

Wood has long been the favourite material for dining tables. Wood may be natural, bleached, stained in many finishes and colours, or antiqued and distressed for special effects of age and wear. Tabletops of plasticized wood have additional protection. Wood veneers impregnated with plastic material and sheets of plastic laminated to plywood are highly resistant to heat, fire, liquids, scratching and chemical action. Polished marble often used for smaller, usually round dining tables is preferred for it varicoloured streaked surface, luxurious coolness, and durability. However, it can be stained by fruit juice and liquid curries. Glass is also used as table tops because of its distinctive brilliance, transparency and resistance to stain but its hard surface can shatter china and glassware, and the edges need careful protection. Plastic tables, whether clear, tinted, or in solid colours, offer easy maintenance and can be made in almost any shape and colour and combined with metal and wood. When choosing dining chairs, comfort is the prime consideration, provided by good back support and soft or contoured seats. Apart from wood, imaginative dining chairs are also made of fibre, metal (wrought iron) plastic, bamboo, wicker and rattan. A cosy informal atmosphere can be attained by using seats other than chairs for dining. Benches without arms or backs are easy to slide under the table when not in use. Other space saving seating includes a variety of colourful, informal chairs such as portable deck chairs with canvas or leather sling backs and seats, and collapsible chairs.

Today a home with one large room devoted exclusively to dining is indeed a luxury due to space shortage. The character of the furnishings for the dining room depend somewhat upon where the room is located. Dining room furniture arrangements are likely to be rather obvious since it is usually most convenient to place the dining table in the middle of the room. However, with the changing trends, the dining table can be placed anywhere in the dining room along with the chairs comfortably. If it is a separate room, the traditional dining room furniture consisting of the table, chairs, a buffet, chests, cabinets, sink and mirror, trolley, and serving table may be used; but if the room is a dining alcove, adjoining the living room, or is actually a part of the living room, it is more interesting to use less usual pieces – a highboy or a chest of drawers, for example, and a table that does not look too much like the conventional dining table etc. The dining area in the living room might, when necessary, be isolated by a sliding curtain on wall tracks or by a screen that rolls up vertically or by a low bookshelf partition, or by some other device. The style, scale and arrangement of dining furniture should fit the dimensions of the room and enhance the desired mood.

The average person wishes to have a restful time at meals, and if there are many objects and much pattern in the room, it will seem confusing, when a large number of pieces of silver, china and glass making up the table service are brought in, they increase the unrest. The first thing to do, then, in order to obtain a restful dining room is to keep the background simple; second, to display only a few objects; and third, to place the table appointments in an orderly arrangement at meals.

One has to feel content with the space provided in the drawing-dining, living, kitchen-cum-dining, or even in a lobby or passage. However, a well-thought selection and arrangement of furniture for the dining area, can provide a satisfactory service to meet the demands of the family members. In some rooms the table can be placed against the windows with the diners facing the garden, which can be lighted to make it attractive at night. Although plenty of wall space is more important than wall furniture in a small dining room, it is sometimes well to put one tall, interesting piece of furniture across the room from the windows for balance and also to create a center of interest.

A dining room serves other purposes also in certain cases – a study, sewing room, sitting room, music room, etc. and in such cases it should be furnished as an extension of living room. A dining alcove or an area in the living room is sometimes located behind a low shelf partition or in a bay window. Thus between meals there need be no evidence that the place is used for dining. A drop-leaf table standing against a wall can also be pulled out for meals.

A spacious dining room might have two eating places – a permanent oversize table for occasions when there is a large group and a small table for one or two, for extras, or for children. The small table might fold up into a wall depression. A small family need not be confined to any particular spot for meals but can eat before the fireplace, beside a sunny window, on the porch, outdoor barbecues or wherever it is most interesting or convenient.

Bedroom

A bedroom expresses the occupant mere intimately than any other room in the house. A bedroom is distinctly a personal one. The function of a bedroom is to promote rest, and so the first requisite of a good furniture item is a comfortable bed. A good bed is an investment for comfort and health – we spend one third of our lives there. As we begin and end each day in the bedroom, the area should certainly be furnished to suit individual tastes and needs. It should hold the body in any sleeping position, taking over the muscles as they relax. A very hard one leaves the body unsupported between shoulder and hip, causing aches and pains.

It is difficult to relax in a room that seems crowded with furnishings or with bold patterns, and therefore, a bedroom should be arranged in a simple way. At the same time a note of vitality is also required which expresses the owner's individuality and taste. The use to which the room will be put helps in determining the type of bedroom furniture to choose. For instance, a bedroom may also function as a study or work room or even as a living room; as in the case of children's room. Generally speaking, the bedrooms can be divided into the following classification; master bedroom, bedroom – sitting room (suite), children's bedroom and guest bedroom.

A very flexible arrangement is necessary for comfort in a bed-room, as the furniture should be readily movable to suit changing conditions. For example, a study table in a bedroom should not be away from the window or in a dark corner. Good ventilation usually requires a different location of beds in summer as compared to winter times. Fresh air without a draught over the bed is always desirable under such circumstances.

The usual arrangement of one or two beds is to extend them into the middle of the room from the center of a wall. Whereas this is sometimes necessary, it is often possible to place the beds parallel to the walls in corners, leaving the middle of the room free. When two single beds are needed in a small room, it is sometimes advisable to buy a pair, one of which is made to fit underneath the other one when not in use. A bunker bed may be used if such arrangement is required in children's room. A well-lighted place near or in front of a window is the best location for a dressing or a study table. In a combination sitting bedroom, a desk can be placed under a mirror and used as a dressing table, with all the equipments concealed in its drawers.

The remaining furniture is located for convenience. A bench or a chest for blankets at the foot of the bed is useful. An easy chair with a reading lamp, an end table or a sewing stand can also be placed in a bed room near a window. A straight chair is convenient beside a desk or table. A bedside table should be large with a drawer that can be opened by the person in bed.

Although furnishings should be chosen for comfort, convenience and beauty, it is the background-colours, pattern and texture in walls, floors, window treatment, bedspreads and upholstery that does the most to create a feeling of comfort and tranquility.

Children's Room

Planning for children's rooms begins with the decisions on the uses to which the room will be put. Apart from providing the furniture needs of a boy, girl or a small child, the room should be much more than just a place to sleep and change clothes. There should be a place to study without interruption, some room to play, pursue their hobbies and to entertain their friends. Furniture must be placed so that one can move around easily in the room. Further, the furniture, furnishings, floor coverings and accessories should be decorative, practical and economical so that bedrooms should literally grow with children.

Guest Room

The guest room is the one type of bedroom which should be impersonal, since it is a room in which anyone should feel at home. There should be no personal photographs in the room, and the pictures used should have general appeal. Since time immemorial guests have been given an important place in a home, the spirit of hospitality is one of the finest attributes of an Indian home.

Every person wishes to show his or her desire to make the guests comfortable by anticipating their wants while furnishing the guest room. Therefore, the guest room, should, if possible, contain the following things, in addition to such necessities as a comfortable bed, a well-lighted mirror and sufficient storage space, a desk or a table on which one may write; a waste basket, a bedside table with a good reading lamp, books of general interest and a comfortable chair. Since a room used for occasional guests is an impractical use of space, specially when family needs are great, the guest room can become, instead, the best multipurpose room in the house.

Entrance Hall

The entrance hall is a transition between the exterior and interior of the house. It is an important place because it gives the first and the last impression of the home to the person who is arriving or leaving. Since empty space is desirable for facilitating traffic, in a hall the pieces of furniture should be few. A small hall needs only a chair or a bench and a mirror. However, a small table is also useful. In a large hall, such pieces as a table, console, chest of drawers, low chest, bench or sofa with end tables are suitable. Pairs of things are sometimes used to give a formal effect. As a hall should not invite one to linger on pictures and very comfortable chairs are usually out of place in it.

Garden Rooms and Gardens

Garden rooms are summer rooms, winter rooms or spring rooms, the rooms that add excitement to what might have been an unused porch area or a dark corridor. Formerly neglected space can be converted to all-purpose rooms with glass walls, brick floors, and lot of flowers and green plants. By recovering outdoor furniture for indoor use, a whole new concept of summer-winter entertaining and family living room evolves.

When planning a garden room, an essential element is glass. The glass may take the form of window walls or of multipaned windows; it may be in fixed panels or in sliding doors; it may be on one, two or three sides of the room; or a delightful garden room may be devised from an interior space by constructing a skylight in the roof. Today a plenty of well-designed garden furniture are available, much of it as handsome as indoor furniture. Pieces constructed of wrought iron, redwood or rattan, for instance, are attractive enough to be used inside the rooms during the winter months (along with the perishable plants) where they will be quite at home along the indoor greenery.

Furnishing Small Rooms

Where space is limited, a wall-storage unit can be of significant help, particularly if it incorporates a fold-down bed. Unit-storage boxes can be added to as required. Fold-down tables, often used in the garden, could be a space saver in a small room. Many kinds of this are available, some painted, or some stained with bright colors or with a natural wood finish. Tubular furniture in a small but modern room creates an illusion of space if the table-tops and seats are of transparent material.

Basic Furniture Measurements

Accurate measurements of every piece of furniture to be used in a decorating scheme will facilitate furniture arrangement and save time, effort and costly mistakes. It is necessary to take measurements along when shopping.

FACTORS INFLUENCING THE SELECTION OF FURNITURE

A house is considered to be well-furnished only when it is suitably arranged with appropriate furniture items. All rooms which are equipped with a large number of furniture need not necessarily be called functional. Some families may find them sufficient and others, either over-furnished or under-furnished. In addition to being functional and in required numbers, the furniture items should also meet the aesthetic needs of the family members. Besides, it is also not necessary that only expensive furniture items are satisfactory. All these requirements can be met only when the family members consider and analyze their needs, the room size and the budget. Once these factors are clear in their minds, they should visit the local markets and make a survey of the kinds and types of furniture available there. A brief market survey would yield a general idea about the availability of the furniture items and their cost. However, the final decision and selection of a specific furniture item should be made only after analyzing the following factors in detail.

Table 11.2: Approximate Furniture Sizes*

Living Room		Dining Room	
Sofa (3-seater)	3' × 6'	Rectangular table	3'6" × 5'
Love Seat	3'× 4'	Round table	5' diameter
Overstuffed Chair	2'6" × 3'6"	Side Chair	1'6" × 1'6"
Wing Chair	2'6" × 2'9"	Arm Chair	2' × 2'
Bridge Chair	1'6" × 1'6"	Buffet	2' × 2'
Flat-top desk	2'6" × 4'6"	China cupboard	1'6" × 4'
Winthrop desk	2'3" × 3'9"	Serving table	1'6" × 3'
Secretary	2' × 3'6"	Chest	1'6" × 3"
End table	1' × 1'8"	**Bed Room**	
Coffee table	3' × 5'	Twin Bed	3' × 6'
Console table	1'6" × 3'	Double Bed	6' × 6'
Round lamp table	2' diameter	Bed table	2' × 2'
Round coffee table	4' diameter	Dresser	1'6" × 3'6"
Round table	3' diameter	Chest	1'6" × 4'
Grand piano	5' × 7'	Dressing table	1'6" × 3'
Upright piano	2' × 5'	Easy Chair	2'6" × 2'6"
		Side Chair	1'6" × 1'6"

* 1' = 12"   1" = 2.54 cms

Comfort

The most important consideration while selecting a piece of furniture is comfort. The livability of a home depends largely on its comfortable furniture items and furnishings. The ready-made furniture available in the market is according to certain standard measurement. Thus a standard easy chair has a seat depth of 22 to 24 inches and is 17 inches high in front and a little lower at the back. An occasional chair is 19 inches deep and 18 inches high. Arm rests are above 7 inches above the seat. Seat backs are 17 to 19 inches high (Figure 11.4).

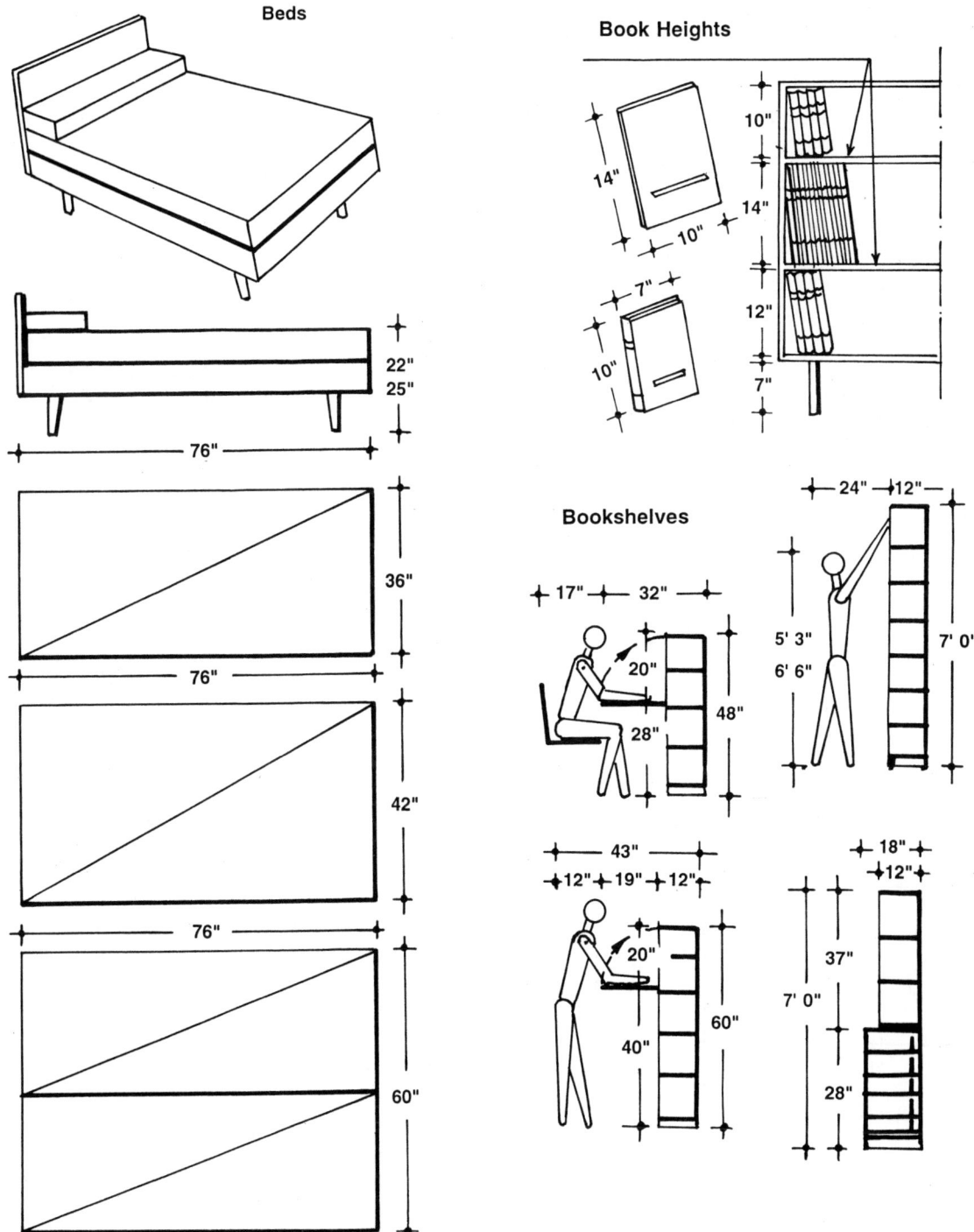

Beds

Book Heights

Bookshelves

Figure 11.4. Basic Furniture Measurements (Contd.)

Low chair

Desk or host chair

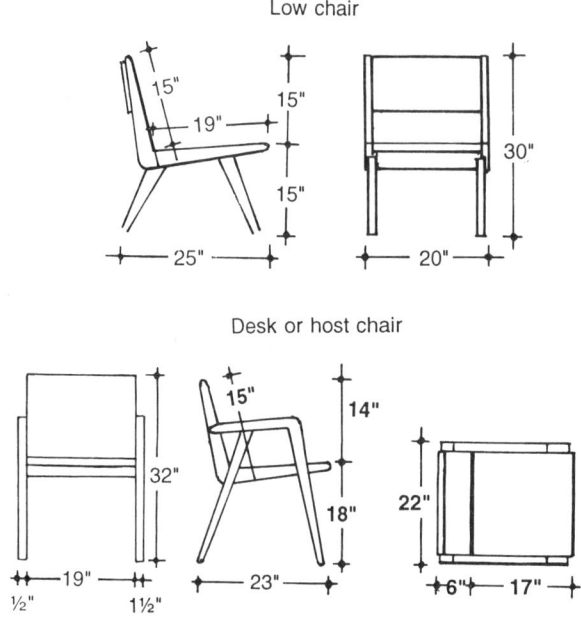

Proportional heights

Side chair

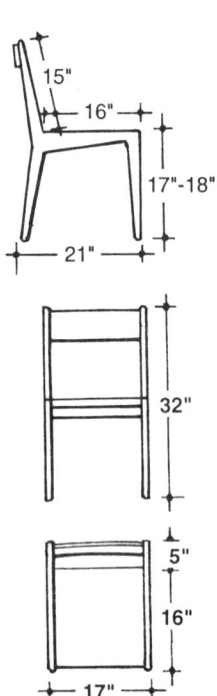

Figure 11.4. Basic Furniture Measurements

PLATE 25

Figure 11.3. Living Room Furniture

Figure 11.6. Decorative Furniture Designs

PLATE 26

Figure 11.7. Upholstered furniture

Figure 11.8. Moulded Plastic Furniture

However, furniture that does not conform to standard measurements is also procurable and can be made-to-order. While procuring such items, one should always keep in mind certain features of comfort. For example – a chair should be deep enough to reach the back of the knees of the seated person. The chair back should have a comfortable slant and height, and should support the shoulder blades well, or even the head, if that is desired. The children's furniture should suit their measurements. Some furniture is made with the adjustable legs that can be lengthened as the child grows. This type of flexibility makes the furniture usable in the long run.

The weight of furniture and its mobility are the other features that also affect comfort of the seated person. Some very light chairs or stools that can be moved easily are needed in living rooms. The most mobile and useful lawn and porch chairs and tables have wheels in place of back legs. Thus it is not necessary that furniture be costly to be comfortable, but it must be designed to fit the body, and make the user to feel comfortable.

Expressiveness

The theme of the room limits the choice of furniture. For example, a cottage style expresses informality, comfort and simplicity and calls for furniture of the same characteristics. The kind of wood, the shape of the article, the style and the colour, all are the elements that help to create the mood or expressiveness desired.

The following are some of the ideas that are expressed in homes consciously or unconsciously – repose, naturalness, sophistication, intimacy, formality, coolness, delicacy, strength, freshness, antiquity etc. For instance, a room having glass topped furniture indicates delicacy whereas heavy furniture items make a room to appear strong and sturdy. An informal drawing room may include a rocking chair and a few other furniture items while a formal room may confine itself just with a sitting area near a window. A room with potted plants and large natural flower arrangements may express freshness and a room in the absence of these may look barren and dry.

Style

Sometimes furniture are bought because they represent certain styles. The three possible choices of style can generally be grouped under as period/cottage or modern/abstract. If the choice is period or cottage, specific periods or types must be selected, like some furniture items representing Moghul period or British style. The style of construction of the house should also be kept in mind while selecting the furniture style. If the house is constructed in a cottage style, the furniture items used in such a house should also be of the same style. Similarly, the modern construction features of a house also demands the kind of furniture with features of the same style. Some houses may be filled with furniture that are characteristically Indian. For example, *divan* has become an important piece of furniture with bolsters and cushions and one or two *piris* or low stools adorning the living room. Along with Indian colour schemes, these houses present a decor-typically Indian. The Rajasthani and Gujarati furniture style presenting a rare style and beauty are superb for such purposes.

Beauty

It was earlier mentioned in the first chapter that every individual is born with some kind of taste for beauty. Any item placed in a house should possess some aesthetic value. While selecting one furniture piece between two similar items, it is the beauty of the piece which influences the final choice. One general rule to follow as a guide in the selection of furniture is that the simple and plain things, are the better choice in the long run. Simplicity should be the thumb rule in the selection of beautiful furniture items. Simple furniture items not only appear to be beautiful, but are also, functional, easy to maintain and eligible to fit into any corner.

Utility

All furniture pieces are bought with the main intention of having some use. Therefore, unless an article is useful, it should not be given a space in the home, regardless of its beauty or sentimental association. This aspect becomes more important in the Indian urban households where space is a limiting factor and flats are made with the bare minimum area because of the cost factor. Therefore, the furniture requirements of every room should be carefully studied and planned. Before making the final purchase, their utility should be calculated and analyzed in terms of the space available and function of the rooms. Sometimes double purposes make furniture more valuable, as table for music system that is also an end side table, the low bookshelves to be used as a seat when a large group is to be accommodated for a short time or a double bed with a box underneath or a sofa-cum-bed that can be used when guests are to be accommodated occasionally.

A manufacturer of unit furniture works to a standard basic measurements/module, so that different pieces can be joined together in a variety of ways to suit individual requirements. Because unit furniture is usually built against a wall, it makes the most economical use of space within a room and simplifies floor cleaning. It is not as easy to rearrange as individual pieces of furniture, but it can be dismantled and moved to a new position.

Storage compartments in furniture designed mainly for another purpose can be great space savers. For instances beds may have drawers underneath, useful for storing blankets, linen/other articles. It is important, that while buying a bed with drawers, it must stand in a position where the drawers can be opened. Convertible bed settees often have space in the base for bedding to be stored. Armchairs and matching stools are available with storage space below both seats. Occasionally tables often include space for magazines, and books storage, and there are designs with lift-up tops and interiors that can be used for drinks, toys, needlework or papers.

The most utilitarian furniture procurable for small quarters is the modern *unit furniture* consisting of cupboards, shelves, tables, radio cabinets, desks and chairs designed to fit into compact, contiguous groups. Unit furniture helps to make a room serve several purposes since it conceals miscellaneous equipment. The effect is the architectural features with the pleasing continuity of line.

Balance and scale

It is always unwise to buy furniture and accessories simply because they are in fashion, without paying attention to their size, proportions and relationship to each other and to the space in the room allotted to them. The actual dimensions of a piece of furniture are basic in determining its scale and proportions, but the way it is designed can sometimes have subtle effects. A long, open-back, cane-seat settee, for instance, can occupy a great deal of space and still give an impression of lightness – out of scale with a heavy, squat armchair occupying less space.

Construction

A well built furniture is always an asset because it gives a long and satisfactory service. A good furniture will give many years of service and satisfaction and for that reason, should be bought deliberately. Every piece of furniture should be studied from the point of view of its construction. A *complete examination* of a furniture and acquiring information about it is a must for every consumer and decorator. The purchaser should look at the back, bottom and inside of each piece as well as at the front.

Drawers should be taken out, doors opened, drop-leaves examined and surfaces, edges and joints studied. *Good workmanship* is indicated by complete finish on backs and undersides and by the use of screws and not nails. Such furniture will have a smooth finish without spoiling the grains of the wood that is used in its construction.

Firmness and rigidity

A well built furniture can be tested by judging its firmness and rigidity as it is being used i.e. sitting in a chair or sofa, lying flat on a bed and so on. Firmness and rigidity under pressure are very important features of a good construction. Firmness depends largely on how the different parts are joined. The legs and frame should be fastened together with glue and also with the dowels, screws, steel clips, and corner metal plates. These joints should not break or make noises when the furniture is put to use.

Joints and joining

'Joinery' may be defined as the trade in wood work in which skilled labour is required to render the wooden members capable of framing together. It is an art of cutting, preparing and jointing the individual pieces of timber, so as to form the frame of the desired shape, size and finish (Figure 11.5).

The joints in wood are of various types. Mortise and tenon or dowel are the two common types of joints that are widely used in furniture construction. The various type of joints used in wood framing can be divided into the following categories:

Butt Joint: As can be seen in figure 11.5, a butt is a simple joining made by nailing or glueing the two ends together. It will not withstand much strain and furniture made of such joint is strengthened by the use of nailed or screwed angle iron.

Dove-tailed joint: In this method of fastening wood-work wedge-shaped or flaring shaped pieces are cut out of each member (wooden piece) and the joint is formed by hooking the projection of one member into the other one. This joint is specially used in cabinet work. In such places, the interlocking tenons which are dove-tailed are lapped and the junction is strengthened be means of pins which guard against the possibility of the members getting pulled apart.

Mortise and Tenon joint: This type of joint is frequently adopted in framing wood work. It efficiently holds the pieces together and is simple in construction. This joint is formed by cutting one of the members, so as to form a projection termed as tongue or tenon which fits into a slot known as mortise, out into the other member. The mortise should be of sufficient length, breadth and depth. Generally, the depth of mortise and tenon is 1/3rd the thickness of the members. The joint is strengthened by inserting wooden wedges from back or by inserting dowel pins from the face. At times the two pieces are put together by smearing the mortise area with a synthetic glue.

Tongue and groove joint: This type of joint is similar to the mortise and tenon joint except that the tongue and the groove extend the width of the boards. Therefore, a projection on one edge fits into a matching groove on the other edge. It is generally used on drawers, and contributes to free movement when put into use.

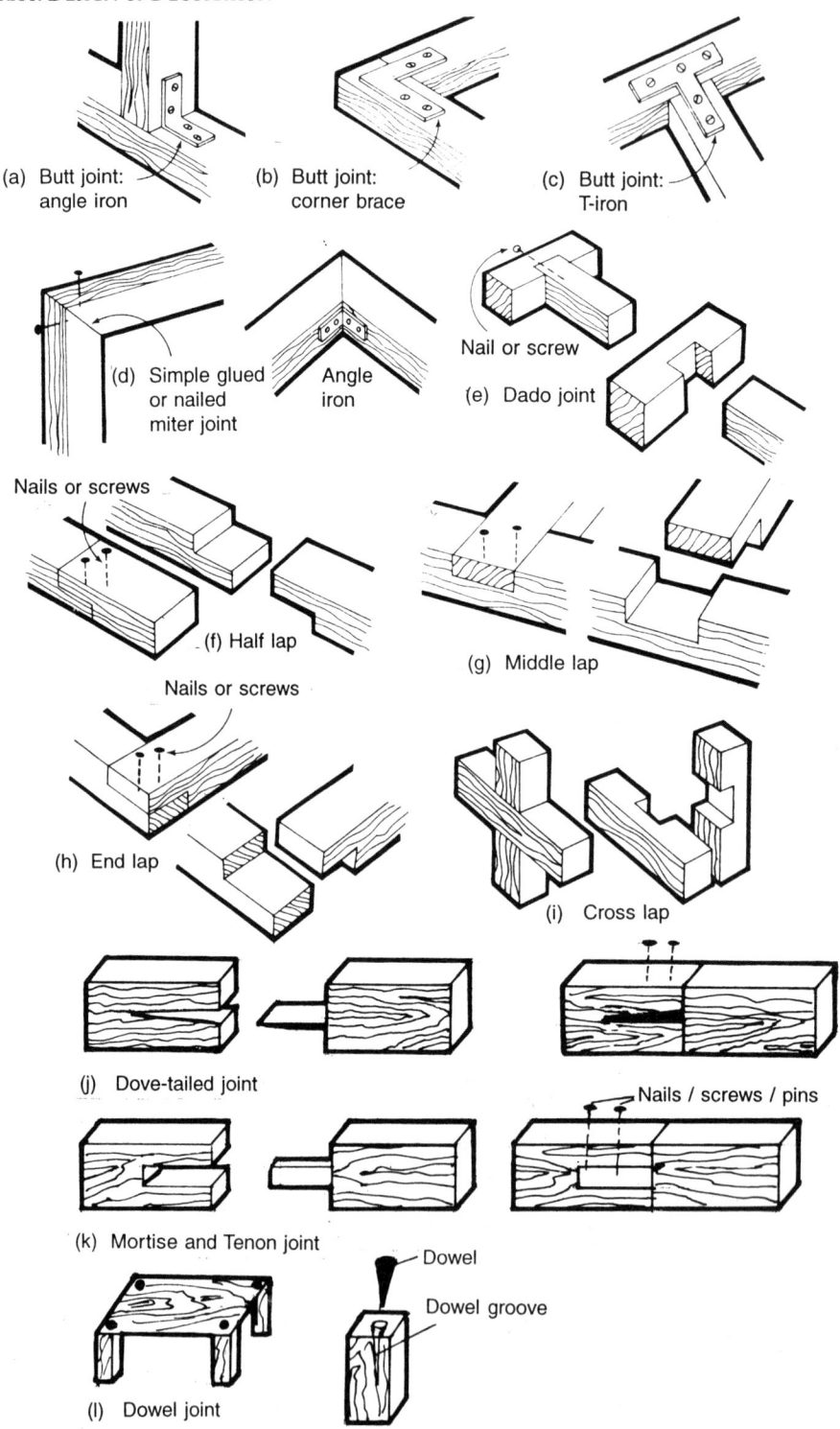

(a) Butt joint: angle iron

(b) Butt joint: corner brace

(c) Butt joint: T-iron

(d) Simple glued or nailed miter joint

Angle iron

Nail or screw

(e) Dado joint

Nails or screws

(f) Half lap

(g) Middle lap

Nails or screws

(h) End lap

(i) Cross lap

(j) Dove-tailed joint

Nails / screws / pins

(k) Mortise and Tenon joint

Dowel

Dowel groove

(l) Dowel joint

Figure 11.5. Furniture Joints

Dado joint

This joint is somewhat similar to mortise and tenon joint in technique. Mortise like slot is made on one piece of wood. The size of this slot is made exactly to the size of the other piece of wood which is going to be placed on the slot. Both the wood pieces can be held together strongly by means of mails or screws (Figure 11.5).

Dowel joint: In this kind of furniture joints, a small peg of wood (dowel) is used to join two edges. Usually the dowels are shaped like a nail with a head and a sharp edge on the other side. The dowel pins are used for various types of joining on chairs, frames, for upholstered pieces and so on. Double dowels provide added stability. Sometimes dowels are grooved so that air can escape when the dowels are driven into grooves. Such kind of joints keep the grains of the wood in tact.

Lap joint: This is the simplest form of lengthening joint and is formed by lapping the end of one member over that of the other and fastening them together by bolting or by using connections. Here two pieces have equal sized grooves so that they are clubbed when placed together. Such joints are used in the case of sliding drawers.

Miter joint: As can be observed, square corners are often mitered. Each edge is out on a 45° angle and the two are held together with glue or nails. Mitering is generally used in photo frames.

The various sections of a piece of furniture should be joined firmly and securely. Careful joining is an art that is of utmost importance to the construction; yet most of it is hidden from view in the finished piece. Nails, screws and glue are also used to hold sections together at points of strain. Nails are the least desirable, because they weaken the grains of the wood. However they are commonly used to provide quick and cheap furniture. Instead of nails, screws and bolts are desirable. They are frequently used for added security. A metal washer under the screw protects the framework of chairs, tables, drawers and so on. In high quality furniture the joinings are as nearly and perfectly matched as possible, and are also smooth and tight. Good quality glue is also used to hold the surface together. One should be aware of crevices and gaps that have been filled in with glue and other fillers which reflect a low standard workmanship.

Since a customer cannot see which type of joint is used she should request that the sales slip should state the type of joint and also the kind of glue used. *Medium and high-priced furniture should have phenol-resin glue,* which is very strong and is heat and water-resistant. Vegetable glue and milk product glue are less desirable and less costly. Old fashioned glues would eventually dry out but new developments have produced glues that are firm and durable bonding agents. They are resistant to ordinary hazards of use such as heat and moisture. Some common glues that are used in the construction of wooden furniture are vegetable or synthetic glue. Vegetable glues are prepared by cooking starch and does not provide good hold. However synthetic glues like fevicol, vamicol etc. are also used and make a firm and good hold in furniture joints.

Other steps/techniques in construction

The style and quality of any particular piece will determine how many steps are necessary from its production till its end. Naturally, the more labour involved, the higher the cost and it gets reflected in the retail price. Furniture construction is a complicated process. It is unnecessary to go into the detail of techniques but there are a few points that should be discussed because they influence the quality and cost.

(a) *Shaping:* For whole pieces that will be made of solid wood or for some part such as pedestals and legs, the timber is cut to the desired size by saws. A plane may be used to shape the edges. If no decorative edge are required the next step may be sanding. Much of this can be done by machine but some areas still require hand finishing.

(b) *Carving:* Certain types of decorative cutting can also be done by machine but the results are somewhat crude. Machine carving is used on mass produced inexpensive furniture only. For better quality initial work may be done by machine but hand-labour is used for finishing. Hand carving is done only on expensive furniture because the process is slow, laborious and time-consuming. It must be done by skilled craftsmen who are trained in the art. Some carved effects are achieved by the use of wood compound moulded to desired shape. The motif, often a beaded moulding is often glued in place on the piece of furniture (Figure 11.6 – see colour plate 25).

(c) *Turning:* Legs, base etc. maybe shaped by a turning lathe (machine) which cuts symmetrical indentation to form a design. The effect of a twisted rope is achieved when the block of wood is moved slowly along the cutting machine.

(d) *Fluting:* Lengthwise grooves may be cut into post pedestals, legs etc. and this process is called fluting.

(e) *Reeding:* This term refers to the decorative process of applying parallel rows of beaded mountings that project from surface. It is opposite of fluting and is also used on legs.

UPHOLSTERING

Today the presence of comfortably upholstered sofas and chairs is almost always taken for granted and yet, upholstered furniture is a comparatively recent development in furniture design. The term upholstery refers to the materials and techniques used in the covering and stuffing of furniture.

Construction of upholstered furniture

A wise buyer should see good workmanship, construction and quality of material beneath the outside covering, as well as in the positions that are easily seen. There are several parts to examine in case of upholstered furniture. They are frame, seat, welting, seam, webbing, springs, stuffing/filling, muslin, fabric, gimp and braiding, tacks, nails and pins (Figure 11.7 – see colour plate 26).

1. *Frame:* A important part of any upholstered piece is the frame, since it both carries the weight and determines the shape. The frame should be securely joined for strength and support.

   It must be constructed to withstand stress and strain over a long period of time. A hard wood that is free from imperfections and has been kiln – dried properly is desirable and durable. Oak, teak, maple etc. are the commonly used ones as they take glue and finish well. In case of cheap furniture items less expensive wood is being used. Wood should also hold nails and screws securely. It should be properly seasoned to prevent it from termites. The frame should also be strengthened by corner block of hard word. Besides this, the joints like 'dowel' and 'mortise and tenon' make the best furniture. For added security glue is used. The legs should be made with one piece of wood, and free from cracks and joints.

2. (a) *Seat:* The comfort is a desirable feature in any furniture and thus the seat should be comfortable. Generally, loose cushions are used on most of the seating furniture of today. Coir, and cotton material make poor cushions and generally synthetic foams make a more comfortable seat.

   (b) *Welting:* Welting is a length of fabric, usually tubular in shape, used to reinforce and disguise seams on upholstery, slip-covers and bedspreads. Welting is frequently used on the seams of upholstered furniture items for added strength and durability. While buying the upholstered furniture, judge it properly to see its neatly fitted corner.

(c) *Seam:* Seams should be well-finished, smooth, even and strong. A firm, clean finish with no projecting ends, shows good workmanship. In cheap quality furniture seams start coming out after a short span of time.

(d) *Webbing:* Webbing stretched across the seat frame is the basis of all traditional upholstery. In the case of chairs and sofas, webbing is closely interlaced, springs are fastened to webbing and to the frame securely. Webbing attached to the frame provides support for the springs and other stuffing materials. The size of the frame opening and the weight to be supported determine the quantity of webbing to be used. Seats require more webbing than backs. The best-quality webbing is a twill weave of pure flax. Webbing can also be made of jute and cotton. The cheapest webbing, quite suitable for most purposes, is brown and made of jute. Jute is used because it has little elasticity and when properly stretched and nailed it will keep its shape for a long time.

3. *Spring:* The coil spring is the best one and is generally used in good quality furniture and is used closely to each-other. Proper turning of spring is important to keep it tight. The type of spring also varies. They are larger on big seats. Correctly tied, springs add strength and hold them in proper place. The spring should have strength and enough elasticity so that it comes out when pressure is released. Galvanized Steel springs make better furniture as compared to aluminium or iron springs.

4. *Stuffing:* Well-made furniture has ample amount of filling selected, so that it will retain its shape for a long period of time whereas poor workmanship will be evident after a relatively short period – the filling will shift and shape gets distorted. If the workmanship is good, the filling material is distributed evenly over the surface of the furniture item, packed firmly and held in place to prevent shifting. The fibres most commonly used as fillings or padding are coir or coconut fibre, Algerian fibre, made from palm leaves and horsehair. Horse hair is most resilient but expensive. It is sometimes mixed with other animal hairs. Usually a fibre is used for the first filling and horsehair or flock for the second. The fillings from existing pads can often be used again.

5. *Muslin:* A layer of muslin is added and covered with the final upholstery fabric. Black calico is used to cover the underside of the seat.

6. *Fabric:* The furniture that will receive hard use should be covered with a fabric that resists both soil and wear. Closely woven fabric is a good choice for the upholstered furniture that will receive hardwear. It should be durable also and according to purpose or use. Now a days leather and rexine are also used to make the seat fabric. Velvets, blended fabrics and polyester materials are also popularly used these days. Each one of them has its own merits and demerits and the buyer should analyse these before opting for any one of them.

7. *Gimp and Braiding:* Tacks and the raw edges of fabric on chairs are concealed under banding or gimp. Bands of the cover fabric may be laid over the tack heads and fixed with decorative nails. Gimp is a narrow braid used for the same purpose. Another traditional finish is leather boarding fixed with decorative nails.

8. *Tacks, nails and pins:* Tacks made of steel should always be used for upholstery. Fine tacks have small heads. Decorative nails, used to conceal raw edges, are made in several finishes.

MATERIALS USED IN FURNITURE ITEMS

Wood

Wood is used for the construction of most of our furniture. The general availability, beauty in itself, flexibility, not hot or cold feeling, noiseless under impact are some of the features that add to its popularity.

Seasoning of wood

Seasoning or drying the lumber of its moisture content (till it is from 5 to 8 per cent) is essential for furniture making. The process involves natural drying (reducing the moisture content to 15 per cent) and then kiln drying. The word of the merchant has to be relied upon as the improper or inadequate drying is not evident when furniture is purchased. However in the case where the seasoning has not been done properly, warping and shrinkage presently show.

Nearly all woods can be roughly grouped as hard and soft wood. Some of the popular wood, their group, source, uses weight and other characteristics are mentioned here. This division has been established by long usage and is not in accordance with the relative hardness of woods (as certain softwoods are harder than some hardwoods) but is concerned with the specific species of the trees. Hardwood trees include the teak, oak, maple and walnut. These trees cast their leaves in the fall autumn. They form a clan of 'broad leaf' trees. Softwood trees are like the pine and spruce. They have needle like leaves which they retain during the winter. Thus softwoods are a group which is confined to conifers which are evergreen trees.

Walnut: Much of the present day furniture is made of walnut or walnut veneers. The properties of walnut are that it is workable, durable and beautiful which makes it almost perfect for furniture making.

Plywood: Several layers of wood are glued over one-another with adjacent layers at right angles and placed under heavy pressure to result in plywood or laminated wood. Plywood is used for large surfaces and for curved planes. Plywood is more economical, serviceable and beautiful and hence its use has increased in today's world than the solid wood. Moreover, it also offers much better resistance to changes caused by dry air.

Veneer: The interesting wood used for outside layers of plywood are the thin logs which are cut in a special way. The freak logs and stumps having eccentric figure and grain are cut up as thin as possible and used as veneering.

These veneers may be produced by:

1. Sawing and slicing
2. Rotary cutting.

In this process the logs are first steamed or boiled and then turned against a knife, a continuous sheet of very thin wood being produced which is dried by air or heat.

Solid wood: The use of solid wood has certain advantages as well as certain disadvantages. The characteristics that it does not peel or blister, no danger from poor workmanship and that it can be chipped or worn or planed down without showing other wood underneath makes it quite popular. However, the solid wood may check or split from the lack of humidity in the heated homes. This can be minimized by making a part like a table top of three adjoining strips and sealing all the surface of the wood.

Teak: Furniture designers, working in the modern style have favoured teak, a wood native to India, Burma and the surrounding areas. It is dense, durable, moderately hard and easily worked. The wood darkens with age but it is often finished in a deep brown tone that is almost black.

Mahogany: It is an excellent hardwood and favoured for fine furniture. The master cabinet makers of the past often preferred to use mahogany because of its strength, its variety of beautiful grain figures and its workability. It has a uniform texture that is adaptable to many interesting finishes.

Birch: It is one of the strongest furniture woods. It takes and retains finish well over its fine grain, and it can be made to imitate costlier woods. It is combined with other woods for strength in plywood. Birch is used in the construction of early provincial places and bentwood.

Ash: Ash is a desirable furniture wood as it has a nice grain, fairly strong and hard, easily worked, bends well and does not warp. While, green and black ash all have white sapword, the heartwood of white and green ash is light grayish brown; that of black ash is darker.

Beech: Beech is a plain, strong wood. It lends well to turning and polishing. It is used in medium-priced furniture for frames, curved sections, drawer guides and bent backs in chair.

Chestnut: It resembles oak and is used for outdoor and simple indoor furniture. It is used for core stock and for some less expensive furniture.

Cherry: It is strong and beautiful in grain and colour but is very scarce.

Elm: It is a durable plain wood. It is used for kitchen furniture, interior trim and frames and decorative veneers.

Cotton wood: It is soft and uniform in texture.

American softwoods: Softwoods comprise less than ten per cent of the furniture woods. They are specially valuable because of their flexibility. Western fir, spruce, hemlock, red cedar, pine, white pine and redwood are the examples. Yellow pine is used for common furniture. Eastern red cedar is used largely for chest, because of its fragrance and its resistance to moth larvae.

Metals

Metal furniture has gained entry into the modern homes, offices, waiting rooms, porches, gardens. It is highly utilitarian as it is durable, fireproof, light in weight; dry artificial heat does not affect it; humidity in atmosphere does not cause drawers to stick and to acquire a musty odor. Various metals and their combinations (alloys) are suitable for furniture. The metal furniture can give a beautiful look if the design is an original one and not an imitation of wooden design. Metal for furniture is gaining new heights of popularity. Wrought iron furniture – beds, garden table and chairs, dining tables with glass tops are now popular. Along with wrought iron – plain or enamelled, brass, steel, copper and aluminium are also being used by furniture designers.

Metal furniture is strong, durable and can withstand weather conditions, and is therefore suitable for indoor and outdoor use. Moreover, mass production has not only greatly lowered its cost but also made it a eco-friendly option but repairs are difficult. Steel and aluminium with enamel finish or chromium plating is well known in kitchens, bathroom cabinets, chairs, tables, etc. Low stools made up of brass/copper are popular in Indian furniture. Frequently, these materials may be combined with wood.

Plastics

Synthetics have become a vital part of the home furnishings field in many areas including the actual construction of furniture. Moulded chairs represent for modern design, a complete break with traditional methods. Sturdy, durable, light in weight, interesting in texture, easily cleaned and maintained and relatively inexpensive, the plastics would seem to quality for a major role in furniture construction (Figure 11.8 – see colour plate 26).

Plastics freed furniture forms from the rigid right angles of conventional construction. Now we see fluid furniture forms, sculptural shapes in hard, smooth plastic or soft polyurethane foam. In plastics colour is intrinsic and will not wear or chip off. Plastics are durable, but not all plastics are unbreakable. There is the possibility of eventual dulling or discolouring or pitting or scratching or of breakage. Moreover it cannot be repaired or refinished, and not fire-proof.

Glass

For many years, the top surfaces of tables and chests were protected with glass, but the idea of a protective glass top has become rather old-fashioned. But it is also likely that it may come into fashion once again. Glass has instead branched out into a more glamorous role in home furnishings. Thick plates of glass are used as tops for dining tables designed for both indoor and outdoor use.

A wide range of furniture items are available in the market. There are branded ones, there are made-to-order varieties, and there are locals who come and make them in your homes. Therefore it becomes necessary to equip oneself about these to make wise choices to suit their pocket without loosing aesthetic and comfort properties.

Chapter 12

ART OBJECTS : SELECTION AND USE

While selecting the art objects, and also when using them, the main objective of every individual is to have an aesthetically arranged home environment. Such a home environment calls for energy, acquisition, involvement and improvement. These are highly related to every object that is used in our day-to-day living in home designing, furnishing and decorating. In each of these activities, one works with sizes, shapes, colours and textures, which must be selected and arranged in accordance with the objectives of beauty, expressiveness and functionalism. While some individuals have the ability to find solutions by working with these objectives with ease, others find it to be an **art problem**. Those who are unable to solve their art problems, can easily acquire the knowledge and understanding of the elements and principles of art, and learn to apply them in any problem situation and achieve success in their attempts.

As the selection process itself poses an "art problem", the person who can build the house in the way she likes, and can furnish it in the way she pleases, is fortunate as compared to a person who has to deal with and use the objects she already has. The person who has to select new objects, has a "lot of options" to choose from, whereas the person who already possess them, has "forced options". She has to compromise and make the best use of whatever she has. In the case of a person who has to select new objects, though the options may be plenty, such a situation is also considered to be a problem. It is so because what appears to be "beautiful" and "appropriate" in a shop, may not necessarily be the same in a home environment. Thus, in both the cases, the persons concerned have **an art problem to solve**.

SOLVING AN ART PROBLEM: SELECTION OF ART OBJECTS

The first concern of every individual while buying an art object for a home is that it should contribute toward making her home to appear distinctive and appealing. It is generally thought that all beautiful objects are expensive. However, cost of an object has nothing to do with its distinct qualities always. Good quality and aesthetically appealing objects can easily be selected and bought even at low costs. The main thing that it has to satisfy is that it should appear to be belonging to its owner and to the home where it is going to be used. A distinctive object will not be eccentric, but it will be individualistic and above all, the details in the objects have been selected with the idea that they look well together (Figure 12.1 – see colour plate 27).

A person who is prepared to make suitable designs and to make wise purchases, has to do the following things*.

- In the first instance, she should be able to measure her choices according to the principles of art in order that her selections may have beauty.
- Secondly, she should know enough about the materials and processes used, to be able to judge good workmanship.
- Thirdly, she has a certain store of related information, such as some knowledge of science and economics, which has a more or less direct bearing upon the problem.

By realising the earlier mentioned factors, the person will be able to form a good judgement upon:

  (i) Whether the object should be purchased or made by oneself;

 (ii) The factors which affect good quality;

(iii) The right price to pay in relation to the income and budget, considering the other demands made upon it;

(iv) The time and effort consumed in making the object in the light of the return it yields (this return may be measured in increased skill, or in the satisfaction resulting from the finished product/object);

 (v) The time and effort it will take to maintain it in good condition.

Thus it can be seen that either purchasing or making an art object by oneself, is not a simple task. Only when a problem in purchasing or in designing has been worked out to satisfy all the requirements as mentioned above, it may be called a "art related problem". The ability to apply this related information should give one a sense of relative values and of appropriateness. Furthermore, such a value-added effort will help to bring art into close terms with everyday life.

Thus a close link between art and everyday living can be brought out and it will help a person to develop good taste, thereby promoting living in an aesthetic and comfortable environment.

Steps in solving an art problem

As mentioned earlier, selecting an art object is not a simple task – it can be said that it is more of "solving an art problem". To obtain good results, it is important to develop a plan which would involve a sequential approach. A well-designed plan needs to be worked out and carried out systematically for this purpose. This plan has been suggested here and it requires four important steps. This plan can be applied in any situation to solve all kinds of art problems. Since it is based upon the generally accepted steps in solving a problem, one does well to think through the various stages and note down her conclusions before making or buying anything that has to be lived with and looked at for a length of time. In general, the plan for solving an art problem has the following steps:

Step. 1. Recognizing the problem: This involves setting up a definite aim or a purpose to be accomplished.

Step. 2. Making a plan: A plan is necessary to work out the problem. This involves collecting all the necessary information related to the problem.

---

* Goldstein H. and Goldstein V. (1961) Art in Everyday Life. Mac Millan Co. New York.

Step. 3. Carrying out the plan and making the final selection: This calls for actually taking a final decision and making the necessary purchases.

Step. 4. Testing the results and making a final judgement: This is helpful in judging the success or failure of the plan before accepting it or discording it to make another plan.

While furnishing a home, a person need to make a number of choices. She has to choose curtains, floor coverings, table mats, suitable lighting arrangement/lamp shades, accessories, decorative pieces, wall clock, wall pictures, flower pot, furniture etc. In all these cases, the selection process will go through all the steps as mentioned earlier. Initially one has to understand the problem, prepare the plan, then carry out the plan and finally make a final satisfactory judgement. Though the first three steps are actually involved in solving the problem, the final step alone would indicate the real success of the plan. The final step, therefore, is the most important one, because this would help a person to realise whether she has analysed the problem properly or not. If success is achieved, she would take similar decisions in her selections in the future or vice versa. A few selections as examples, are worked out in this chapter to give an idea about the details in solving the art problems in selected situations.

Selection of a floor covering

One of the difficult tasks for many of us is to select a floor covering for our homes. For some of the items like a dress, utensil, consumer durable, soft furnishings, etc., one has a somewhat sufficient knowledge, since these are either easily observable, or bought more often. However, floor coverings are not very often bought or selected for a home. There are also not many retail outlets which sell floor coverings. The sale of these items are limited only to a selected specialized shops, and that too, only in highly sophisticated commercial areas. Keeping these factors in mind, the selection of a floor covering is considered as a typical problem for a home-maker/designer. The details of the plan in this case too, would involve the four steps as mentioned earlier. The selection details would be filled in somewhat in the manner described in the following paragraphs.

Step 1. Recognizing the problem: The aim is to select a suitable floor covering for the drawing room. The drawing room has already been furnished with curtains, furniture, showcase etc. Only the floor is bare with a mosaic finish.

Step. 2. Making the plan: Prior to the actual selection of a satisfactory floor covering, the following factors are taken into consideration:

- *The room:* its main features such as its purpose, size and shape, the additional features like wall treatment, soft furnishings and furniture, finishes used, and finally, the people (family members) who are going to use the room – their taste, requirement, habits etc.

- *Material/textile information:* the quality related factors of the floor covering, such as the properties of the fibres used, the yarn and weave construction, the thickness/height of the pile, the amount of wear and tear the floor covering is likely to experience or exposed to, the suitable texture, and other textile related information such as its utility in the room, the type of the other furnishings used in the room, etc.

- *Art information:* standards/requirements for good structural and decorative designs in a floor covering, use of colour and its implications with the other furnishings and art objects in the room.

- *Economic aspects:* the budget allocation/money available for the purchase of floor covering, time and effort required to maintain and frequency of cleaning, wise purchasing of floor covering in terms of reduction in the investment cost (purchase at the time of discount sale) and maintenance costs, durability, effects of buying good materials, colours and patterns, a reliable manufacturer, and shops which offer a good amount of variety in floor coverings and a warranty.

Step 3.1. Considerations for carrying out the plan: For carrying out the plan, the following details are considered and analysed.

- *The room:* Let us assume that the size of the drawing room (purpose) is 6 m × 4 m. The furniture items that are placed in this room are made of rose wood with a similar polished finish. The polyester furnishing (curtains) cloth with self prints has a tone of the colour of rose wood. The upholstery material for the sofa is soft velvet in maroon shade. Accents in the room – wall picture, flower arrangement, cushion covers, sofa backs, etc. are crimson red, saffron yellow, mustard shade, black and olive green. the family consists of two adults and two children, aged 16 and 12 years. The family has simple taste, and a moderate living style. These family characteristics are somewhat reflected in the kind of furniture and furnishings used in their drawing room. The family seem to enjoy beauty resulting from good line, good construction and simple decorations.

- *Textile information:* The main considerations in the choice of materials are appearance, quality and economy. The market survey has shown that the fibres used for preparing floor coverings are wool, linen, cotton, jute and synthetics. It was also seen that worsted wool and synthetics are durable and last longer. The two principal types of weave construction in floor coverings are the flat weave and the pile weaves, cut/uncut. There are floor coverings which are prepared either by cementing the fibres together or by cementing the fibres on a base fabric. Texture for the floor covering should be sturdy and soft at the same time, with high and dense piles to withstand wear. The sitting area in the drawing room needs a texture that is serviceable as well as good in appearance. The characteristics of furniture and the decorative objects in the room suggest an intermediate texture and a good substantial fabric.

- *Art information:* While considering the structural design requirements in the floor covering, it is necessary that the shape of the floor covering should be in harmony with the shape of room. It is desirable to have a wall-to-wall carpet to cover the entire floor space adequately. Usually the margin of the plain floor space left around the carpet is about 5 to 8 cms. Therefore, in a room of the size 6 m × 4 m, the carpet should measure slightly less than this size to make it suitable within the area of the room. This size should be available in broadloom carpeting or in some ready-made carpeting. The other choice could be a vinyl floor covering sold in lengths and, therefore, can be bought in desired length and breadth. Regarding the decorative aspects of the floor covering, the main consideration is that it should provide a good background for the other furnishings in the room. The effect should be quiet, i.e. it is inconspicuous and appear to stay flat on the floor. The type of design should be in harmony with the furnishings of the room. These considerations call for a plain carpet or a well-backed geometrical or woven pattern in a conventional style. The colour of the floor covering should be such that it is an intermediate shade between light and dark colour for its practical and good background qualities.

- *Economic aspects:* The family, in this case, has already kept aside sufficient amount for the purchase. However, they may not like to spend money more often on drycleaning. They would also expect the floor covering to last longer so that they need not have to replace it at an earlier date. Therefore, the carpet should be made of a fibre and weave which will wear well, and of a pattern and colour which will not show dirt easily. It requires a lot of expertise to judge the qualities of a carpet/floor covering and so it is desirable to select a manufacturer and a shop/retailer who are dependable, or enjoy a good reputation. Such an information can easily be collected from magazines, friends, relatives or neighbours who already have a floor covering in their homes.

Step. 3.2. Carrying out the plan: In this step, the final selection of a desirable floor covering is made by carrying out the plan which is made earlier. Since the necessary information concerning the factors listed earlier has already been acquired and applied to the particular problem, it is possible to make a satisfactory selection. The family will be able to select a floor covering with the following features:

- *Size* of carpet - 5.8m × 3.8m (6m length of a floor covering having 4m width-cut, finished and fixed to the floor)
- Woollen *fibre*
- Worsted *yarn* (long/twisted)
- Closely *woven* loop pile with a strong thick base fabric to make the piles secure and to provide a firm body and a cushioning texture.
- *design* of the floor covering is an allover self-woven geometrical or floral pattern in a single colour (tints and shades) rather than a multi-coloured pattern, since the rest of the soft furnishings used in the room are all of plain coloured fabric.
- *Colour* of the floor covering can be either grey or mustard brown or olive green (medium value) to tone with the accents in the room.

Step. 4. The final judgement: In this case the final judgement can be made only after using the floor covering for some time. The success of the choice of this floor covering will be measured by the degree of satisfaction received from the actual floor covering during the period of its use.

However, the family can assume that it is likely to get the expected satisfaction, since it had all measures to go through the necessary steps and analysed the pros and cons of selecting a floor covering/carpet. It is worthwhile to mention here that the family has conducted a market survey, collected all the necessary information from others, and purchased the one that has been manufactured by a reputed firm and sold in an authorised show room/retail shop. The success of selecting a suitable floor covering – the solution to the art problem – can be expected.

It does not seem unreasonable to predict that the owner of the carpet would find that it is a well designed floor covering for the purpose and fits well in the drawing room. It is going to withstand the wear factor too as enough care was taken to choose a durable one from a reliable manufacturer/retailer in the selection process. Thus it is also fulfilling the problem of maintenance and also not to go in for another one in the near future, thereby making the investment as well as its maintenance at a nominal cost. This last consideration is, after all, of utmost importance in **solving the art problem.**

SELECTION OF ART OBJECTS AND FURNISHING OF A HOME

It has been seen earlier how one might plan the choice of an expensive and an important item like a floor covering for the drawing room, and carry out the plan as an exercise in solving an art problem. Whatever may be the costs involved, it is equally important to prepare a systematic plan and follow a proper procedure in selecting each and every object in a home. It may be a simple object like a lamp shade or a place mat or an elaborate bed cover or a decorative chandelier.

Irrespective of the nature of the object, or the other costs involved, the steps involved in planning for the selection of these objects, are going to be similar. Only the lesser details under each category may vary. For better understanding and application, details for the selection of the other household items such as table lamp, wall picture, bed cover, cushion cover, table napkin and a decorative accessory object are given in Table 12.1. One can take benefits from these details while purchasing or selecting these objects for their use in their homes.

When a person wants to select functionally effective and aesthetically satisfying art objects, he has to think and follow certain guidelines to achieve the best results. She has to systematically carry out her selection process under four main steps as mentioned earlier in this chapter. However, this is not a simple task, but a complex process, which everyone has to practise and master to learn the art of selecting the best art objects.

Personality expressed through choices

As mentioned earlier, when we choose an object for our house, we are satisfying some need or desire. Besides meeting our needs, we are also expressing unconsciously our personality to others through the qualities possessed by this object. The way we live in our homes, the way we wear our clothes, and the way we furnish our rooms, all mutely proclaim to the world just what sort of a person we actually are. All these things tell our interests, prove or disprove our sincerity, display our imagination, or lack of it. It is, therefore, important to make an effort to utilize our knowledge and express it in all our choices, which will in turn reveal our real personality (Figure 12.2 – see colour plate 27 and Figure 12.3 – see colour plate 28).

All belongings have a tremendous influence in forming character. Strong character and high ideals are often the reflections of the home surroundings. Artificiality in the form of wall paper and metal to appear like wood, or shiny fabrics to imitate costly silks and damasks, can be avoided if their significance is understood. Unfortunately, quality in things is more or less intangible, and difficult to define as personality in an individual, but the outstanding features can be recognized and classified. With a clear mind and the eyes open, one can easily reach a point where every picture, every piece of furniture, or drapery pattern, speaks its note of social grace or friendly domesticity, vigour or fineness. Similarly, one might feel, after spending some time with some people surrounded by their own things, that it is better to spend the time with them in a hotel or in any other impersonal setting!

Those people, who have not experienced such feelings as stated above, can realise them by thinking about the impressions they got while watching a theatre show. When the curtain raises, by seeing the settings on the stage one can get a definite idea about the people who are likely to be present there. If a stage setting shows a living room with glaring lights, florid wall paper and rugs, showy lace curtains, and overdecorated lamps, one expects the people who live there to come on the stage in flashy clothes expressing a gaudy taste. On the otherhand, if the setting shows a room with soft and mellow lights, walls and rugs with subdued and harmonious colouring, thin white glass curtains, well-designed furniture, with comfortable chairs, lots of books and flowers, a few good pictures and decorative objects that catch the light and create points of interest, the audience would expect the people who live in this room to be tastefully dressed, well-bred, and charming.

However, for an ordinary person to be aware of the fact that her home reflects her personality, the answers to the following questions can provide the guidelines:

- Have the lamps, vases, candlelight holders, and other decorative objects been chosen

PLATE 27

Figure 12.1. Good quality and aesthetic art objects

Figure 12.2. Personality expressed through choices and arrangement of objects

PLATE 28

Figure 12.3. Personality expressed through neat arrangements

Figure 12.4. This place needs to be re-arranged to provide an orderly look

Table 12.1 Selection of Art Objects (Solving Art Problems)

S. No.	Art Object	Step. 1 The Problem	Step 2 The Plan	Step 3 The Action	Step 4 The Judgement
1.	Table lamp	Selection of a table lamp for a study table	*The table:* Its purpose, size, shape, chair, the type of polish for the wood and furnishing materials in the room and the likes of the child who is going to use it.	*The Table:* Size, tone, and colour of the wood and its polish, the other furnishings in the room, choice and taste of the user are considered while choosing the lamp.	The lamp shade is suitable for the table size and shape. The colour of the shade is neutral and so blends well with the colour scheme of the room. The lamp shade is not very expensive and is easy to maintain and also, dust-proof, thus economical at the time of purchase as well as for maintenance. The user taste is also taken care of and the lighting is suitable for the purpose of reading and does not produce glare. Thus the lighting arrangement is suitable for reading purposes, and according to the initial plan.
			Material Information: The materials used in making the lamp and its shade, their properties and texture, its suitability for the purpose.	*Material information:* Material should be easy to maintain, and be dust/dirt proof. Medium texture and colour in harmony with other furnishings in the room.	
			Art information: Structural and decorative features, colour, balance and proportion of the lamp in relation to the table and between the lamp and its shade.	*Art information:* Stable base for the lamp with a shade of wider circumference at the bottom, must give both direct and diffused lighting, not too decorative and be simple.	
			Economic Aspects: Initial and maintenance costs, availability of replacements and a reliable shop/market source for making the purchase.	*Economics:* Wooden or hard metal/pottery/plastic base, sturdy and not very heavy, with a plain silk or polyester cloth lamp shade of beige colour, and is easy to maintain and does not attract dirt/dust.	
2.	Wall picture	Selection of a wall picture for a drawing room	*The room:* Its purpose, size, shape, wall treatment, furnishings used, and the family taste, style of furniture setting.	*The room:* Size is 6m × 5m, somewhat rectangular shape, colour plan of the room is an analogous colour scheme with blue as the dominating colour	The wall picture choice was between a painting or an embroidered one. An embroidered piece with mirror work in a traditional style was

(Table 12.1 Contd.)

S. No.	Art Object	Step 1 The Problem	Step 2 The Plan	Step 3 The Action	Step 4 The Judgement
			Picture Information: Either a canvas painting or a cloth pattern with block printing or embroidery work. *Art information:* Modern or traditional pictures or a natural scenery, colour combinations in the picture, structural and decorative design characteristics. *Economic aspects:* Initial cost of mounting, laminating and maintenance. Available in a government emporium/art exhibitions/trade fair at a fairly reasonable cost, or during discount sales.	in a traditional setting. The family has a traditional taste and living style. *Material information:* Material is chosen for appearance and economy. The picture should be mounted, but at the same time, should not heavy, should be easy to clean and maintain. *Art information:* The shape of the picture should harmonise with the shape of the room, and furniture items, kind of picture – abstract or a natural scenary to match with the room setting. *Economics:* A canvas painting of a famous painter/artist may be a bit expensive, and so, a fabric painting or an embroidery work in a traditional style and is of moderate cost.	chosen and purchased because it was going well with the taste of the family and the style of total furnishing in the room. The shape of the mounted picture is rectangular i.e. 100 cm × 40 cm., going well with the shape of the room. The picture can be mounted on a platform with a laminated finish at a nominal cost and therefore, is easy to maintain. The colours used are orange, rust, mustard olive green and black to provide an accent to the colour scheme of the room. The background colour of the cloth is off-white/beige.
3.	Bed cover	To select a suitable bed cover to be used in the master bed room	*The room:* Purpose, size, shape, orientation, wall treatment and the furnishing plan, and the taste of the couple who are going to use the room.	*The room:* Master bed room, dimension – 6m × 4m facing south-east with a secondary triad colour scheme, bed is finished with a natural wood colour polish and a laminate, the couple has a simple taste, but highly social, fond of nature and natural objects.	After visiting a number of shops and analysing the various factors, a double bed cover of 8' × 8' size is bought made of a cotton and polyester blend. Twill weave with a floral pattern to suit the taste of the couple. An overall pattern with a border

(Table 12.1 Contd.)

S. No.	Art Object	Step 1 The Problem	Step 2 The Plan	Step 3 The Action	Step 4 The Judgement
			Textile information: Weave construction, fibre properties, suitable texture for a bed covering and the use of the room, type of mattress and pillow covers, foot mat, floor covering etc. *Art information:* Standards of structural, decorative designs/patterns used on the bed cover, colour scheme, etc. *Economic aspects:* Initial cost of the bed cover, its maintenance cost (home laundry or dry cleaning costs), wear and tear, wise purchase through a market survey, availability from a reliable mills' show room, or a reliable shop selling such materials.	*Textile information:* Material is chosen for pattern and its colour scheme, fabric is blended with silk/cotton/rayon and polyester, plain or twill weave, fine and smooth texture, that gives a good service and look, and will not attract dust/dirt. *Art information:* Shape and size of the cover to fit the shape and size of the bed – a double bed of 6' × 6' size. The size of the cover slightly larger than this size. Floral pattern or designs on the bed to suit the likes of the couple and the other objects in the room, and of a matching colour. *Economics:* Moderate cost initially and also during maintenance – home wash-able, need not be sent for dry cleaning, easy to maintain, fast colour, strongly woven and fibres with good strength – long lasting.	on all four sides-block-printing using natural dyes to give a natural appearance. Colour combination is of cool green, saffron yellow, and mauve. A smooth and soft finish that will not attract dust. Not very heavy; easy to wash and maintain. Easy to replace and change, if necessary. Medium priced-within the budget of the couple.
4.	Cushion covers	To select a set of 3-4 suitable cushion covers for a child's room	*The room:* Purpose, size, shape, wall treatment, furnishing plan, and age and sex of the child.	*The room:* Child's room of 4m × 3m size, walls are painted with light colour, bed at the floor level, primary triad colour scheme for room furnishings against white/off-white background. The child is active and likes bright colours.	After finding out the cost of ready-made cushions and the material cost, the decision was to make them at home. the home-maker had sufficient time and polyester blend fabric in bright red, blue, green and yellow

(Table 12.1 Contd.)

S. No.	Art Object	Step 1 The Problem	Step 2 The Plan	Step 3 The Action	Step 4 The Judgement
			Textile information: Weave construction, fibre properties, texture suitable for cushion covers, use of the room and the type of the other furnishings used.	*Textile information:* Material chosen for colour, appearance and pattern. The fibres available are cotton, satin, rayon and polyester blends. Plain, twill or satin weave. Fine and smooth texture. Strong and dust resistant.	coloured material was bought. Three cushions – red coloured apple, yellow mango and blue cherry shaped cushions were prepared with green stems and leaves. These colours were in harmony with the colour scheme of the child's room. As home-made cushions, they were more satisfying by saving money, and the personal touch. The bright colour fruit shaped cushions proved to match the requirements of a child's room.
			Art information: Standard sizes and characteristics suitable for good structural and decorative designs for cushions, and colour scheme of the room, texture of the material.	*Art information:* The shape of cushions can be like apple, pear, mango etc, in suitable colours and patch work. Type of design in harmony with the patterns of other furnishings in the room. Bright red, yellow and blue colours are preferred.	
			Economic aspects: Amount allotted in the budget, time and effort costs of maintenance and frequent washings, availability and access to a nearby shop or a manufacturer's show-room – either ready-to-use cushions or available materials for stitching them at home.	*Economic aspects:* In this case, there is enough money. The home-maker can also cut and sew.Therefore, the choice was between readymade cushions and made-at-home cushions. Good polyester blended fabric can be bought for their strength, fast colour and a smooth texture with a satin finish.	
5.	Table napkins	To select a set of 6 table-napkins for a dining room	*The Table:* Purpose of the room, table size (4 or 6 seater), shape, wall treatment, table finish and the	*The table:* Size of the table 6' × 3', rectangular, finishing with a wallnut coloured laminate, simple furnishings with	Market survey is conducted. A set of 6 table napkins, is purchased from a well-known store, selling household

(Table 12.1 Contd.)

S. No.	Art Object	Step 1 The Problem	Step 2 The Plan	Step 3 The Action	Step 4 The Judgement
			other furnishings in the room. Taste of the family members—simple and practical-oriented. *Textile information:* Materials available in the market, and their properties. Fibres used are cotton, rayon, synthetics etc. Weave construction, texture of the material, other furnishings used in the room, the crockery and cutlery accompaniments etc. *Art information:* Standard sizes available in the market, possibilities of patterns/designs — both structural and decorative aspects, colour combi-nations, and a suitable colour scheme in relation to the other colours used in the room. *Economic aspects:* Money available or the budget allocation, time and effort, costs of washing and maintaining the mats, choice and variety available and a comparison of costs quoted in reliable or reputed stores/show rooms.	beige background and curtains with geometrical patterns. *Textile information:* Material is to be chosen for appearance and maintenance, availability of matching table napkins — good cotton casement fabric woven in plain weaves, not of very soft texture, durable fabric to withstand wear and tear on account of frequent washings. *Art information:* Simple prints — block printed or embroidered, either grey or beige coloured base, good workmanship and consistent texture. *Economics:* Amount of money, time and effort needed in buying and maintaining the mats, whether it can be cleaned with ease and at no extra cost. Durability of the material to provide a good service.	furnishings. The pattern is either embroidered or block printed to match the geometrical patterns found in the curtains. Casement fabric (of cotton) is bought which is easy to wash and maintain. Napkins are made of plain woven poplin cotton fabric, suited to the taste of the family and the informal settings in the room.

(Table 12.1 Contd.)

S. No.	Art Object	Step 1 The Problem	Step 2 The Plan	Step 3 The Action	Step 4 The Judgement
6.	Decorative Accessory (Pottery)	To select a decorative accessory for the drawing room	*The room:* Its purpose, size, wall treatment, and the furnishing plan, other decorative accessories in the room, the living style and taste of family members. *Material Information:* Materials used in the making of pottery are clay, bronze, brass, copper and wood, their finishes, their properties, and its use in the room. *Art Information:* Details, on structural and decorative designs and features of the decorative accessory (pottery) finish of the object, traditional piece, sculpture/pitcher/pottery. *Economic aspect:* Money available for meeting the cost of the object and its maintenance, effect of buying good material, reliable shop and a manufacturer.	*The room:* Drawing room of size 6m × 5m. The walls are off-white (cream) colour, the colour scheme of the room-mono-chromatic with brown and dull-orange. The accent is of saffron-colour. The family has a taste for traditional/cottage style. *Material information:* Material chosen for beauty and appearance, to match with the traditional setting of the room, a pottery made of clay of 3' height with a dark brown finish (wallnut colour) requiring less care than that of a metal finish. *Art information:* Beautiful carving on traditional pottery, well-balanced, having a good structural design, colour in tone with the colour scheme of the room. *Economics:* Not very expensive, cheaper than metal ones, maintenance is simple and inexpensive, A good finish, can be used as a decorative pottery as well as a flower or plant holder.	Market survey and the other collected information reveal the availability of good traditional potteries in State Emporium. Clay pottery found to be suitable for the room setting, not very heavy, but sturdy, good appearance, goes well with the other art objects in the room. Finished with a thin coat of an enamel paint, and so is easy to clean and maintain. The patterns are carved and formed part of the basic design of the pottery, and keyed to the colour scheme of the room.

for their beautiful shape and colour and refinement of decoration, or are they over-
ornamented?

- Are these decorative objects placed where they are needed – to relieve a bare spot, to create
interesting shapes and spots of colour, or to balance some other object? Or, on the other
hand, they are put up for show purposes?
- Are these decorative objects so numerous that they do not enhance one another or the object
on which they are placed, but add to the confusion of an overdecorated room?
- Does the furniture express the kind of person its owner would like to be?

Feminine Vs Masculine furnishings

Wall hangings, pictures, decorative objects and furniture may suggest either a masculine quality, a
feminine quality or they may be intermediate. While furnishing a room for a girl, a man or a guest
room, same things cannot be chosen. A feminine quality will result from the selection of a little
lighter type of furnishings with a slightly smaller, finer pattern in the drapery material, and a little
more grace in the lines of the furniture and other objects. The colours in a woman's room should be
somewhat different from those in a man's room, and be higher, and the textures finer. A man's room
need to be more solid than a woman's room, and somewhere a forceful note of contrasting colour
can be introduced. A guest room should be intermediate, i.e. between the feminine and the
masculine, so that either a man or a woman will feel at home in it. A transitional quality should be
present which could be achieved by selecting furnishings neither distinctly light nor heavy, a
pattern neither very small nor very large, and colours neither dainty nor heavy.

Domestic, social and impersonal qualities

Besides expressing a feminine or a masculine quality, objects may give a social or a domestic feeling
or, be impersonal. Domestic quality is the outward expression of love and affection of a home and
family and is usually informal. The social idea is usually expressed more formally than the
domestic. When the term "social" is used to define one of these group expressions, it must be
understood that the more limited sense of the word is intended, as referring to the characteristics
resulting from an interest in the conventions of a formal society. This expression varies according
to the social standards of the individual. If she is a person of taste, if she is sincere and his standards
are high grace, charm and fine quality will be reflected in the choice and arrangement of her
furnishings. Such finer qualities will be missing if she has the taste of being insincere and pretentious (Figure 12.3 – see colour plate).

At this point, it is important to understand that questions of expense, of good or bad taste, of
richer or poorer materials never enter into these attributes of objects – the social or domestic, the
feminine or masculine. It is simply the individuality of an object, just as it is individuality which
gives a distinct character to an individual. Although the actual furnishings of the houses of the
individuals may change with their changes of fortune and their acquisition of taste, the essential
quality would always remain the same. The possessions of each individual will reflect his personality because he could not help surrounding himself with things that reflected him.

MAKING THE BEST USE OF ONE'S POSSESSION

When a person is able to build the kind of house she likes, and furnish and beautify it in the way
she pleases, she is lucky. But it may not be the same with some others. For many couples, when they

start a home, they receive a few art objects as gifts or as inheritance. In that case, they are forced to compromise with whatever they receive and make the best of them. The problem becomes acute when they make unwise or impulsive purchases in addition. Then they realise that they have invested much money and that they may not really be able to make use of all of them. In the case of those couples who often have to shift houses, the problem of using the objects or not needing the objects, because of the changing environmental factors or climatic conditions, is felt more. Whatever may be the situation as mentioned above, the couples can think of using the following remedies to meet these **decorative mishaps**. These are:

- Elimination
- Re-arrangement, and
- Concealment.

Elimination

Elimination is the first measure suggested for such situations. Each object should be judged critically and impersonally, regardless of its sentimental value. For some, an interest in antiques sometimes leads to the belief that anything old is automatically beautiful. Any gift from a close friend or a relative is felt as precious one to be preserved forever. Any inherited object is also considered "valuable" because of the "memoirs" that it may bring in. All these factors are not the ones to judge whether an object is really or aesthetically beautiful or not. The real test of any design, is the quality of its structure and its decoration. If any object, does not add to the beauty or to the comfort of the people in the room, it need not be retained any more, and, therefore, need to be discarded.

It is, therefore, necessary for everyone to collect all the objects, analyse them from their practical utility, beauty and comfort at frequent intervals. If an object is found to be out of place, or does not contribute to the decorative features of the room, such an object need to be removed or "eliminated".

Besides the ones that a person owns, she may also keep adding or additions may "just happen". This may add to the already existing ones, and too many objects for the size of a room may call for an additional sifting. Therefore, it is necessary that everyone analyses the objects one owns once in a while, and a "spring cleaning" may suggest "elimination" of unwanted objects.

Let us now take the example of a family who is shifting to a new house. In the house where they were living earlier, there were no built in cup-boards. They had to purchase a show-case for the drawing and an almirah for the bed room to keep their accessories and clothes, respectively. The new house where they are moving in now, has a good wall-to-wall show case-cum-book case-cum-TV top. Under such circumstance, the show case can easily be disposed off because it is not only required in the new house, but also may occupy space in the drawing room where there is space only for keeping a sofa set, a centre table, and perhaps, two side tables and a settee. Here the options would be between two show cases and one settee. The family can easily "eliminate" one of the show cases and retain the settee since it can provide more sitting places. Similarly this family can also eliminate the almirah which they had bought earlier for storing clothes. Since the new house has enough built-in storage for clothes, they can "eliminate" this almirah. Instead, they can either buy a study table or keep a T.V. stand for the bed room, which may have better utility than the clothes almirah. Many families may face similar situations. One has to think and analyse the objects one possess, and use only those which are required, and eliminate the unwanted things.

Re-arrangement

Everyone of us is tired of seeing the same thing in the same place again and again. 'Change' is the spice of life. Same rule applies to the placement of art objects too. When a person is unable to eliminate certain objects, she has to think of an alternative to make the best use of them. For this purpose, re-arrangement is the next step (Figure 12.4 – see colour plate 28).

Order is the first requirement for beauty and functionalism and what can be achieved through orderly arrangement is termed as "shape harmony". In any arrangement where a number of shapes are used, there should always be an effect of organisation, or, in other words, of orderly arrange-ment. If a sense of order is to result, shape and harmony must be followed. To achieve this effect, large objects or masses should be placed to follow the boundary lines of the enclosing shape, and only the small objects should vary from the general directions. To give variety, some of the small objects may be placed at slightly varied angles, but too many angles should also be avoided.

Though orderly arrangements can contribute beauty and functionalism, it is pleasant to see an occasional change in the appearance of a room, and very simple re-arrangement of the furnishings or art objects are often sufficient to arouse interest. More often, it is only a matter of moving an object from a more obscure position to one of prominence in order to give an impression of having something fresh and new in a room. It is good to experiment with the various objects by positioning them in different ways. By doing so, one will also realise the best way of arranging things. At the same time, variety in arrangement is produced to reduce monotony or boredom of seeing the same objects in the same places.

It is not enough that objects be grouped in order to gain the impression of unity. They must be placed in such a way that their lines will carry the eye along the paths staying well within the boundaries of the arrangement. For example, a simple way of turning a teapot so that its spout faces towards the centre of a shelf, rather than away from it, may hold an entire arrangement together. Such re-arrangements produce beauty and variety thereby making the room to look "different" and "appealing".

Concealment

Similar to elimination and re-arrangement, concealment can also help to solve some of the prob-lems related to "art mishaps", though it is the last measure suggested for such problem situations. After all the unessentials have eliminated, and the room has been well arranged, some unsightly objects, necessary for their comfort and functionalism, may need to remain. Then the problem of hiding their deficiencies can be taken care of. Slip covers or sofa backs are the examples of this kind for chairs and sofa. Where a wall finish is not desirable, and is required to be concealed, a big wall picture can be hung to hide such defects. There may be many places or parts of objects which may require such treatment, and similar measures can help to conceal unwanted sights or objects.

Some problems may necessitate all three measures – elimination, rearrangement and conceal-ment. Let us now take the example of a room in an old house or apartment. There is so much heavy wood work as compared with the size of the room that the amount of wood becomes oppressive. There may be architectural pillars and beams which is likely to spoil the appearance of a room. There may be other mistakes besides the architectural detects in this room. Too may objects may be present or disorderly arrangement of objects may provide an ugly sight in this room. They may not be placed in an orderly manner, thereby necessitating a re-arrangement. Some of them need to be eliminated or concealed. There may be too much emphasis on many objects, though the background may be simple. In all these cases, some remedial measures need to be taken to produce a simple, restful and unifying effect in the room.

Besides elimination, rearrangment and concealment, one can manipulate lines and colours in such a way that remarkable changes can be effected. Windows and Rooms which are too square may be made to resemble oblongs through the use of a decided line movement (vertical) in one direction. The oblong windows which are too long may be made to appear shorter by using lines (horizontal) that repeat, and thus de-emphasize the short side. Objects may be emphasized or suppressed as desired, by means of the colours used in the background against which they are seen, and in the objects around them. Some objects can be highlighted or concealed by focussing light or by keeping these objects away from the light source.

Thus it can be seen that it is not difficult to make an environment in the home the way one would like to have. A thorough knowledge of the elements of art and the principles of design would aid a person to apply them in various problem situations. When one must make the best use of what one has, the ability to use colour and other art elements to merge well, and to apply the principles of art and design in all the arrangements are of immense value for those who lack imagination, the suggestion is to observe good arrangements in the shops show rooms, homes of their friends and relatives, or wherever they go, and use them in their own home surroundings.

Some tips for Do's and Don'ts for decorative arrangements

Do's

- Add beauty and home-like quality to the rooms through interesting decorative arrangements.
- Any patterned wall paper or hanging that is to serve as a background should be unobtrusive.
- Decorative objects should be placed against backgrounds that will show them to advantage.
- Use simple decorative objects in functional rooms.
- The objects placed in open cupboards should have decorative quality and should be arranged to create and attractive design.
- Use slip covers to conceal unattractive sofas/chairs or to harmonise unrelated pieces of furniture.
- Place ill-shaped furniture against backgrounds that do not create contrast or at places where there is diffused light only.
- Let your decorative objects supply accents of bright colour in rooms in which a quiet colour scheme is used.

Don'ts

- Do not make the room too stiff or barren that they look cheerless and unlivable.
- Do not forget to change the decorative arrangements occasionally.
- Do not put decorative objects of intricate design against a strikingly figured background.
- Do not use cupboards as a storage place for piles of dishes.
- Do not cluster all objects of intricate designs in one place.
- Do not put objects that are very small in scale near clumsy or over-sized furniture.
- Do not focus/use direct lights on objects/furniture that are ill-shaped.
- Do not let your rooms lacking in interesting colour and patterns.
- Do not leave too much space between the objects that the arrangement lacks the effect of unity.
- Do not hang wall pictures, textile or mirrors so high above a shelf or furniture that they fail to form a part of a group.

(Contd.)

Do's	Don'ts

Do's

- Arrange the objects in a group so that they will appear to be unified rather than scattered.

- Hang a picture or a wall hanging or a textile pattern or a mirror close to a shelf or a piece of furniture so that it becomes a part of a group.

- Hang small pictures at short distances so that they form a part of a decorative group.

- Use decorative objects according to the space available – smaller objects in small shelter and larger objects in large rooms.

- Group a number of small objects so that they will appear to be in scale with larger objects, or with large wall space.

- Secure rhythmic effect in decorative arrangements.
  - by means of connecting lines of a rectangular tray or plate, a plant, flowers etc.
  - by the use of a horizontal shape between two vertical objects.
  - by using varied heights and sizes in one arrangement.

- Place pictures or objects that the lines of the composition will lead to the centre of interest.

- Choose one object or a group of objects to stand out as a principal centre of interest.

- Use borders to set off objects or areas you want to emphasize or to create a series of related shapes to give a room unity. Print borders break the monotony of continuous areas of solid colours.

- While furnishing, use both plain and patterns to create interest. In any room either one-fourth or one-third can be patterned and the remaining portion, plain.

Don'ts

- Do not hang single wall pictures of smaller size alone to avoid the impression of being isolated from the other group.

- Avoid making decorative arrangements that appear to be scanty and inadequate.

- Do not use too many contrasting colour combinations in a single arrangement.

- Do not place emphasis on a number of objects that may spoil the focus on a single one.

- Do not place a picture or an object so that its lines carry the eye away from the centre of interest.

- Do not leave larger gap either in the centre or in between the objects in a decorative arrangement.

- Do not use two objects so equal in importance that they compete for the centre of interest.

- Do not use too much of solid colours in most of the areas.

- Do not use big motifs or designs for border areas. The size of the motif should not be big in small rooms.

- Do not use exclusively plain or patterned objects and furnishings in a room.

For those who find decoration as a problem or unable to decide on use of the art objects in their homes, a basic knowledge on art elements and design principles would help them to overcome such difficulties. Some practical tips, the steps to be followed in solving an art problem and the ways and means of making the best use of one's possession as elaborated in this chapter, can help them to make artistic and aesthetic arrangements in their homes and make their living environment a pleasure.

INTERIOR DECORATION

Accessories

Flower Arrangement

Rangoli: The Art of Floor Decoration

Interior Decoration: Trends in India

We have shown high esteem for any society that has shown sensitivity to beauty in all aspects of its daily living. We admire the people who appreciate and exhibit beautiful things in the areas that permeates into our daily living. Perceptions and the ability to enjoy the expressions of beauty, depend to a large extent, on the degree of sensitivity we developed towards people, ideas and our physical surroundings. The home atmosphere in which we live, is an important factor in determining our values and ideas on beauty. This is the place where we develop our sensitivity towards beautiful objects and surroundings that results in good taste.

Little touches in a room help to provide big effects in decorating. Just the necessary items and furnishings alone do not provide the relevant atmosphere in a room. It is seen that some objects and the other accessories supply the much needed sparkle and liveliness to an uninteresting and drab room. Just as an ornament/jewellery adds to the dressing up of a human being, accessories play a big role in decorating a room. A house, without accessories, looks barren and incomplete. Besides, an object meant to be an accessory, may also be functional in its own way. Thus, accessories not only add to the beauty of a room, but also reflects the function or the comforts in a room.

Today's generation is extremely conscious of the nature's role in the interior of a home. The use of potted plants as well as the cut flowers in a bowl are regarded as essentials in decorating a room. The wide variety of flowers available in the markets of today, speaks of the interest people show in them. Though the Indian homes are traditionally familiar with the use of fresh flowers, the art of arranging them has acquired a professional status now. Further, arranging flowers – right from their selection to the final arrangement – satisfies the creative impulses and needs of a human being. Flowers and their artistic arrangements have thus become a part of our lives today.

Similar to the art of arranging flowers in a vase, rangoli is an important traditional Indian art used to decorate the courtyards and walls of a house. Besides the use of flowers as fillers in a rangoli design, many more materials like saw dust, pebbles, marble clups, coloured powders etc are commercially exploited. It has become a distinct art and has acquired a prominent place in the curriculum of many courses involving creative arts.

Designing home interiors is considered as much an art as other artistic forms like painting, weaving, handicrafts etc. Beautifying the human dwellings is an integral part of everyday living in India, as Indians are firm believers in creating things of enduring value. The varied strands of our culture are woven into the fabric of our day-to-day life. It is no wonder that life in our country is moulded in an aesthetic environment. Interior decoration is an art which makes a setting appropriate for a congenial home. It aims at providing comfort, pleasant working conditions, relaxation, a warm and friendly atmosphere along with aesthetic surroundings and factors which provide satisfaction to its inmates.

Chapter 13

ACCESSORIES

L ittle touches in a room can add upto big effects in decorating. They are likely to supply the sparkle and liveliness needed to give a drab room a highly imaginative air and their effectiveness bears little relation to their costs. There is no doubt about the importance of accessories in decorating a room. Accessories are the elements that add charm, individuality and vitality to a room. They complete the decorative aspects of a room so that it is truly alive and can be fully appreciated. They play a subtle but a vital role by offering pleasant clues to the personalities of the owners and making the room as unique as the tastes and interests of the people who live there.

Accessories are usually acquired slowly over the years. They may be changed around, or replaced occasionally to provide some variety. *Pictures, calendars, lamps, mirrors, wall hangings, statues, collection of books, flower arrangements, clocks, curios, murals, handicrafts, artifacts earthen-ware and ceramic pots and textiles* are some of the commonest accessories seen in a home. In the recent times, potted plants are also added as accessories in home interiors.

Interesting accessories can make or mar a room's decorative scheme. The most carefully arranged and decorated room can also be bland and unimpressive if the personalities of its occupants are not reflected in the choice of its furnishings and other elements of decoration. For this reason, accessories deserve as much attention as any other element in the decor of a room. Often they are the most memorable features of a room. The decorator's touch with accessories can make it a reflection of the occupant's tastes and interests and increase the beauty, expressiveness, function and comfort of the room (Figure 13.1 – see colour plate 29).

It is important to mention here that an object meant as an accessory may also serve a function like that of a clock in a living room or a decorative lamp on the bed side. The accessories can either be purely decorative like art objects, antique implements, paintings and so on, or be functional like lamps, pillows, mirrors, flower containers, room dividers, bowls and the ubiquitous ashtray. Imagination, inspiration and good taste must go into the selection of accessories. Sometimes a person may accidentally come across an object and know immediately that it is just what she needed for a certain spot in the home. In general, one can say that an item added as an accessory in a room meets all the general objectives of fulfilling beauty, expressiveness and functionalism in decorating a room.

SELECTION AND ARRANGEMENT OF ACCESSORIES

Accessories must suit the overall picture of a room if they are to be effective. The size, shape, colour, theme, texture and purpose of an accessory are to be considered in the first place. Moreover, they should blend harmoniously with the colour scheme that is already established in the room. If the room furnishings are primarily of one period (traditional/contemporary etc.), the accessories should keep to the same period or be in harmony with it. One of the important rule to follow in order to give the accessories their proper setting is to mix formal with formal and informal with informal. A careless mixture never produces successful or the desired results. In all, accessories should meet the basic requirements such as beauty, functionalism, expressiveness, colour, line and form.

Beauty

Most of the accessories are chosen for their beauty alone. It provides the aesthetic pleasure to the members of the family and to those who visit the home. It further reinforces the development of visual tastes and an appreciation of all that is pleasing to the eye.

Function

The homes of today must provide maximum comfort, service and pleasure in return for the minimum care. The accessories chosen may be selected for serving dual purposes of beauty as well as functionalism, for example, books kept in a neat row not only add a pleasing sight but also encourage value for reading and knowledge thereby adding to the intellectual, spiritual as well as the emotional growth of the family members.

Expressiveness

Accessories should express the same idea as the home itself and should harmonize with it. The mood of the furnishing should be extended with the accessory being used. The personality, hobbies, interest should reflect in the selection and arrangement of accessories by the family. Accessories like books, plants, periodicals and hobbies such as painting, stamp-collection, photography, etc. generally reflect the taste of the family.

Line and form

In line and form, the accessories should conform to the general design feeling of the home. Period, non-period/modern furnishings should include accessories having the same general line and form as the furniture. However, deviations from the general style are more acceptable in accessories than in other furnishings.

Colour

An accessory should be carefully related in colour to the furnishings of the room where it is to be placed. Although accessories often provide the brilliant colour accents in a room, it is good to choose some accessories that have less compelling colour than others.

PLACEMENT OF ACCESSORIES

The accessories are placed in an interior which has an already existing mood. The accessories thus incorporated should enhance the beauty of the already set trend of the room. The most important

PLATE 29

Figure 13.1. Accessories in a room can add upto big effects in decorating

Figure 13.2. Living Room Accessories

PLATE 30

Figure 13.3. Stuffed toys, books, paintings are interesting and cheerful accessories in a child's room

Figure 13.4. Room divider, Side lamp, Indoor plant, Wall pictures, flower arrangement etc. act as accessories in a room

accessories in any room should be at the center of interest. The secondary centers should have less important accessories.

One should try to incorporate new ideas to perfect the art of decoration. The advantage of being original is that one can improve and evolve the product constantly unlike those who depend on borrowed ideas. Articles that are placed together should be made of materials that agree with each-other. A beautiful piece of sculpture might be placed in the center with a lesser object on each side of the mantel. A large low coffee table may hold a pile of two or three new periodicals, an ash tray, and a plant/cut flowers. A piano should have nothing of these on it (Figure 13.2 – see colour plate 29).

One can find unusual articles in importing shops, decorator's shops, antique shops, studios of craftsmen and artists and out-of-the way shops. Auctions, junk yards and storage warehouse sales are places where one sometimes can find old but useful articles.

HOME ACCESSORIES

A large variety of accessories are used in a home. The most common ones are curios, books, clocks, statues, etc. There are other accessories too which are used by many families. Some articles like screens, accent furniture, dining table mats and similar items add variety to the kinds of accessories used in a home. Some of them are listed in the following paragraphs.

Mirrors

Mirrors work magic, as there are an endless number of visual coups one can pull off in their rooms. When a room is too long and narrow, it can be widened by covering one of the long walls with mirror. If a dressing room is tiny, hanging big mirrors on adjacent walls can get a double playback of space. To get more mileage out of the candle light in a dining room, placing the candles against a mirror will get twice the candle power. Use of a mirror on a blank wall facing a land scape or a scenary would bring this visual beauty into the interior of a room in a fascinating way. The partitions that create a cramped look can also be made to disappear with the use of a mirror. While wanting to decorate a pillar or the top of a chimney without using a painting or a picture, it can be done with a mirror which would not distract the beauty of the room or the fire place. Thus a mirror can be used in a number of ways to brighten up any space or decoration.

Mirrors have been used in decoration for well over three thousand years. The mirror of the ancient Egyptians, Greek and Romans was a thin disc of metal, usually bronze, slightly convex and polished on one side for viewing. In hand mirrors, the reverse side was often engraved with line drawings, usually of mythological characters. Many of these metal mirrors, can still be seen in museums. It was not until the fourteenth century that the technology was advanced enough to produce glass mirrors in any significant quantity, but then they were only available to the very rich. It was in Venice in the fifteenth century that the making of a flat glass mirror began on a commercial scale. In France, plate glass was invented in 1691, thereby opening the possibilities of a variety of size and far greater availability.

Along with mirrors, their frames can also provide variety and beauty in a room. Painted mirror frame is an important colour accent in any room. Similar effect can also be produced by the use of a shiny material like silver or stainless steel frames. Sometimes colour plastics can also be used as frames for the mirrors to create variety. A wooden carved antique mirror frame can be a focal point in any decoration.

Mirrors, when used architecturally to cover walls or partitions, are part of the surface decoration of a room, similar to paints, wallpapers or paneling. The style of the frame is as much an individual element of period detail as any other piece of furniture such as a chair or a table. These days when a mix of styles or of periods is considered lively and attractive, one can choose a mirror style either to match or contrast the furniture style. Having a general knowledge of styles can be of help in making final choices.

Though mirrors may not appear as accessories on their face value, they can successfully be used in rooms as additional accessories. The mirrors help in the reflection of the person, personality, interests, nature as well as the mood. Mirror frames are often more elaborate than picture frames and like an important painting, can grace a wall alone. A mirror may be placed in a frame of wood, metal or a composition material and is equipped for hanging. Mirrors add life to a room. They should be free of distortion and may be used to make a room much larger.

More than any other element in decoration, the mirror produces a feeling of spaciousness. Entire walls of mirror are the most effective and are not so costly. Permanent reflections in mirrors should be pleasing and would add to the beauty of the room. Mirror tiles may bring both spaciousness and an incomparable gay glitter wherever used.

Books

In interior design, books are classified as decorative accessories. The combination of books and the other objects thoughtfully arranged can be the focal point of a decorative scheme. Books add warmth and character to a room and give a lived-in look and feeling. In any room of any style, book-shelves and book cases can serve as centres of interest. Whether the home library is small or big, one can always find ideas for adopting it to one's needs and decor.

It is not surprising that books and fireplaces go together. It is easier to fill in the spaces between the fireplace and the walls with a combination of storage units below and bookshelves above. This will add stature to the fireplace itself and also make the room seem bigger, because the entire wall space is used. Books and fire place combine to give a home a feeling of warmth and welcome that is hard to duplicate. Bookshelves over the mantel, can, no doubt, provide the ornamentation needed on a fireplace wall.

Books can be used any where in a home. A desk and bookshelves make a natural combination, particularly if the desk is for study or for a home office for obtaining reference books immediately. One good idea is to angle the desk out from the wall, and it will serve as a room divider, and to place the shelves on the wall next to it. Another way is to place the desk against the wall, flanking it with ceiling-to-floor bookcases and cabinets that hide files and home records, The walls in a dining room can also become a combination library and display area for fine china. A study room is the natural place for the books along with a pleasant, comfortable place to sit with good lighting. In a living room, an entertainment centre can be created. This area can be designed to accommodate books, record player, television set, tape recorder, and storage space for tapes and records.

Hall ways, the place underneath the stairs, or the passages can be used to keep bookshelves of 8 to 12 inches to accommodate books or some decorative accessories. In a bed room, a wall of book cases can add a bit of live-in appearance. In a guest room, a small book case can have an enclosed lower section for storage of guest linens and a pull-out counter to be used as a writing desk. Space under an attic ceiling is perfect for a read-and-listen corner.

To some people, books are almost sacred things; to others, they are just objects of certain sizes, weights and colours. To most people, however, books are the sources of pleasure, instruction, or diversion, for almost everybody own some. For avid readers, their books get piled up everywhere in the home. Whether one has a large collection or only a few books, it is good to have fresh look at them, and consider them not so much as loved possessions alone, but also a part of one's decorative scheme. Books are an indication of the interests of a person. However, they also contribute colour, pattern and interest to a room by serving as accessories. One may bind the books in colourful coverings and arrange them in an orderly manner. The books lend a warm and lived-in look to a room apart from serving as a favourite time-pass for many of us.

Books along one of the short walls are probably enough in a living room unless there is a definite reason why the occupant would feature books. Built-in bookshelves are preferable to movable cases because they conform better to the lines of the room. Shelves should be adjustable to fit the height of the books, periodicals, phonograph records holders. Bookshelves should not be left empty, and they can be filled temporarily with some second-hand books.

Books should be arranged with consideration for their size and colour. For reasons of stability as well as good form, largest books should be arranged on the lower shelves. The darkest book look well near the bottom and along the ends of the shelves. Pottery, figurines, carved book-ends may serve to break the monotony if the number of books is very large (Figure 13.2 – see colour plate 29).

Murals

A mural is a painting or a decorative surface (such as wall paper) applied to a wall. A mural is distinguished from a fresco in that the latter is part of the wall rather than an applied surface.

Murals can be used on walls for colour, glamour, and excitement. Murals are like exclamation points to be used when one has something important to say. They create an intense decorating effect, as large in size, with bold colours and dramatic design. Murals are not for the timid because no other decorating accessory can express so much value for the taste, flair, and creativity one invests in selection and display. One may choose to splash bright colour in bold abstract forms across a dining room wall or paint a whimsical tree, complete with birds and butterflies, on bedroom wall and make obvious that one is approaching decorating with vigour and a light heart. One the other hand, if a delicate classic Japanese screen is chosen to cover one wall the entrance foyer, a mood of elegance and quiet restraint is being established, even before guests have entered the living room.

Because they are surfaces placed or painted upon walls, murals should be chosen with as much care as painting or paint. Accessories may be placed on a painted wall with adhesive or attach mural – size posters, prints or photostats. Therefore, murals can be devised from almost anything, as long as they are large and effective and choose a mural that complements the room's style.

Clocks

Clocks can be considered as functional accessories. Clocks have been a necessity from time immemorial and served as an indicator of the advancement in technology. The earliest measurement of time was done by the sundial. Then came the hourglass followed by the invention of pendulum. The early pendulum required a tall case, and to this day we enhance our homes with 'grandfather' clocks. These clocks are available in style ranging from elaborate period cases of heirloom quality to modern sculptures of Plexiglas. Thus a large variety of clocks are available to add beauty to a room, besides serving a function of indicating the passage of time.

Mantel and wall clocks are generally intriguing in shape and materials. The period pieces serve decoration as well as the functional aspects. A clock with plain lines and little decoration is usually the best choice for almost any room.

Accent Furniture

The accent furniture creates a centre of interest in any setting. It makes the eye rest on it and thus invites the guests to come and appreciate it. The accent furniture may include small pieces such as a corner basin stand, wedge/handkerchief table, reading table, pier cabinet, curio cabinet, coffee table, wall cabinet, table desk, flip-top console table, oriental chest, wall screen, etc. These pieces apart from being beautiful themselves are able to display other art treasures as well. Period furniture, finely carved ones and rocking chair arranged at focal points act as accessories.

Decorative Hardware

Decorative hardware can transform ordinary furniture into distinctive accent pieces. The furniture pieces may have knobs made of wood but could be replaced by ivory and metal knobs which make an ordinary piece look expensive and stylish.

Wall Unit

The most remarkable aid to living room storage is the wall unit. Instead of appearing as a conglomeration of diverse items, the wall unit by means of consolidation and organisation, produces a calming and cohesive effect, as well as an important focal point in a living room. Every room has a wall that can be used this way. A wall unit is very individualistic. Its base may be a solid series of chests and cabinets with shelves rising above. Gaps may be allowed in the base furniture for a television set, desk, chair/sofa and shelves may be open for storage of books and ornaments shuttered and doored for storage.

Screens

A Screen is used for privacy, separation between two functional areas, may conceal poor architectural features, create balance for a door/window, or provide protection from glare, etc.

A screen should be decorative, should suit its surroundings and should be large enough to be really useful. Screens are available which are decorated with greatly enlarged photographs of plants scenery. Wooden frames can be made with the center of cloth, matting, reed, bamboo, painted hardware cloth, fabric, plywood, composition board or translucent composition over screen wire.

One can also create dramatic and practical effects with colourful curtains of beads. Bead hangings make decorating effects in countless number of ways. They can create a dramatic division and at the same time, can keep space free-flowing. They can be used to curtain a window, disguise a non-view, and at the same time, allow light and air to filter through. As a divider, they take almost no space, yet give a feeling of semi-privacy in each of the separated areas. Along with beauty and versality, there is another factor in their favour – beads require no dry cleaning laundering or pressing. They are only to be wiped with a damp cloth to restore them to their original appearance.

Many kinds of beads are available. There are plastic beads that come in kits of plain or mixed colours. Wooden beads in several wood finishes and metal ones, in silver and gold finishes are also popular along with ceramics, crystal and golden tones. They also come in various shapes and forms – round, oval, multi-faceted, or random shapes. All one need to do is to fasten them together in a string to the desired length. Bead hangings can be designed to solve a number of decorating problems in a home. They can be used as window curtains to reduce light glare and give some day time privacy without obstructing the view. They can also be used to make a bright centre of interest as a wall-hanging in a room. Bead hangings not only provide partition in a living room, but also give a background for furniture arrangement. A bathroom can be given a decorative treatment by using a bead hangings over a plain shower curtain. A bedroom can also be dressed up by trimming ready-made draperies with bead hangings that blend in colour and style of the room.

Pillows and Hassocks

Both these can provide interest through shape, colour and design accents. The furniture pieces such as the sofa, chairs etc., can accommodate pillows, cushions of varied pattern and colour with rich fabrics. Contrasting colours may be used to emphasize.

Quilted, patchwork and furpillows are a few items to mention which add the desired quality of aesthetics. Hassocks serve not only the decorative purpose but also serve as seating and storage. Ruffles, tassels and cording trim these useful and decorative accessories (Figure 13.3 – see colour plate 30).

Dining Room Accessories

In a dining room, the same set of china, tableware centre piece and place mats day after day, no matter how beautiful, can become boring and dull. Introducing a change in any one of these items in rotation would add interest to a room.

Place Mats may be oval, round or oblong. They can be made of linen, organdy, lace with smooth or rough texture. Patchwork print quilted mats are casual and practical, vinyl mats are easy to clean with a damp cloth. Designs should be bright and vivid. A few sets of mats can vary the colour scheme and mood of the table setting. Napkins may blend or add contrast. A lot of variety in texture, material and colour for tablecloths are available and add interest in a dining room.

Centrepieces may include bowls of china, glass, copper, brass, stainless steel, enamel, silver/ pewter may hold leaves, flowers, fruits and vegetables. Candle-sticks of any material are an old-time favourite, which can be used on special occasions at dinner times.

Sideboards: Various articles may be artistically decorated on the sideboards. A silver service, cut glass decanters, glass sculptures are all delightful and appropriate to bring in variety and beauty in a room.

Artifacts

The various kinds of artifacts – relics from the past or products of primitive societies – can play a significant role in decorating the home and its immediate surroundings. Among the artifacts more usually found in the home are primitive tools, ancient coins, and pottery fragments. These are often displayed in combination, especially as wall decorations. Modern reproductions of ancient statuary – especially busts of members of royal dynasties such as the Egyptian queen Nefertiti – are used successfully as accessories to modern furniture. The larger pieces may be placed alongside drive-ways, or conveniently settled in a garden. Artifacts – original or reproductions – reflect a time in history when using these antiquities as accessories, a balance in style may be tried.

Antiques

Anything – art objects, furniture, porcelain, and commonplace items such as boxes or paper-weights-can become an antique if it stays around long enough as, an antique is something from another time. Purists maintain that only items made before 1830 are true antiques. After that time, many handcrafted items were produced in factories using machines.

Antiques for decoration can be classified in two categories – furniture and accessories, but sometimes some may fall into both categories. Often reproductions of antiques can be used in place of real thing as accessories. To gain authenticity, these can be blended with a few really old pieces. However, it is necessary to consider scale and colour to make a perfect blend.

There are many antique items that adapt themselves to be used in living and other common rooms to add a decorative touch in the arrangements. Large ironstone pitchers that, before the days of modern plumbing, stood with the wash bowl on top of the wash stand, are good containers for the long-stemmed flowers, or can be wired for lamp bases. Old-fashioned washstands can also be refined and re-finished if necessary, to be used as an extra storage space. They can even serve to house stereo components and records. There are a number of antiques like candlestands, bowls, coffee and tea services, and many other items in brass, copper, silver or clay that date from earlier times. All these antiques can be re-furbished and put to good use. It they are in need of more than home cleaning and polishing, there are experts who specialise in mending, cleaning, buffing and lacquering to restore these decorative treasures to their original beauty and finished form.

A few factors should be considered while buying antiques to serve as accessories. If a family's taste run towards accessories of yester years, it should look for curios such as old molds, skillets, flat irons, copper pans, clay jars, match safes and cooky containers. These are often found in auctions or fairs. Old school bells, clocks, cradles, picture frames, bird cages, and earthen wares can also make interesting decorative items for creating accents. Usually all these items need thorough and careful cleaning, and this may reveal that they also need painting, polishing, or re-plating. If these pieces are expected to add interest to the furnishings, they should be selected wisely, and restored to bring out their natural beauty. However, the final impact will depend upon the way they are arranged tastefully.

While searching for genuine antiques and art objects at sales, it is necessary to realise that these are very scarce and highly priced with keen competition. These pieces must be examined and studied thoroughly. It will also pay to study books on antiques and to check the price lists of antiques deals to check what similar articles are worth on the current market. It is also important to remember not to buy too many of the same kind. Variety is the spice, as far the accessories are concerned. In all the cases, it is wise to show restraint and fewer accessories for creating interests and accents, when used sparingly. With patience and a great deal of time and effort to pursue this hobby, a person can be fortunate enough to acquire these wonderful treasures and use them to their best advantage.

Baskets

For those with a limited budget, baskets offer as inexpensive accessories. They are the smart, unstereotyped accessories, which can be placed at many places with extra advantages. Often of wood tone, they serve as colour neutrals, blending well with woods in any room. Occasionally they come in bright colours to be used as accents. As one of the traditional objects, they have now become so popular that almost every city in the country has shops that sell baskets. Department stores, gift shops and florist shops often have special basket sections, with different styles, shapes, sizes and weaves for meeting every one's needs.

Baskets are the inexpensive, useful and handsome accessories for every room in the house and for outdoors as well. They provide both storage and beauty. Baskets can be made from the slimmest of twigs, thin strips of wood and bamboo or any other flexible material woven together. They range in size from small serving pieces to enormous storage trunks or carrier baskets. Depending on the type of weave, baskets can go well in classic, country or contemporary rooms.

Even though most baskets come in wood tones – dark or light – they can also be colourful. As can be seen in their variety available in the market, there is a growing interest in painted or dyed ones as decorative accents. Some baskets are available in tones that could fit into any decorating scheme by picking up major colour or by providing accent or emphasis in a room. When needed, they can be painted with any colour one wants to add. Since these can easily be painted at home, one can also make quick and easy change to some other colour.

Baskets are versatile and provide varied kinds of uses. A small entrance room or a hall can be furnished with a shallow chest or table above which can be hung a mirror or a picture grouping. On top of this, a lamp and a basket can be combined to hold mail, gloves or small note books/scrap books that a hall table can hold. Care may be taken to relate the size and shape of the basket to the scale of the table. Baskets can be used to hold fireplace wood, to take the place of a magazine stand, to serve as a toy chest in a child's room or as a clothes hamper in a bath room. For a mobile family who change their apartments often, they can buy easily movable furniture or storage, baskets can provide an inexpensive furnishings or store trunks/baskets for them. A large hamper with a lid can serve both as table and storage.

Similar to living room requirements, baskets can effectively be used in a dining room too as centre pieces or serving bowls for cookies, nuts, toffees, deserts etc., besides being a flower pot. Baskets can also be converted to shades for lighting fixtures or to drawers, and massed together for accessory or storage groupings. They serve as handy pickup and take along containers for sewing and knitting or for gardening tools. Potted plants in a range of bright colours can take on unity and importance when they are grouped together with baskets as pot cover ups. Thus baskets can create a variety or produce a delicate effect, besides being practical ones.

Glass, silver and china accessories

Glassware as accessories plays just as important role in home interiors and decoration as silver, china, linens and the other art objects. Just as a simple dish is much more appealing on a colourful plate than on a plain white paper, so does a glass of ordinary drink gain importance in a graceful stemmed goblet.

Glass consists of sand plus various bases. These include potash, time or lead oxide. Crystal is an especially durable form of glass noted for its brilliance and clear bell-like ring when struck. Crystal is made of flint and lead oxide. Ordinary glass contains no lead therefore, has no ring and little sparkle by comparison. Cut glass is crystal that has been cut in faceted patterns that reflect light and add highlights and brilliance. Pressed glass is made by pouring molten glass into moulds where it is pressed into patterns. Pressed glass is made in a variety of colours, and is, therefore, used widely for many purposes. Hand-blown and pressed, or moulded glass are both produced now-a-days and each variety is valuable for a particular purpose. Hand-blown crystal is precious, expensive and breakable and is therefore, usually not used every day.

There are countless shapes and sizes of glassware – many more than one family would normally desire or have the space to store. A glass bowl meant as a container for desserts or any other serving dish or for making an attractive floral arrangement or a finger bowl can be used to make a floating flower arrangement with a few single blossoms. Thus the glasswares, varied in size, shape and style, can represent just about every purpose to which glass is put – from a formal table ware to a dessert dish to a delicate cut – crystal glass accessories in the show-case of a drawing room.

Music system

Music systems like hi-fi and stereo are a good leisure resource, a means by which music previously difficult to access and performers who once had limited audiences have been made widely available. Fine music and musicians, once heard only by aristocrats at court performances, or in auditoriums and folk songs once brought only to local towns by strolling country artisans and musicians, can now be heard almost anywhere and at almost anytime by anyone relatively at a low price of an audio or video tape. Home libraries of the finest music can provide a good collection of popular hits, poetry readings, and speech of history-making importance in small packs in the form of an audio or video cassette discs.

The number of people with a taste for pre-recorded entertainment has increased greatly as the quality of the equipment has improved and as prices have dropped to a level many can afford. Today millions of people buy the needed equipments and spend many hours enjoying it. Home record parties, evening get-togethers, musicals and dances, celebrations of birthdays, anniversaries, and similar events, call for the use of the electronic equipment, and tape libraries are needed for these activities, and are therefore, important new factors in home decorating.

While including musical equipments as a part of the interior decoration, they, must be housed with an eye to its appearance and an ear to its acoustical needs. The cassette player and the tapes/ compact discs must be stored comfortably and a favourable listening environment must be created. Fortunately, the audio/video players, speakers, amplifiers, and tuners lend themselves to a number of satisfactory arrangements. They may be treated as separate components and placed on open shelves or concealed in cabinets. They may be concentrated in one room or scattered throughout a number of rooms.

Hi-fi and stereo installations are highly flexible and can be incorporated into traditional or eclectic rooms, and into large or small spaces, creating a listening environment for dining, for conversation, or for concentration on the music. A conventional living room, too small to accommodate an extra piece of furniture, can have a music system added without sacrifice of comfort or loss of charm into the room's decorating scheme. In a large family room, the music system components can be an integral decorating element, displayed like works of art, when the components are wall-mounted or set on shelves, each placed for aesthetic interest and ease in handling.

It is generally seen that the modern musical equipments are finely machined and assembled and has an elegance of its own that is as decorative as it is functional. The proud owner of a good quality equipment takes pleasure in the beauty of the fine equipment and enjoy having it on display as part of his room decor. An alternate solution to housing a music system is to enclose all of it, or part of it in a cabinet, since the cabinet has the advantage of protecting the delicate equipment from accidental damage and dust. The design and finish of the cabinet or the shelf for the musical system can be blended with the decorating scheme of the room, in which the equipment components are out of place. If there are small and curious children with exploring fingers in the house, or the location is unusually laden with dirt and dust in the air, a cabinet is a wise choice.

Fine electronic equipment can be a major investment as an accessory in any decorating scheme. It, therefore, requires much research and planning on the part of the owner, who may justly be proud of the results of an aesthetic arrangement. Almost as important as the equipment is its placement in the room. It is important to have the turntable easily accessible to change the cassette and tapes, or to alter the tuners and volumes. Finally, all parts of the system, turntable, tuner, amplifiers, speakers and tapes, must be attractively displayed.

Hobbies and collections

Hobbies, whether they involve making things/handicrafts or collecting them, growing things or engaging in sports, help to keep us interested in life and act as safety valves for relieving stress and tensions. Since hobbies enable us to express our individuality and reveal our personalities, one of the best ways to make sure our homes bear our individual stamp, is to decorate and plan our rooms with hobbies in mind. Many people like to paint, weave, make things of art, frame pictures or bind books or engage in many other innumerable possible ways of spending time fruitfully.

When carried out sincerely, the results of the hobby are suitable for display, and so can be placed any where in the house. Walls, shelves, cabinets, bookcases, any other part of the house can be used to exhibit the handicrafts or the collections. In short, hobbies can be a decorative asset in a home. Some hobbies can even develop from an object that was obtained in the first place as a decorative accessory. Many aquarium and plants that are purchased for pursuing a hobby can be used as a pretty accessory for the home. Many wealthy collectors have gathered assortments of paintings, furniture, or other treasures which can easily become a part of a decorative scheme in a home. For others, collecting things as a hobby help them to acquire things at relatively very little expense. It may range from seashells or stones smoothered and shaped by ocean or stream to a diverse patterns of pressed glass or from a metal paperweight to soft toys or from old coin bank to ceramic pots. Any of these things that are collected can be used to enliven the home. However, all these collections of identical items should be kept together in one place for maximum impact. If the items of collection are tiny or fragile, displaying them behind or under a glass cabinet, or a collector's table with a glass top, or glass-enclosed shelves is suggested. Some of the items can also be hung on one wall of the room.

House Plants

Contemporary architecture is partly responsible for the growing popularity of indoor plants as accessories in interiors. Even those people who have overlooked the decorative, sculptural qualities of plants, have started exploring the use of plants. Both the formal geometric interiors of most office buildings and the informal modern houses, benefit a great deal from the colour, softness and the graceful shapes of plants. Thus, plants are the tools for the interior decorator to exhibit his creativity. Their decorative usefulness extends from a single small plant on a table to a grouping of these plants to provide the focal point in a room.

Plants, when placed cleverly, can hide, or atleast minimise unfortunate room features like pipes or obstrusive beams or hanging cable wires. In some cases, they can camouflage an ugly radiator or an unattractive air cooler. A mass of greenery can serve as a temporary substitute for a piece of furniture which has gone for some repairs, or can even fill the space in a drawing or living room with its own kind of beauty (Figure 13.4 – see colour plate 30).

As home accessories, plants can give a fresh look to any interior. This is particularly true in the summertime, when the greenery of the plants, and their colorful flowers can give any area a gardenlike look and a feeling of coolness. Plants are inexpensive items when compared to many other accessories. One can also choose those plants that require little care, to avoid unnecessary trouble in maintaining them. It is not necessary to change the plants after a short span of time. There are many house plants which thrive for many years. A professional attitude toward beauty as display must be developed by the person who wants to use plants effectively in the home.

Plants can be placed anywhere in the home and many will thrive as well in artificial light (fluorescent is best, but incandescent will do for some) as they do in daylight, but this depends upon the kind of plant. the plants that one select will depend on where one lives, the time one can denote to them, and the kind of effect one wants to create. To thrive in the house, a plant needs approximately the same moisture, light and soil as it had in its natural habitat. The needs of a plant are usually simple. Leafy plants need light, but little direct sun; flowering plants need sun in varying degrees; a few plants need special soil; and all, of course, need moisture, but care must be taken not to over water them.

Plants can be used as accessories in every room. Nature's colours are easy to work with, from the many shades of green that most people find pleasing to the vivid colours of flowers that enliven a whole room. Plants thrive in a bath room because of humidity, and they contribute a great deal to soften the usual porcelain look of the ceramic tiles. Though one may think of a plant taking some amount of oxygen, the amount actually used by many plants in a room would be inconsequential. Thus they can safely be used even in bed rooms.

The kitchen becomes a happier place to work in when there are plants, and an entrance hall becomes more inviting when decorated with growing flowers or greenery. Thus there is no place in a home which cannot contain green plants. These plants can be placed either singly or in groups. Large plants, however, especially the tall ones, can stand alone quite successfully.

Not only flowering plants provide variety, but also their containers. Some simple household articles can be converted to serve as plant holders. Asparagus, ferns, money plants etc. are among the leafy plants that grow well in hanging pots. Each potted plant should have some kind of saucer or dish under it to catch the water that runs through. Containers for hanging plants should usually be light in weight. Variety of baskets can also be used for this purpose. While clay pots admit air to the soil better than plastic pots, the latter are a good choice for hanging plants. A good way to water these hard-to-reach plants is with a watering can that has a long, curved spout.

Pictures and wall Hangings

Pictures reveal the stage of aesthetic development of their owners more clearly than any other articles of furnishing. Pictures should not only say something, they should also do something to an individual. A good picture should stimulate the imagination in such way that the observer is permitted to share an artist's mood and see the inner meaning of the subject he interpreted. What a picture has to say depends both upon what the artist has put down and the personality and understanding of the observer brings to it.

The appeal of a picture

Pictures make their appeal in various ways-form, colour, pattern, subject – each has the power to stir the imagination. A good picture is an interpretation rather than a literal representation of a subject. In selecting a picture, one's first concern should be for the individuality.

All pictures have a composition or design. Sometime the plan of the composition stands out prominently, like a framework but often it is so observed that the casual glance does not reveal the outline upon which the picture was built. If the framework is badly designed, the picture appears to lack organisation, while if it is so concealed that it cannot be traced easily, the picture seems structureless. Colour, is the quality most universally enjoyed in pictures. Quiet, gay, restful and stimulating colours make their special appeal to the individual. When the theme of a picture is interpreted with destination, it adds character to the other elements like line, form or pattern. The pattern of the dark shapes is beautiful against the lighter background tones. Such sensitive variation is felt in the lights and darks in the picture that colour is suggested even though it is not there.

Oil Paintings

An oil painting should not be considered an impossible luxury by a family of modest means. A young artist of ability who has not yet attained recognition is usually willing to sell an original oil painting that has merit for a price of about 2 to 3 thousand rupees. Such pictures should not be confused with the commercial type of oil paintings sometimes available in ordinary department stores.

Oil paintings are usually on canvas, and linen is prepared. Oil paints are made from pigments of mineral or vegetable matter which are well ground and mixed with fine oil. Naturally oil paintings are more durable than water colours. Most of the serious painting, exclusive of fresco, is done with oil paint because of its *permanence* and *flexibility*. It dries slowly; therefore it is the best medium for subjects that require deliberation. the painter has time to analyze his work, and, if he is not pleased, he can scrape off the paint in part or all of his picture. Even if the paint is dry he can remove it with paint remover. He can then repaint the picture until he is satisfied with it. Oil paintings often have a studied quality due to the consideration they have received. On the other hand, some oil paintings are done quickly and freely by artists who have an emotional approach.

A wide *variety of techniques* is possible with oil paints. Good and bad pictures are painted in all techniques. The uninitiated should not make the mistake of thinking that a smooth surface is preferable to others. Some artists employ *thin oil paint*; others prefer a consistency similar to that of *house paint,* which can be applied in several ways. Some who use the thick paint just as it comes from the tubes apply it to the canvas so that it is 1/8 inch thick in places. Fairly thick paint may be applied all over a canvas or only in those foreground parts that receive emphasis. The size and shape and the handling of the brush are also important factors in technique. Because of its flexibility, oil paint encourages experimentation in texture, which is now receiving more attention than before. Texture as well as technique are only means to an end, however, and are unimportant in oil paintings as they help artists to express their ideas.

Water-Colour Painting

Original water-colour paintings should be used much more generally than they are, as they are reasonable in price and also desirable. Unframed water colours of merit are sometimes sold for the price of a reproduction. Water-colour pictures are painted on paper with water-soluble paints that come in tubes or cakes. A somewhat rough paper is desired by most painters, for it adds textural interest to the otherwise flat paint. The white of the paper is often left unpainted in places throughout the picture, thereby contributing a sparkling quality.

A good water-colour painting is usually *highly subjective* and requires a very different approach from the carefully considered oil painting. *Emotional intensity* is necessary for painting a water colour. The artist has to work at top speed because otherwise the paint dries so quickly as may interfere with his plans. Different artists handle water colours differently, however. Some use wet paper, and some dry. Others feature such characteristics as large wash areas, many separate brush strokes, delicate blotted areas, graded planes, crisp dry accents, or beautiful brush strokes.

All water-colour paintings should be spontaneous and fresh. Usually they are treated broadly so that there is no niggling detail in them. Water colours are as a rule somewhat sketchy so that the observer has to exercise his own imagination in completing them. One charm of water-colour painting is that the accidental effects often are better than those that are planned.

Water-colour paintings are limited in size because the medium is difficult to control. A 20"×24" painting is about the maximum size for home use. Water colours are most effective for quick, sparkling sketches, which are shorthand notes on nature.

Development of sense for selection of pictures

It is every difficult to select pictures that have aesthetic quality unless a person has seriously studied pictures. However, constant association with fine pictures might develop a sure taste. It is possible that a dilettante could visit art exhibitions for a lifetime and still not know what to look for in pictures. The sense of a good painting can be developed by reading a comprehensive history of painting, a good recent book on the aesthetics of painting, and also the contemporary art periodicals. A person can join a course in appreciation and analysis of pictures which are generally provided in museums. No one can thoroughly appreciate pictures, however, unless she has had some good instruction and practice in painting in both oil and water colour. Painting can be used as a medium whereby one can express or communicate an idea or feelings.

Some handicrafts of India

The imaginative skill of the craftsmen of India is limitless. With their expertise and skill, they continued to produce priceless artifacts which helped to preserve a heritage and tradition dating back to many thousands of years. The various handicrafts – whether *metalware, pottery, mats or wood-work* are generally made to serve a need in the daily lives of the people. At the same time they provide them with an opportunity for artistic self-expression.

Metal work in India goes back to the third millennium B.C. The earliest bronze figures found at Mohenjo-Daro reveal a high degree of skill in this art nearly 5000 years ago. Craftsmen use a variety of techniques and styles. Engraved brassware with or without a covering of enamel is made mainly in Moradabad and Varanasi in U.P. and in Jaipur, Rajasthan. Popular items are traditional *wine jugs, flower vases and fruit bowls, table tops, gongs, bells, candle stands, ashtrays, beer mugs,* and other decorative items. From Bidar and Hyderabad come the beautiful *bidri work* – lustrous silver or gold inlay done on a striking jet black oxidised background. *Wall plaques, cigarette boxes, ash-trays, paper cutters and fruit dishes* are often ornamented in this way.

Metal sculptures, for which the South of India (Madras, Madurai and Bangalore) is famous are often in the finest Chola tradition. They represent gods and goddesses and may be made in brass, bell metal or oxidised to give a black finish. Both solid and hollow images are cast. The tribal people of Eastern and Central India produce *brass and bell-metal images* which have a character of their own. This *Dhokra work* shows a primitive style which appeals to the tribal folk.

Wood carving is an important traditional industry in India. *Images and panels, architectural elements* (like doors and carved pillars) *utensils* and *various decorative* pieces are made. Modern workers make *furniture* of new designs – tables, low stools and chairs in Punjab. In Kashmir, rich, variegated designs in raised or engraved patterns are carved out in walnut wood. *Inlay work* in wood is carried on at a number of centres. The inlay may be of bone, ivory, metals or even plastic. Sankheda in Gujarat is renowned for the production of lacquered wooden furniture. Rosewood (now difficult to obtain) and sandalwood are used in South India for the manufacture of decorative items as well as furniture. Similarly, workmanship on bamboo is another artwork of the people of Kerala.

Stone carving has been known in this country for centuries and a wide range of styles may be seen, especially in temples all over India. Stone carvers now adapt these designs and motifs to make decorative pieces for our homes.

Pottery as a craft is widespread. In Rajasthan, blue and turquoise glazes obtained from cobalt and copper oxide are famous. Other important centres in the North are Khurja, Azamgarh and Aligarh, where a type of interesting black pottery is produced. In South India, Bengal and Maharshtra *terracotta* work has long been popular. The Bankura horse of Bengal has now become known all over the world.

Basketry is primarily a folk craft. Bamboo, cane, grasses and reeds as well as the leaf of the coconut and date palms are used. Mats and baskets, boxes and trays, toys, dolls, costume jewellery and wall hangings are produced. Such work is to be seen in Assam, West Bengal, Bihar, Orissa, Uttar Pradesh, Tamil Nadu, and Kerala.

Indian textiles have now captured a world market. The art of fine weaving, varied process of painting and dyeing, hand and loom embroidery have been perfected over the ages. *Silks* of many kinds ranging from Banaras brocades, Gujarat patolas and Kancheepuram silk are too well-known to need discussion here. *Tie and dye* techniques are practised in many states – Sarurashtra, Jaipur and Madurai being well-known for this art. In addition cotton textiles – fine mulmull and cambric, Chanderi and Kota materials to mention only a few – present an endless variety to choose from.

A house without accessories would look barren and unpleasant. Moreover, accessories reflect a person's personality and quality of the home. An accessory may be a small piece as a curio on a side-table or as big as a wall-to-wall carpeting. Therefore, it is necessary to analyze the requirements and needs of the home and make wise choices to make the house more attractive and functional.

Chapter 14

FLOWER ARRANGEMENT

"In the beauty of the flower is revealed the hand of its creator"

Using fresh flowers as a decorative accent in today's homes is an important aspect of interior decoration. There is a growing enthusiasm for this practical as well as visual art form that everyone enjoys. It is undoubtedly true that flowers bring beauty and fresh life in a home and provide a touch of graciousness to the entire room settings. Further, arranging flowers – right from the selection of an arrangement and its corresponding materials in the making of it, to the final touch in the creation a design of one's own—satisfies the creative impulse of the human beings.

The art of arranging flowers comes as an inborn gift to some. However, it may not be difficult for the others to learn it. Any one who handles and cares for flowers and the other natural things and is prepared to take a little time and effort to collect them, can easily acquire the necessary skills and techniques of this art. Anyone can enjoy using the plant materials in the way they are seen in nature, along with a little bit of one's own imagination.

There may be a minimum arrangement, using a single bloom or an extravagant floral fantasies, intended for special functions like that of a wedding celebrations. Between these extremes, is a huge range of shapes and forms, including the ones like that of a sculptured effects, surreal new flower "species" made by using foliages and flowers from different plants, and the exotic combination of wild and cultivated plant materials.

In a systematic creation of a flower design, the application of principles of design and art is vital to get a scientifically finished effect. By doing so, one will find that flower design is an intensely practical and enjoyable activity. It shows as to how to get the most out of the flowers by handling them carefully so that they work like the coloured pieces of a jigsaw puzzle, slotting into place within each arrangement and within the room setting one would like to select.

The ideas for designing arrangements are infinite. Any one can take inspiration from any natural source or plant-related materials—whether exotic or an ordinary one. Picture postcards, photographs, wall papers, fabrics and china designs of flowers and gardens, are also good sources for obtaining ideas. One can also get inspirations and ideas from the still-life paintings exhibited in the art museums and galleries.

DESIGN RULES

The art of arranging flowers depends on certain basic design rules. Various elements make up a flower arrangement the shape, form, colour, texture etc. Scale of the flowers and foliage in relation to the container is also equally important. In order to utilize these elements, one needs to develop an understanding of the natural attributes of the materials one is working with. Some of the important aspects one has to keep in mind are:

- Shape
- Colour
- Texture
- Harmony
- Fragrance
- Proportion
- Balance

Shape

This is a key aspect of flower arranging because it provides the structure for the whole arrangement and hence makes a major impact on the overall impression. There are two principal shapes – facing arrangements and round arrangements. The round arrangements are mass arrangements which are intended to be viewed from all sides whereas the facing arrangement which is seen primarily from the front and is therefore, primarily line arrangement Table 14.1 and Figure 14.1).

Table 14.1: Basic shapes and their types for flower arrangements

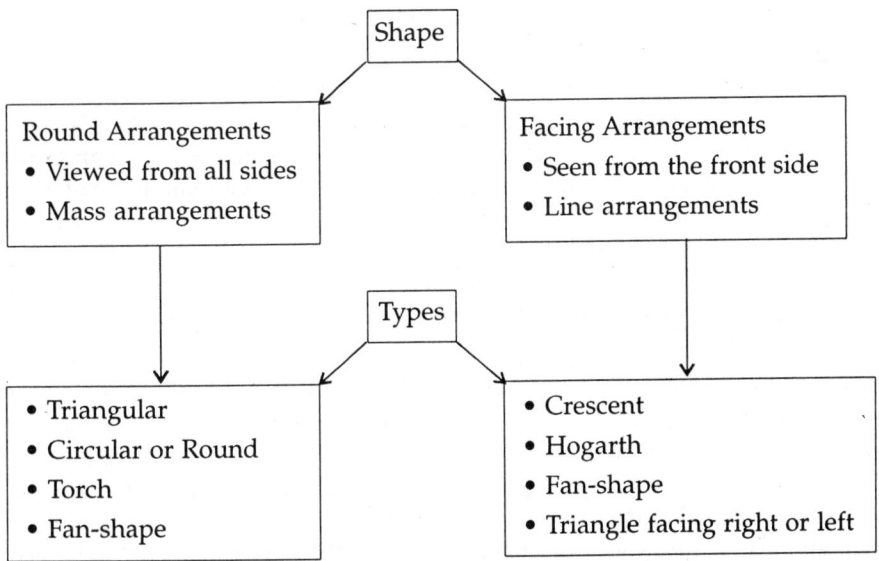

Round arrangement is an arrangement that is to be placed in a central position – in the middle of a table, for example – needs to be created in a mass or round form so that it looks equally good from all sides. It maintains a balance of flowers and foliage all around and leaves no obvious gaps or holes. The most successful round arrangements for central positioning are reasonably symmetrical, but not uniformly rounded. The form can vary from a low cluster to a tall dome shape and the basic shapes under this could be triangular, circular or torch arrangements.

(i) **The triangular shape** is a popular basic shape for symmetrical arrangements and it lends itself to many variations of height and width. After establishing the height and width usually with flowers or foliage of finer form or paler colour, a focal point of interest with larger or darker coloured flowers at the centre and just above the rim of the flower vase is established. Then the arrangement is filled up with flowers and foliage. The equilateral triangle has been one of the most popular designs in flower arranging.

(ii) **The circular shape** inspires its form from the naturally occuring circular shape of majority of flowers. The circular shape being repetitive is satisfying to the viewer's eye. By inter-spreading foliage with flowers, one can avoid monotony of round flowers in a circular design.

(iii) **The fan-shape** round arrangement is actually a dome-shaped mass arrangement. It is symmetrical and can be viewed from all sides. It may be arranged in such a way that radiating lines emerge to form an interesting fan-shape. When designing flowers for the centre of the table, it forms an attractive centerpiece. When it is kept low it will not interfere with across-the-table talk.

(iv) **A torch arrangement** tall and commanding can also be a long mass arrangement in a tall vase. It is a vertical, symmetrical arrangement usually used in large spaces as a point of emphasis and centre of attraction. Long flowers with elongated stems such as gladioli are especially suitable for this type of arrangement. The torch arrangement is mainly used for height effect because of its vertical design.

The most successful round arrangements for central positioning, whatever be the shape, are reasonably symmetrical, but not uniformly rounded. It is realized that composition and proportion are more important than symmetry to achieve a good balance of foliage and flower forms to prevent it from looking lopsided.

The **facing arrangements** allow an arrangement to be placed against a backdrop, such as a wall. Unlike a round arrangement which is viewed from all sides, a facing arrangement is seen principally from the front side. The shape can vary from a very wide fan to a tall and narrow one, but in all cases the flower arrangement assembles the display with one primary viewpoint in mind. A facing arrangement does not need even coverage on all sides since the back of the arrangement is never seen once it is in position and so it requires fewer stems than a round arrangement. But it could occasionally be seen from the sides and therefore both the sides need to be balanced.

The various facing arrangements which are to be viewed from the front are fan-shape, triangular arrangement's facing left or right, crescent, hogarth and torch arrangements.

The **crescent arrangement** is an appealing facing arrangement, which draws its inspiration from the not fully visible shapely line of moon. It is to be viewed from one-side only. The crescent line arrangement is a formal design, which is asymmetrical as it has one-sided height. It is usually done in a shallow bowl and is dependent on a large base for its balance requirement to support its one-sided height. If it is the longest branch which gives the height, then the big sized flowers are arranged in a mass style at the focal point just above left side of the vase, then the crescent line is filled up with flowers, leaves and other foliage so that the graceful curve-line is achieved. It is important to make sure that the stems of the plant materials that are needed are pliable enough to

PLATE 31

Figure 14.2. Some Colourful Arrangements
with fresh flowers

PLATE 32

Figure 14.3. Arrangements with dry leaves, twigs and fruits

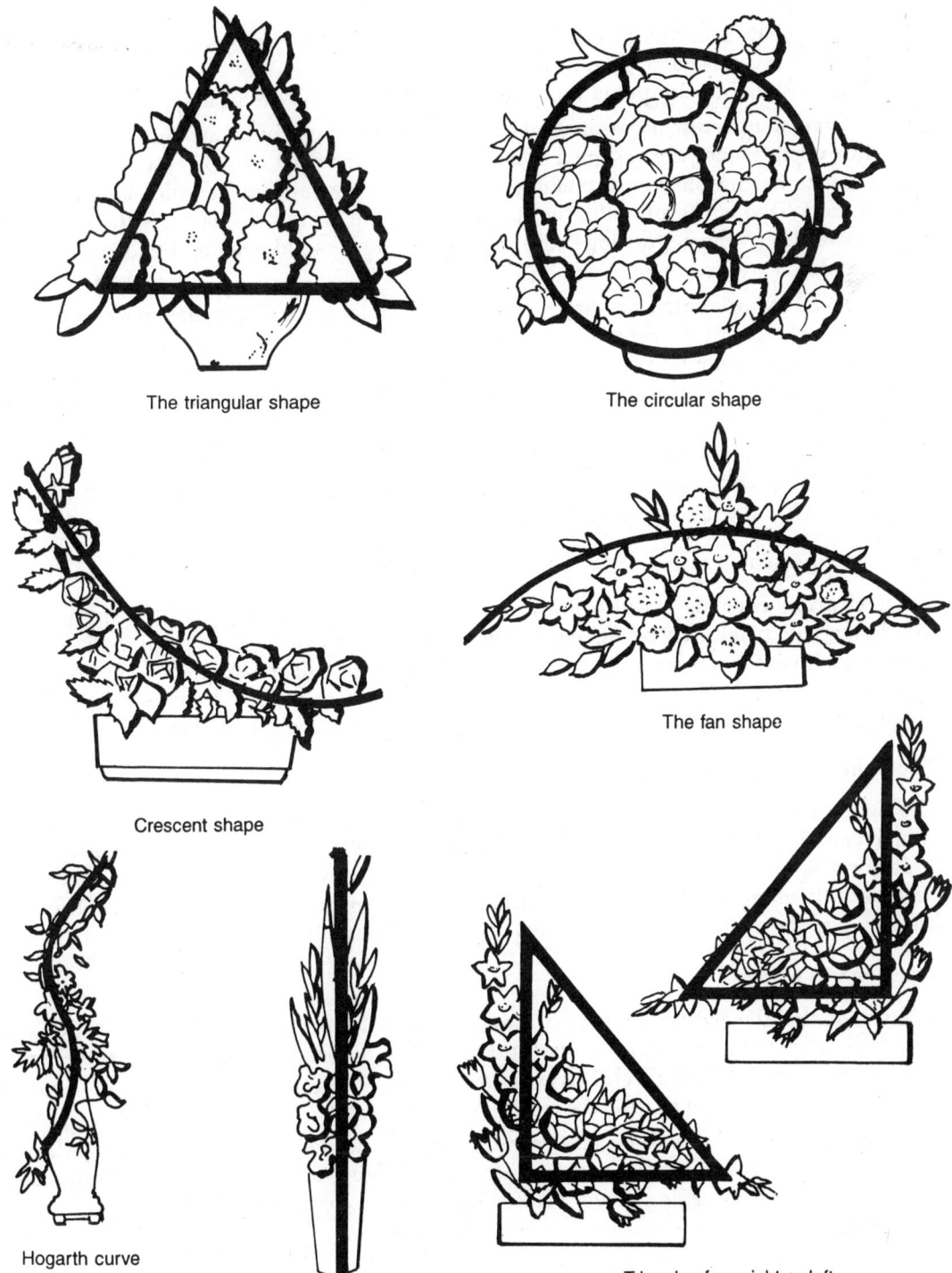

The triangular shape

The circular shape

Crescent shape

The fan shape

Hogarth curve

Triangles face right or left

Figure 14.1. Basic shapes for flower arrangements

permit manipulation because the brittle stems will probably snap before they can be bent to the desired curve. It can be kept, in mind that its execution requires a lot of skill and experience on the part of the designer as compared to many of the other basic styles of the flower arrangement.

The **fan-shape crescent arrangement** is also called a **horizontal arrangement**. These usually have little depth as they are designed to be viewed from the front only. They are therefore, ideal for displaying on narrow shelves or mantelpieces. It is a line arrangement which is symmetrical. It is usually the foliage which provides the skeleton structure for this display. It is then filled out with vivid flowers in lively contrasting colours and if the flowers fade before the foliage withers, they can be replaced with something similar.

For a more **horizontal fan shape**, which can be suitable for displaying on top of a dresser, for example, the height of the central part of the arrangement may be lowered and the length of the stems at the sides may be extended. It is important to make sure that the proportions are well balanced. For instance, for a low fan shape, the arrangement should be roughly three times as wide as it is tall.

With tube-roses or gladoli flowered stems, a beautiful fan-shape arrangement can be made for a corner with gladoli forming radiating lines. A palm leaf can also be used at the back of the flowers, as it has the natural **radiating shape**.

Torch arrangement

The torch arrangement is a vertical line arrangement, which is especially effective for lending itself to a design with height. Therefore, it could be ideal when one has limited display space. It is made in a tall, long vase and many tall plant materials such as gladoli are adaptable. The arrangement is made by placing the tallest stems at the centre and then others on the sides facing the front. The use of tender foliage at the sides may soften the stern lines of this arrangement.

The right-angled or left angled triangular arrangement

The right angled triangle is nothing but a squared – off cousin of the crescent. Only its lines are straighter and its focal point precisely at the ninety degrees angle. Triangles could face right or left depending upon location. These arrangements with asymmetrical triangles are very appealing because of their attractive lines and are most effective in low rectangular containers. These flowers are very popular with modern decorators because of their attractive lines.

The Hogarth Curve

It is named after the eighteenth-century painter artist William Hogarth, who once added a palette with an S-curve to his signature on a self-portrait, with the words "the line of beauty". From that incident, the name of this graceful style of line arrangement is derived. Its shape is beautiful and rhythmic though difficult to produce requiring some practice.

The vase used needs to be tall, graceful, sturdy and well balanced. The Hogarth curve is achieved with pliable stems or curved branches and putting some flowers in the centre and some flowers and foliage sweeping gently below the rim of the container.

The element of colour

Of all the attributes of flowers and foliage, colour excites the most attention and makes the most immediate impact. The art of flower arranging makes the home more attractive and beautiful mainly because of the element of colour in flower arrangements. The use of colour contributes to the decorative assets of flowers. For example, the splash of vivid colour in an arrangement can light up a neutral colour scheme of the room while white or pastel will enliven a darker one.

Colour harmony in flower arranging may be achieved by using principles of colour wheel, the classification of colours and the colour schemes. It is important to note that in nature colours are never pure. In a single flower, red may shade off to purple and orange to yellow. It is not possible to match flower colours to decorating schemes, yet we can learn a great deal about the use of colour from the colour wheel. To create interesting and striking colour combinations, knowledge of principles of colour theory is essential.

Guidelines in the use of coloured flowers

Some guidelines to using flowers, organized by colour are given here. The dark-coloured flowers recede, light colours highlight, and areas of colour intensity or very pale colour produce a compositional focal point. Whites and greens are the colours of the more retiring flowers and leaves. Some of the common white and green flowers and leaves that are frequently used in flower arranging are white roses, gladioli, lily, carnations, sweet peas, chrysanthemum, tulips, cow parsley, anemone, geranium leaves, thuja, asparagus, ferns, touch-me-not, etc. These can be used to highlight other materials that have greater depths of colour. Used on their own, without the addition of other colours, they can produce striking near-mono-chromatic designs.

From the warm end of the colour spectrum, come the yellows and oranges. Some of the common yellow and orange flowers used frequently in flower arranging are orange zinnias, yellow and red chrysanthemums, (tiny and large) yellow snapdragons, carnations, lily, narcissus, marigolds, polyanthus, carnations, marigold rose (yellow rose), florist's rose (red rose), sunflower, etc. These yellow and oranges combine well with blues and are also effective when used to create strong colour clashes. One warm colour only may be used if a gentle harmonizing design is to be achieved – for example, one yellow flower with single or multiple whites and greens.

Warm coloured flowers create a feeling of warmth and welcome. Seasonal berries, fruits and golden foliage may also be used to contribute variety to arrangements that focus on warmth. Most of these warm colours blend happily with each other, but may lose the overall feeling of warmth if pure white or pastel shades are added to the arrangement.

Containers for warm displays should share some tones with warm part of the spectrum. Wicker baskets, for example are ideal, as are earthenware jugs, wooden buckets, bamboo and terracotta pots. Copper and brass buckets and jugs are also suitable and add a contrasting shiny surface.

The reds and pinks are found in the hottest part of the colour spectrum. A selection of common red and pink flowers used frequently in flower arranging are red rose, geranium, sweet peas, anemone, red campion, peony, camellia, hyacinth, gladioli, tulip, orchids, willow, etc. Reds and pinks that have some blue in them harmonize well with blue and mauve flowers. Reds and pinks which have yellow in them harmonize with yellow flowers. Combinations of bluish-pinks, mauves, and white are effective, as are bolder combinations of reds, blues and purples.

Mauves and blues come from the cool end of the spectrum. A selection of mauve and blue flowers used frequently in flower arranging consists of rosemary, sweet-pea, clematis, polyanthus, lilac, phlox, anemone, iris, hyacinth, pansy, monkshood, delphinium, blue bell, cranesbill, catmint, lavender, earkspur, etc. The mauves and blues are cool colours and combine well with other colours, either creating contrasts – such as blues with oranges – or harmonies – blues with pinks for example. The blue colour tends to recede and the addition of white or pale yellow to an arrangement containing blue boosts contrast and enhances the impact of blue. Arrangements that make use of blue colours evoke a mood of peace and restfulness.

Colours for unity

To give an arrangement a unified look is to pick a limited palette of colours from the same section of the colour spectrum – say, yellows, greens, and blues, or pinks, mauves, and blues – and then the use of deeper and paler shades of these colours to shade from light to dark within the arrangement will give the same effect. It is important to be careful not to create too great a contrast between light and dark shades, or the arrangement will not appear harmonious. White, cream flowers or foliage may be used to lighten the overall effect.

Flower colours to be in harmony with their surroundings

Flower arrangements have to be considered in relation to the colour schemes of rooms whether it is made in the house, commercial spaces, or any other surrounding to use flower arrangements as decorative assets. For example, in a monochromatic colour scheme, flowers that repeat one or more of the room colours will be most dramatic in enhancing the decor. Using flowers according to colour schemes add more to the beauty of the arrangement in relation to its surroundings rather than just putting flowers in a riot of colour (Figure 14.2 – see colour plate 31).

Texture as an element in flower arrangement

Texture is the surface quality of the materials and it contributes a great deal to the success of an arrangement. Paying special attention to the texture helps to create a feeling of substance in an arrangements. Contrasts of different textures, such as smooth and rough, matte and glossy, sharp and spiky, add immeasureably to the final quality of the arrangement. The ferns and slender grasses create a delicate and airy texture in an arrangement (Figure 14.3 – see colour plate 32).

For more interesting arrangements, contrasting textures, which are easy to secure, should be included. It is worth looking for a variety of flowers, leaves and fruits including berry textures for different seasons by, just taking advantage of nature's bounty. It is easy to combine rough with smooth, dark with light and round with spear. For example, in a hypothetical arrangement, slender stems (of dogwood), spiky mahonia and holly leaves, soft-petalled chrysanthemums and feathery fronds of cypress or thuja would create an arrangement with exciting textural contrasts. This arrangement further points to a festive look by combining the contrasting colours like rich red and deep green. Particularly in winter and autumn when there are fewer flowers to choose from, the contrasts of texture and form help to create interesting arrangements. Without textural variations, a flower arrangement can look bland and uninteresting.

Sometimes, the flowers and leaves in an arrangement may all have a similar overall texture – soft and smooth as it helps to unify the arrangement. It can be a useful tip for dealing with a large number of different materials. Similar sized flowers along with limiting the colour palette also helps to pull the arrangement together. Since it is important to include some contrasting elements, differently shaped and textured flowers can be used to give an extra edge to an attractive flower arrangement. This helps us to conclude that texture could be used in a bold or subtle way depending on the type of arrangement. Texture may not be the first element that one notices in an arrangement but it definitely plays an important role in creating an impact.

Harmony in flower arrangement

Harmony or unity in flower arrangement is achieved by proper selection of plant materials, containers, artificial flowers or plants, accessories, and the background setting for the arrangement. In other words, it is achieved with a perfect blend of the various ingredients of the flower arrangement. All these together form a good design and will constitute a harmonious arrangement.

Harmony requires the contribution of judicial use of colour, form, texture, type of flowers, containers, and other attributes such as smell or scent along with accessories used and final setting of the arrangement for an ultimate unified and satisfying arrangement.

Avoid monotony by contrasting texture, colour and form

For more interesting arrangements it is important to include contrast. Combining rough with smooth, dark with light, round with spear will create contrasts. *Colour contrast* within a flower arrangement is gained by combining hues of greater and lesser values. Pale hues have less value than deep shades. Dark colours look best low in an arrangement, as they appear heavier to the eye. For help in choosing flowers to make the right colour contrast within a room scheme, refer to chapter on colours for more information.

Achieve harmony with a proper blend of bloom and foliage

Harmony or unity is always the final goal in arranging flowers. It is a result of making a skilful selection of plant materials, containers, accessories, and setting so that they will appear to belong together. If all these elements have blended effectively, the outcome will be a satisfying design and will constitute a harmonious arrangement.

Contrasting textures in an arrangement are easy to secure. Often what one needs is to take advantage of the contrasts of nature's soft velvety flower petals with shiny, glossy foliage; or coarse, ruffled petals with sleek leaves. When the plant material that is being used, has no 'built-in' contrast between flowers and foliage, use imagination to make combinations that provide good textural contrast. *Contrasting forms*, like the rounded bloom and pointed leaf, enhance each other when placed together. Deeply cut leaves are more interesting if combined with solid-looking flower heads. Often, the flower's own foliage is contrast enough. If not, search for others that give the desired contrast.

The element of Fragrance

Who has not been exposed to the heady fragrance of hyacinths, the sweet perfume of lilac, the spiciness of herbs, astringent-like fragrance of mint, slow but all-pervading scent of flowers of neem tree, the sweet smell of jasmine flowers in the garden especially at night! A variety of fragrant flowers and aromatic herbs that both look and smell good can be included in an aromatic display of flowers.

Scent is a quality found in both flowers and foliages, though the type of scents vary enormously. Liking or disliking a particular scent is entirely a matter of personal taste. But it is important to make sure that fragrances do not clash or cancel each other out in an arrangement using several scented plants. Some of the scented flowers are various varieties of rose, hyacinth, lilac, "Star-gazer" lily, daffodils, sweet peas, carnations, honeysuckle, jasmine, geraniums, narcissus, flowers of various trees, etc.

Herbs have been valued for centuries for their aromatic and healing properties. Herbs make a wonderful contribution to flower arrangements, either grouped together in an old-fashioned posy or combined with other flowers to add scent and spice to an arrangement. Some popular and useful herbs, which may be included in flower arrangements are rosemary, sage, thyme, mint, lavender, fennel, geranium etc. The scents of different herbs have remarkable therapeutic properties: lavender is generally regarded as calming, rosemary as invigorating, for example. A scented herb arrangement gives any room a welcoming feel for dining and kitchen areas. Sometimes herbal scent might interfere with the aroma of the cooked food to be served to the family members and guests. For such occasions apart from using flowers and leaves, fruits may also be used effectively.

Scent of plants, invisible to the eye, is that valuable ingredient that gives an extra dimension to any flower arrangement. When using herbs in an arrangement, it is useful to make sure to condition them first as the soft stemmed forms are inclined to wilt rapidly. Conditioning of flowers and stems is thus an important aspect of any arrangement. There is a word of caution to the decorators while making arrangement. Though largest profusion of scented flowers occurs in spring, the sweet and light, strong and heady, spicy and musky scents of flowers and leaves cannot be ignored in the other seasons.

Proportion in flower arrangement

A good proportion in flower arrangement is when the flowers are in the right size for its container. The best rule for proportioning an arrangement was realised much earlier by the Japanese who had mentioned them in their rule books centuries ago. According to these rules, the flowers should be atleast one and a half times the height of the container and for low bowls/vases the breadth or diameter of the container is taken into account. Rules may be broken or ignored by experts so long as the container complements the arrangement for the most pleasing effect. When these rules are ignored in an arrangement, it may give a restless effect.

The flower arrangement is in good proportion when it seems the right size for its container. If it is a tall vase, a safe general rule is to have the height of the flower materials which extend above the rim to equal 1½ to 2 times the height of the vase. The standard height rule of arrangements in low containers is that the tallest stem should equal 1½ to 2 times the length or diameter of the bowl. Experts in all styles ignore these 'rules' as their skill and sense of proportion become well developed through practice. It would become second nature gradually for the others to examine the proportions of their displays.

Balance in flower arrangement

For that solid-looking stable feeling, arrangements have to be balanced. It should not appear to be lopsided, rather it should be well-suited to its container and the arrangement should be well settled in its design. Balance can be obtained in two ways – symmetrical and asymmetrical. The symmetrical balance has two halves which are identical on both sides, whereas the asymmetrical balance has two halves which are not actually equal or identical, but appear to have equal weight or importance to the eyes. The symmetrical balance is easy to secure but the asymmetrical balance can be challenging creatively and requiring practice though satisfying. Asymmetrical balance may also provide variety.

Conditioning Plant Materials

When an arrangement is made, it is equally important to keep it as long as possible. Certain general guidelines are therefore outlined here to suggest several ways of conditioning the various types of lowers and stems to prolong their life. It is natural that some flowers can survive longer than the others when cut but these basic conditioning rules help all kinds of arrangements to last as long as possible.

- Sufficient water should be provided to the arrangement and fresh water added to the container whenever necessary especially in hot weather. A solution of ammonia and sugar (florist's shop also stock these) when added to the water also helps to prolong the life of the flowers.

- Make sure that the water in the container is clean. Rotting leaves in the water should be removed as it causes tainted water, which causes flowers to droop rapidly.

- A moderate steady temperature is important for the arrangement. It should never be positioned close to a radiator or direct sunlight. Similarly, never place an arrangement where it will be exposed to the draught from a fan or a window.

- A variety of stem treatments can be used in order to improve their ability to take up water –

  (i) *removing leaves:* Gently strip off any leaves that will be below the water line when flowers have been placed in their container.

  (ii) Plants with woody stems (such as roses) or fibrous stems have difficulty in taking water through their stems. The last one inch of the plant's woody stem is *crushed with a hammer* before plunging in water. Making criss cross cuts or cutting them at an angle rather than a flat cut to have a largest possible cut surface will be exposed to the water to keep the flowers fresh for longer periods. For fine-stemmed flowers, wrapping them in ½" or 3/4" squares of papers and then squeezing the wrapped stems down between the needles of a pin holder will be helpful.

  (iii) *Singeing stems:* Using a lit candle or slow burning match to briefly singe the cut stem will seal the stem and prevents the sap and the nutrients from escaping out and contaminating the water. But it will still allow water to be taken up into the stem.

  (iv) *Removing dead flower heads:* Flowers that have several heads on a stem last longer if the blossoms that have faded are removed. This encourages those flowers that are still in a bud form to open fully.

  (v) *Removing stamens:* Flowers that have prominent stamens such as lilies, last longer if the stamens are removed. Either they could be pinched out with the fingers or snipped of with sharp scissors. This technique also reduces pollen as well as pollen stains.

  (vi) *Trimming of flowers:* Trimming is done for reasons such as shortening the stems to fit a smaller container; wish to extend the life of flowers that are slightly past their best by cutting them down in size and placing them in a small bowl; and stripping of dying foliage. In all trimmings, use a pair of sharp scissors and take care not to damage the stems.

  (vi) *Straightening stems:* It is a useful way of getting flowers to behave as one wants. Stems of flowers that are inclined to bend over or crooked can be straightened by wrapping them tightly in a few sheets of newspaper. Another method to shape the flower stems is to soak them in a bucket/kitchen sink/bath tub or plastic waste baskets. Allow the stem to soak for a few minutes and then, working under water, shape the stem a little bit at a time with a controlled gentle pressure until the achievement of the desired shape.

(vii) *Wiring stems:* Flowers with weak or floppy stems can be stiffened by wiring. A medium-gauge florist's wire (wire wrapped with green paper or painted green) is pushed through the centre of the flower head and taken out at the base. About five cm of wire is left protruding out of the head and this short end is looped into a hook. The wire protruding out of the base of the flower is gently twisted around the flower stem and any surplus wire is trimmed off. Another method is to wire the weak-spined stem to another stiffer stem, and the weak one will maintain its posture.

(viii) *De-thorning:* Carefully snip off any thorns on the lower part of the flower stem with a sharp scissors. Thorns make it difficult to insert rose stems into oasis.

(ix) *Removing damaged petals:* For improving the look of flowers, say roses, gently remove any damaged or faded petals before arranging them for maximum impact.

(x) *Making stems less bulky:* The large number of leaves, say on a plantain plant stems, can make it difficult to display them attractively. Some of the leaves may be stripped off to make the stem less bulky while taking care not to bend the stem.

(xi) *Opening flower buds:* Some flowers with strong, waxy petals can be persuaded to open out more fully. Gently pull the petals outward, using your finger and thumb. Once pulled out they will stay in position.

(xii) *Handling heavy-headed flowers:* Handling heavy-headed flowers requires some patience and skill. Candle wax may be put at the base of the heavy flower head keeps bottom petals from falling off the flower like chrysanthemums. To repair the bent stem of a heavy headed bloom, a thin stem or tooth-pick may be pushed through the centre and into the stem.

(xiii) *Leafy tips:* Leaves with brown or decayed edges may be trimmed away along its margins and get a shape resembling its original proportions to gain a neat and healthy look to the leaves and flower arrangement.

It is therefore very important to keep in mind a number of factors to retain the flowers in an arrangement as long as possible. Procuring the flowers from the plants early in the morning helps to keep them fresh and to avoid withering easily. It is wiser to pluck those buds which are about to open rather than a raw bud or a fully blossomed flower. While plucking the flowers, it is good to use a sharp scissors or a knife and not to pluck them with finger nails. Using finger nails damages both the stems of the flowers as well as those of the plant. Immediately after plucking, the flowers should be placed in water. If the flowers are to be transported to a long distance, they should be wrapped in a wet soft cloth. While carrying in hands, the flower part of the stem should face the ground to facilitate water from the stem to flow towards the flower. Sprinkling water on fresh flowers as well as after an arrangement, will also aid in keeping them moist and fresh for long. A pinch of salt or an asprin to the water in the container is also an useful tip for this purpose. A few precautions and measures in the care of flowers as they are collected from the plants can help an arrangement to appear fresh and last long.

Tips on cutting and preparing flowers

Flower when to cut and how to treat

Anemone	*½ to fully open. Likes deep water.*
Aster	*¾ to fully open. Scrape stems.*
Azalea	*Bud to fully open. Scrape and crush stems.*

Atriplex	*boil stem ends.*
Buddleia	*¾ to fully open. Scrape stems and slice open.*
Calendula	*Fully open. Scrape stems.*
Carnation	*Fully open; snap or break from plant. Scrape stems.*
Canna	*½ to full open.*
Chrysanthemum	*Fully open. Scrape stems and slice at base.*
Clematis	*¾ to fully open. Pre-arrangement drink in shallow boiling water.*
Daffodil	*As colour shows in bud. Cut foliage sparingly or buds will not mature.*
Dahlia	*Fully open. Sear stems in flame, or give hot pre-arrangement drink.*
Daisy	*½ to fully open.*
Day Lily	*¾ to fully open. Flowers last just one day.*
Delphinium	*¾ to fully open. Scrape stems; snap off top buds.*
Dicentra	*4 or 5 florets open.*
Geranium	*Fully open. Arrange at once.*
Gerbera	*¾ to fully open. Sear stems in flame.*
Gladiolus	*As second floret opens. Scrape stems; snap off top buds.*
Heliotrope	*¾ to fully open. Shallow water pre-arrangement drink.*
Hollyhock	*¾ to fully open. Float florets, or scrape stems and give hot water drink.*
Hydrangea	*At all stages, boiling water drink.*
Iris	*As first bud opens.*
Larkspur	*¾ to fully open. Scrape stems; snap off top buds.*
Lilac	*½ to fully open. Scrape and crush stems; float wilted branches in 110-degree water for an hour. Remove all leaves.*
Lily	*As first bud opens. Cut no more than ¾ of stem or bulb will not mature.*
Marigold	*½ to fully open.*
Mignonette	*¾ to fully open. Pre-arrangement, shallow water drink.*
Morning glory	*In evening when closed. Wrap each bud in soft paper, boil end of each vine stem; let stand in deep water overnight.*
Narcissus	*As colour shows. Cut foliage sparingly.*
Nasturtium	*½ to fully open. Use with its own foliage.*
Peony	*Bud in colour to fully open. Split stems.*
Poinsettia	*Full colour. Sear stems and points from which leaves have been removed.*
Poppy	*Night before opening. Sear stems; drop of wax in heart of flower keeps it open.*
Rose	*As second petal unfurls. Cut stem just above a five petal leaf or plant will stop blooming. Scrape stems. Pre-arrangement drink in shallow boiling water.*
Snapdragon	*¾ to fully open. Split stems.*
Stock	*¾ to fully open. Scrape stems and split.*
Sweetpea	*¾ to fully open. Cut stem from vine.*
Tulip	*Bud to ½ open. Cut foliage sparingly. Wrap flowers in paper; stand in deep water overnight.*
Violet	*½ to fully open. 'Harden' by soaking in water for short time.*
Water Lily	*Tight bud. Sear stems in boiling water; drop of wax in heart of flower keeps it open.*
Zinnia	*Fully open. Pre-arrangement drink of hot water.*

MATERIALS REQUIRED

There are a number of essential items of equipment for arranging flowers.

Equipment for cutting: Scissors and a sharp kitchen knife are sufficient to cut flower stems. Size and weight are not important, but such types should be choosen with which one feels comfortable while handling. They must be sharp and in good condition.

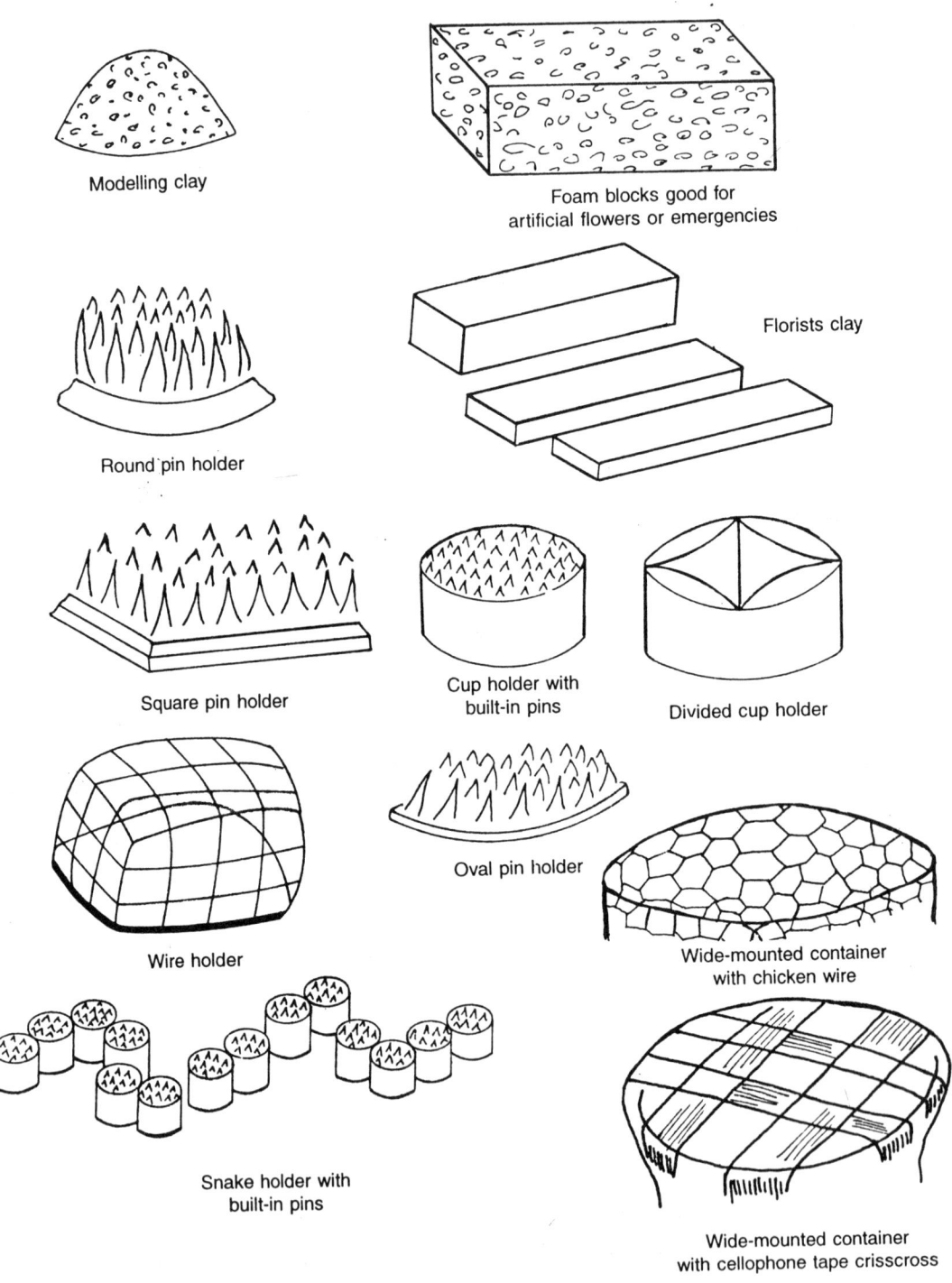

Modelling clay

Foam blocks good for
artificial flowers or emergencies

Florists clay

Round pin holder

Square pin holder

Cup holder with
built-in pins

Divided cup holder

Wire holder

Oval pin holder

Wide-mounted container
with chicken wire

Snake holder with
built-in pins

Wide-mounted container
with cellophone tape crisscross

Figure 14.4. Equipments for foundation in flower arrangements

Equipment for fixing: A wide range of tools exists for fixing foundations to containers or flowers to foundations. Florist's wires in different sizes, string, tape, pins (long and oasis pins), rubber bands are all the other essential tools to arrange flowers in an effective manner.

Equipment for foundation: One must have a solid foundation for the flower arrangement so that the flowers stay in the positions one has chosen for that purpose. The type of foundation one chooses depends on the flowers to be used and on the choice of the container (Figure 14.4).

Foam blocks of "Oasis" (Oasis foam ring, Oasis foam ball, Oasis spike or 'frog') for fresh flowers and styro foam are useful for making artificial flower arrangements. They are extra easy to assemble and keep. A block of artists' modelling clay or plasticine, florists clay, metallic pin-holders (round, square, oval), cup holder with built-in pins, snake holder with built-in pins, divided cup holder, wire holder, chicken wire are some of the holders one may need to possess.

Wet oasis and chicken wire are the most commonly used foundations for flower arrangements. Wet oasis is sold in blocks that need to be soaked in water before the flower stems are inserted. However, oasis is not suitable for very soft and thin stems. Oasis can successfully be used only for such flowers which have thick and firm stems. But there are times when one is using a clear glass vase, for example, you need something less mechanical looking. Clear adhesive tape and glass marbles make effective anchoring materials for such situations. A clear adhesive-tape grid over the neck of the vase can be made.

Containers: As far as containers are concerned, anything that holds water is a container – a cake tin, glass ash tray, jam jar or a shampoo bottle. In other words, one can use almost any type of receptacle as a container for arranging flowers. In addition, China and earthenware containers, glass containers, decorated china containers, informal-looking containers such as wicker and wooden baskest give a good range of containers. The recent innovative idea in this area is that of metal containers including that of stainless steel (Figure 14.4).

After the flowers themselves, the container is the most important part of any arrangement. The right choice of container can enhance beyond measure the whole composition while the wrong one will only look out of place and detract from the beauty of the flowers. Here is a selection of all types of containers for the different styles of arrangements.

Figure 14.5. Containers for flower arrangements.

Equipment for giving finishing touches: Ribbons, raffia, rope and string are all useful for decorating containers and for gift wrapping bouquets. One may keep a few different styles and colours handy and use them on and off. A ribbon bow provides a professional-looking finish to a bouquet or a tall arrangement. One may choose a ribbon that complements the flowers: use either a matching or a contrasting colour that picks up at least one of the shades used in the arrangement. Reversible ribbon is the easiest to work with. The use of accessories with flower arrangement, such as sculpture, figures, pebbles, rocks, etc. can also be used to give special effects. Accessories are generally used to enhance the beauty of the flower arrangement. Sometimes accessories give a theme to an arrangement. For example, a block and a book case near a immature arrangement can represent a study corner.

The ability to recognise beauty in unexpected places is characteristic of the most gifted arrangers. To the uninitiated, a rock is a rock, a shell a shell. But to the artist arranger, each is a treasure that can enhance the beauty of the flowers and foliage with which it is combined. A stroll in the woods or along the edge of the sea has new interest when you are looking for curiously shaped bits of wood, shells, and water-smoothed rocks and pebbles-all things you might once have ignored. Miniature scenes from nature such as the Japanese excel in are eminently suited to 'found art'. Or, as the arrangements pictured suggest, a shell can be a container; rocks and weathered wood can be part of a landscape; or they can be attractive masks for the mechanics of an arrangement and lend a finished look to your compositions. **Autumn Day**, a design which is a happy reminder of an autumn walk. The dry ferns, weathered branch and piece of wood recall quiet still days at summer's end. The ferns and branch are secured on a small pinholder hidden by the wood (Figure 14.6).

Collecting rocks pays when flower arranging is your hobby. Of course you use rocks and pebbles to disguise a pinholder. Do not forget, however, that they can be extremely effective as the dominant feature in an arrangement which seeks to call up the image of some remembered landscape (Figure 14.6).

The infinite variety and appeal of shell is too seldom appreciated by flower arrangers. Employ them as a rich source of intricate patterns, shapes and colours. Use these in your arrangement of flowers and leaves that grow near the water (Figure 14.6).

Steps in preparing a flower arrangement

Any arrangement with flowers must be pleasing to the eyes. When the arrangement is being made, it is possible that one aspect of the arrangement is more satisfactory than the other, or one particular bloom the most beautiful. To arrive at an overall, satisfactory and attractive **arrangement**, following a systematic step-wise arrangement would be helpful. The steps involved in making arrangement is as follows:

Step. 1. Collect all materials required to make an arrangement. The required materials are:

- Container
- Pin/stem holder
- Flowers
- Foliages and fillers
- Water
- Scissors/sharp knife
- Accessories (optional).

Autumn Day design

infinite variety of shells

Collecting rocks pays

Figure 14.6. Accessories in Flower Arrangements

Assemble all these materials on a table in on orderly manner. Do not overcrowd the area.

Step. 2. Cut the stalks of the flowers to the desired length. This length depends the type of container—tall or a flat or a round bowl. It is also good to have a fair idea of the kind of arrangement preferably with a sketch, one would like to make. After cutting the stalk of the flowers, put them back in water.

Step. 3. After assembling the floral materials, anchor a pin/stem holder firmly into place in the flower bowl/container, using florists clay. Make sure that it stays firmly and can hold the flowers without falling.

Step. 4. Start arranging the flowers in the order of their length – the tallest first, the second tallest second and so on. Compare the results with pre-conceived arrangement (Step. 2). If necessary re-arrange the flowers. Some buds can be opened partially, or a stem may be bent slightly under the water.

Step. 5. Once the flowers are arranged, fill in the spaces between the flowers with leaves or fillers. Use extra greenary to hide the pin holder or the container, if necessary, laying foliage flat at the base of the arrangement.

Step. 6. Water is added in the last, after the arrangement is kept in its place, particularly if the container is quite shallow. Add a pinch of salt to the water. Sprinkle a little water on the flowers and foliages, if the weather conditions happen to be hot and too dry. If the pin holder is still visible a few pebbles or marble chips can be used to cover it.

To make a finishing to the arrangement, a few accessories, as suggested earlier can be used to make it attractive and appropriate to its location.

SOME PROFESSIONAL TIPS TO IMPROVE ARRANGEMENTS

- There are tricks to all trades – flower arranging included – that help to make results look more professional. Here are some used regularly by experienced arrangers. They will serve you in good stead also, and will make cut flowers stay fresh longer, speed the assembly of an arrangement, and guarantee a better appearance to the finished design. Study these eighteen tips for better technique. Whenever you are making a flower arrangement, put them into practice and you will find that they work.

- Carry cut flowers from the garden in a heads-down position. Heavy-headed flowers will not snap off.

- Lay flowers flat, wrap in newspaper. Plunge bunch into tepid water for 3 to 5 hours or overnight to condition.

- To revive wilting flowers, snip off a half-inch of stem under water and plunge them in a deep container of water.

- Shape the leaf to resemble its original proportions when you must trim away a brown spot along its margins.

- Never place an arrangement where it will be exposed to the draught from a fan, window, or to full sunlight.

- To repair the bent stem of a heavy-headed bloom, push a toothpick through the centre and into the stem.

- Many flower and foliage stems are pliable and can be shaped by placing the thumbs and applying gentle pressure.

- Preserve or revive woody stems by pounding the bottom two inches before plunging them in water.
- Ensure enough water for woody stems by paring off the bark from the bottom two inches and crosscut stems.
- To reduce underwater decay, strip the stems of all foliage and thorns which fall below the water line.
- Bloom of some flowering shrubs has a tendency to wilt. To reduce the problem, clip off small shoots at the top.
- Dahlias, poppies and flowers with hollow stems should have the stem ends seared to prevent sap escaping.
- Roses and tulips open rapidly in a warm room. Use sticky tape to keep blooms from opening while arranging.
- Prolong the freshness of an arrangement by spraying with a syringe of tepid water morning and night.
- Candle wax at the base of the flower head keeps bottom petals from falling off flower like chrysanthemums.
- To secure the stems of flower arrangements using large branches, stuff the container with chicken wire.
- Cut a 'y' shaped Forked twig and place it at the neck of a tall vase to hold the stems and flowers in an upright position.
- Insert a stick to strengthen the stem of such flowers as delphiniums, but be sure the stems reach water.
- Add salt to water to lengthen the life of an arrangement
- Change or add water after 24 hours to ensure that the container does not become dry.

Leading Styles in Arranging Flowers

There are three leading styles in flower arrangements – Oriental, Traditional and Modern styles. It would be informative to go into the detailed characteristics of the above mentioned styles.

Traditional Style of Flower Arrangement

This style has its roots in the art tradition of England, France and other western world of the sixteenth and seventeenth centuries and also seen in famous paintings of western artists such as Rachel Ruysch, Monnoyer.

Cut flowers have been a source of enjoyment for centuries but originally they were not always put into water but arranged in garlands and strewn over the floors. Gradually they were very often tied together in a bunch and inserted into a vase. The flowers and leaves were arranged in vases of every kind during the Victorian times. Decorative pottery-china and glass-stem holders became generally available. At that time flowers from the hot houses and garden were brought together and arranged in showy containers for house decoration in many homes. In large houses the head gardener was generally responsible for arranging the flowers.

Opulence is the hallmark of the traditional style of flower arranging. Traditional, also called period arrangements are typically full, symmetrical groupings of mixed flowers displayed in a highly decorated ornamental vase resulting in a colourful, expensive looking mass decoration. In such mass arrangements, large blooms and deep hues are usually concentrated at the centre; fine flowers or foliage and pale colours are used to outline the basic shape of the design. The overall form of the traditional arrangement may be pyramidal, round or oval depending upon setting and the plant materials.

The charm of the traditional arrangement lies in its most splendid effects with a mass of bloom. It is the overfull and crowded mass style which is today synonymous with the victorian era. The massed style of flower arranging has always remained popular and inspiring to the flower arrangers all over the world.

Modern Style of Flower Arranging

As in painting and sculpture, modern style is a highly individualistic composition. It follows no fixed rules or formulas for correct proportions and employs unorthodox materials for making flower arrangements. They are representative of modern life and exhibit uncluttered, stark lines reflecting on the simplicity of our furniture and architecture and way of living, with which they are more in keeping than the traditional massed arrangements.

Modern arrangements usually are original, made with plant materials minimally but frequently include an unorthodox container as an integral part of the composition. In the modern and abstract styles, artists (flower arrangers) believe in greater freedom in personal expressions. Arrangers working on such styles generally make the fullest use of very striking, dramatic plant forms often in simple yet distinctive containers. To bring added interest for the visual impact, the modern arrangers may leave voids of different sizes and shapes. The aim is to achieve a strong uncluttered effect with natural plant forms, relying on contrasting textures and shapes for greater impact. Use of accessories may sometimes be essential, not incidental to the design.

Modern and abstract styles are becoming popular because it provides the scope for continuous growth in flower arrangement and embraces new thoughts and methods. Since the style is concerned with personal interpretive expression, their only test is the effect of an arrangement on a sensitive and knowing viewer's eyes. Although arrangements in the modern and abstract styles are original, they owe a lot to Oriental style in concern with their stress on line and a symmetrical balance.

Oriental Style of Flower Arrangement

On examining the pictures of the most ancient art of China or Egypt, it is seen that the flower arranging has been practised with consummate skill for a great many centuries. In the orient, the art of arranging flowers is centuries old.

Ikebana

No information on flower arrangement will be considered complete without a reference to the famous flower arrangement of Japan, "Ikebana". The other forms of this style of flower arrangement, are also equally important to understand this style fully.

When Buddhism was introduced into Japan about the middle of the Sixth Century, A.D., it must have been superimposed upon an earlier naturalistic religion with whose philosophy it had much in common. The symbolism and tradition of the early altar arrangements formed the basis for the art of *Ikebana;* they were later formalized into the first school of *Ikebana,* the Ikenobo School, which still flourishes. Through the centuries the concept of *Ikebana* was gradually modified and expanded from its original status as an entirely religious ceremonial activity to a broader interpretation that permitted its use for non-religious ornamental and decorative purposes, at first only by the nobility, but later by all Japanese.

In spite of the great popularization of *Ikebana* in Japan, a strong overtone of tradition and symbolism still persists, and the religious and philosophical origins of *Ikebana* have by no means been forgotten. An important feature of many oriental religions and philosophies is the sense of the oneness of man with nature; he shares the universe with plants and animals and is not in conflict with them, as, unfortunately, is too often true in Western cultures. Flower arranging to the Japanese is a special way of life, with a traditional background of spiritual and philosophical meaning; to them *Ikebana* harmonizes the laws of nature and of humanity. The traditional use of three branches, flowers, or other objects in some styles of *Ikebana* to symbolize heaven (*shin*), earth (*tai*), and man (*soe*), thus symbolizes also the whole universe.

Flowers arranged with artistry are used as constant decorations in Japanese homes. The flower arrangement is an indispensable part of life, and they are displayed in an alcove called 'Tokonoma' which acts as a frame for the flowers. During the ceremony of serving tea, the guests entering the home are made welcome and then brought to sit and admire the beauty of the flowers for a moment or two of tranquil silence and then the ceremonial serving of tea begins.

Oriental arrangements are more than aesthetic groupings of plant materials. They are symbolic presentations of an ideal harmony which exists between earthly and eternal life. Basic to oriental style is emphasis upon line in every design. In each arrangement there is a triangle. Its tallest line represents heaven: facing and looking to heaven is man; looking to both is earth.

The symbolic philosophy behind each stem is represented by the difference in the heights of stems. Ikebana calls for the observance of rules and proportions. Broadly speaking, all schools in Japan have two fundamental styles: moribana or arrangement in a low shallow bowl style, nageire or arrangement in a tall vase style. Finding the formal, upright style of Ikenobo completely unsuitable for the new plants, a wholly new style that used low, wide containers was developed. These flat, shallow containers permitted a more casual arrangement that soon progressed to the "natural scenery" *moribana* that were to change the whole future of *Ikebana*. *Moribana* means, literally, "piled-up flowers in a flat basin," and represented the first serious attempt to utilize Western flowers in Japanese flower arrangements.

The Basic Forms of Ohara School*

Basically, *Ikebana* of the Ohara school is classified into two primary categories: *moribana* and *heika* – or *nageire*, as it is commonly called.

Moribana arrangements are those made in a low bowl. This may be a small compote, a flat plate, a wide shallow basin, or any similar shape in which water may be contained if needed. Usually a holder – either a needle-point holder (*kenzan*) or a heavy metal holder with separated openings (*shippo*) – is employed.

Heika or **nageire** arrangements are those made in a tall container, a bottle-shaped or cylindrical vase, or any vessel having depth. Usually such containers have a smaller opening than the low bowl, and ordinarily no holder of any kind is employed in the placement of materials.

---

* *Source:* Houn Ohara, "Flower arrangements",

A careful study of materials, to learn to determine the basic form – *moribana* or *heika* – most suitable for their arrangement, and to select an appropriate for their arrangement, and to select an appropriate container, is the most important step to good harmony in arrangements, for the container does much more than merely contain the flowers; it becomes an essential part of the total design.

When flowers are combined with branches, the branch material is used for the primary and secondary stems and the flowers are used for the tertiary placement. This tertiary grouping is known as the *nejime*. **Modern shoka** was developed to meet the requirements of contemporary living and permits great freedom in the choice and combination of both materials and containers. The classical theme is used as a basis for free expression. Chabana styles are created with flowers and vases keeping neatness and simplicity as the main aim in the arrangements.

Rikka meaning "standing flowers" is the oldest studied form of Japanese flower arrangement. It had its origin in the mid-fifteenth century and in its original concept depicted nature in all her glory and grandeur. For several hundred years this style of arrangement reigned supreme and was usually engaged in by the warrior class, the Japanese nobility and the Buddhist priests. The arrangements were usually large, elegant and magnificent, at times being massive in their proportions. They were beautifully balanced and called for considerable skill on the part of the arranger.

THE FIVE BASIC STYLES

The two basic *Ikebana* forms may be interpreted in the lines of five different Ohara styles – the style being the shape outlined by the principal materials. Three main branches – subject, secondary and object – form the foundation or framework in each style. The length and position of these three stems differ in each style, the result being in each case three main points forming a triangle of different shape and dimensions. The proportionate lengths of the principal stems are predetermined to assure a well-balanced structure; however, the container itself is the controlling factor in determining actual measurements and angle of inclination of stems. The position of the arrangement, considered from the viewer's angle, is also an important influence. The first step is always to consider the material itself.

As you will see, each style is designed to bring out a particular aspect of the beauty of materials, by emphasizing their highly individual characteristics. Good judgement in the choice of style must coincide with the proper selection of form and container in order to achieve a successful arrangement. Study and practice, plus experience gained through handling a wide variety of materials, are the requisites for acquiring this kind of judgement.

To complete the design begun with the container and three main branches or stems, additional materials, known as fillers, become a part of the arrangement. Fundamentally, they fill in the outline provided by the main stems. Their purpose may be to accent, to contrast, to provide movement, to emphasize a mood or atmosphere, or to supply the unexpected touch that lifts an arrangement from the ordinary to the artistic, while always playing a supporting role to the principal stems.

Both *moribana* and *heika* forms are classified into five styles:

- Upright style
- Slanting style
- Cascade style

- Heavenly style
- Contrasting style

In both forms, and in all five styles, we find variations and free styles.

THE FIVE METHODS

The next step, after determining the appropriate style for the material, is to decide upon the method. What is the point to be emphasized? Colour? Line? Volume? The various methods are designed to bring out and enhance the characteristics that will produce the desired effect. An atmosphere of feeling may also be expressed by controlling the method of arrangement, as you will learn. The five methods include:

- Natural method
- Colour-scheme method
- Mass-effect method
- Line-scheme method
- Abstract method

These five methods in turn fall into one of two classifications: natural and colour-scheme methods follow the realistic trend; mass-effect, line-scheme, and abstract arrangements are classified as non-realistic. These trends are explained in more detail later on.

Natural

To express this in a fuller sense we might say a "method of presenting nature as it is," this being a translation of the Japanese term for this method. This is believed to have been the fundamental idea behind all *Ikebana* – to present plant-life as it is seen from season to season, to express a love of nature by reproducing, imaginatively, a portion of her beauty within the small sphere of a flower container. It is the essence of this beauty, with nonessentials removed, that is translated by the *Ikebana* artist. To recognize the beauty in every aspect of nature requires daily observation, for it is not to recreate an arrested moment of beauty that we wish, but to imbue the arrangement with the life and movement that exist in nature, without which it will be static and meaningless.

Colour-scheme

The purpose of this method is to emphasize the beauty of colour in materials, by combining colours which harmonize, contrast, or complement each other. To achieve this purpose, the natural growing characteristics and seasonal significance of material are sometimes disregarded; however, good design is never sacrificed for colour effect. Balanced composition is essential. More effective distribution of colour is possible by skillful composition of a design, using the varied lengths and positions of materials to bring out or subdue a colour, to point up a contrast or to blend harmoniously. The colours themselves may be selected to create a soft, quiet beauty, or a joyous gaiety.

Colours are extremely important in all Japanese flower arrangement, having profound psychological influence upon human beings, as well as reflecting the cultural tastes and personality of the arranger.

Mass-effect

Modern *Ikebana* has developed through the need for stronger expression, in keeping with the urgency and stress encountered in daily life. Creative urges have been stimulated by postwar trends and developments found in all phases of our existence.

The mass effect enables the arranger to meet this need for stronger design by utilizing volume to produce emphasis. Massed chrysanthemums, for example, are much more striking in colour as part of a design than individual flowers would be.

Massing is usually done by grouping and tying materials before arranging, rather than attempting to arrange individual stems for the desired appearance.

Line-scheme

The opposite of the mass-effect method, but equally important, the line-effect arrangement uses every line of the material in the production of the desired design. Not only the line, but the space around it, becomes a part of the arrangement, and it is possible to express strong movement artistically by skillfully arranging lines in rhythmic patterns. *Ikebana* has developed along with Japanese painting, and a strong mutual influence is inherent in the two arts; however, the line-effect has developed even more rapidly in modern *Ikebana*. Geometric detail and directional emphasis in line arrangement are the result of the application to *Ikebana* of techniques used in modern abstract painting.

Abstract

The use of materials other than plant-life in *Ikebana* is usually attributed to the newer, modern trends; however, in *rikka*, the oldest known form of *Ikebana*, sand and stones were used in some arrangements. This indicates that there has long been recognition of the unexpected beauty to be found in objects and materials other than flowers. To find this beauty and to enhance it by adapting the material to a design, either alone or in combination with similar materials, or with living plants, requires a keen eye and artistic sense.

As a ceramic artist begins with clay to form a thing of beauty, so the *Ikebana* artist assembles iron, plaster, stones, glass, etc. to express his recognition of the kind of beauty to be found therein by composing an artistic design.

The Trends

In the early days of the Ohara School, when *Ikebana* was only a means of artistically expressing appreciation of the beauty of nature, the natural and colour-scheme methods of arrangement alone existed and were practised in both *moribana* and *heika* forms; the natural to interpret the beauty of scenery, the colour-scheme to emphasize to the fullest the beauty of colourful materials.

Recognition of the individual beauty to be found in colours and shapes of plants and flowers brought about a new trend, using each material as part of a design, rather than as its natural self.

Realistic arrangement

Whether combined in a natural scenery arrangement, or as individual plants, materials arranged realistically are displayed exactly as they appear in life, their beauty enhanced by the idealistic and imaginative touch of the arranger.

Non-realistic arrangement

In non-realistic arrangements, each material becomes an object – a shape, a colour, a line –

completely without regard for the manner in which it grew. The combination of materials into an arrangement becomes the creation of a design. Flowers, branches, leaves and roots may be composed in a pattern altogether contrary to their natural aspect. Startlingly different from the arrangements of the old days, which were imitations of nature, these non-realistic designs bring out a greater depth of loveliness, through stronger emphasis on the forms of materials. For example, the mass-effect arrangement is a more emphasized form of colour-scheme, and belongs to the non-realistic, because it does not conform to the natural growth of the materials.

ARRANGEMENT USING ARTIFICIAL FLOWERS

One of the interesting modern developments in flower arranging is the trend towards the use of artificial flowers. There was a time when artificial flowers were frowned upon because they were poor imitations and gathered heat and dust. Not so today when plastic flowers, foliage and even stems are so graceful and real-looking that even the birds and the bees are fooled by them. In addition, they last for a long time and are washable and can be used over and over again in any arrangement in place of fresh flowers. They are especially good choices for permanent arrangements, especially as an alternate for expensive exotic flowers.

It may be remembered, that fake flowers will never affect sensitive noses or people with allergies. So if you want arrangements that never droop or fade, require no water or other attention, and are available in season and out, artificial flowers are a wise investment. They will increase the range and scope of the art of flower arranging.

Fun with and without flowers

Being creative with the groupings of plants, fruits, leaves, pine cones, the drift woods, the moss, the stones, and the plant parts and branches, even with the roots and the other dry materials may all be carefully utilized and used along with flowers by the people living in the country side. However, people living in the cities may buy carefully potted plants for using flowers only on special occasions.

"All green", "foliage only", "flowerless arrangements" or "dry plant materials" are the other options. There can be arrangements with drift wood pieces, fruits, shells, leaves, berries, gourds, seed heads, grasses and similar ornamental parts of the plants.

Care of indoor plants

The best advice would be not to be ambitious at first as it may be possible to be successful by starting off with a few "easy to grow" plants, leaving the more difficult ones until one gets the practice. It is also important to keep in mind that all the indoor plants are kept away from direct sunlight and draughts. It is good to fork over the top soil now and then to aerate the earth and to remove all dead and yellowing leaves. During spring and summer times, plants need a fortnightly application of fertilizers. The soil should be kept moist. During winter months, the plants resting time, no fertilizer is required. Very little water need to be sprayed then.

Interior decoration is a fast developing discipline. Its importance lies in the fact that the interior use of an enclosed space is changing frequently due to the fast changing life styles of people. A variety of new materials are available today and there is a growing awareness among people to keep their surroundings attractive and appealing. Arranging flowers and keeping potted plants is one activity which is becoming more and more popular to meet this need of the people.

Chapter 15

RANGOLI : THE ART OF FLOOR DECORATION

India has a culture as diverse as her people, climate and size. Indian population is polygenetic and is said to be the melting pot of various races. Inspite of this diversity, surprising that it has a rich cultural heritage and is the home for a variety of art and art objects. One such an important art is that of floor decoration, popularly known as Rangoli.

Rangoli is an important art originating from India, and is used to decorate mainly the floor, and in some areas, the walls too. The roots to rangoli can be traced to the ancient cave and temple paintings. Though it is a distinctive art of India, it has spread to the other neighbouring Asian countries as well.

Anyone, who visits India, will be treated to the aesthetically attractive painted patterns on the floor, or the courtyard or the frontyard of the houses, specially in the states like Tamilnadu, Andhra Pradesh, Kerala, Karnataka and Bengal. The luxurious ornamentation of doors, walls and courtyard with rangoli using patterns of singular charm and graceful variety, can be found almost all over the country. One simply has to step into the house of a family during festival or auspicious days to discover the sparkling beauty of this art.

When people visit Hindu homes on special occasions, they will, in all likelihood be treated to a royal welcome with the kind of decorations done on the floor, walls, and the entrance to the house. The house wife will festoon the main entrance with garlands of flowers and mango leaves, draw elaborate patterns on the floor, and light an oil lamp, all indicators of a warm welcome to the visitors. She will make her guests stand at the threshold of the entrance door, and apply bindhi (a red coloured dot of kumkum (vermilion) on the forehead) and sandal wood paste on the forehead and neck, sprinkle a few yellow rice grains (coloured by turmeric powder) over them as a blessing, and offer sweet item as special treat in their honour.

On various festivals, celebrations and other happy occasions which includes family gatherings, Indian Women with their nimble fingers prepare fine lines with which every prominent corners of the house is decorated. Similarly, Indian farming families also wish for the arrival as well as blessings of Goddess Lakshmi with extraordinary and beautiful floral patterns after the hard days of working in the agricultural fields. After cleaning the house, the villagers decorate their house like a bride, draw attractive and colourful patterns on the floors, light it with diyas (mud lamps), and wait eagerly for the goddess Lakshmi's visit.

Decorating the floors with patterns is considered as important as cooking food or washing the clothes, since the decoration of the house begins with the drawings of rangoli first in the morning in the frontyard of the house. The frontyard is cleaned and then the drawings or patterns of rangoli is made prior to the start of any other chores of the household. Similarly, rangoli patterns are drawn before the start of any of the religious ceremonies, functions or celebrations. Such is the significance given to rangoli which, it is believed by its presence, heralds all good things in life.

Beautiful and artistic designs of rangoli are, therefore, not only confined to the front courtyard or the place of worship, but also to the verandahs of the house, the centre and corners of the rooms, the main entrance etc. In India, where it is considered auspicious to decorate the house and its entrance, sometimes, the entire stretch of the streets are also decorated with rangoli patterns. This is very common during festivals, melas (fairs) and special ceremonies and functions in the temples and religious places.

Thus, it can be observed that rangoli is the art of drawing designs in the form of geometric configurations, birds, flowers or tendrils on to the floor, done normally with a paste of rice flour and water, or rice itself in its powdered form. At times, lime, or chalk powder, water colours or the other indigenous materials like turmeric, red mud, mehandi, flower petals etc. are also used to make and fill the designs and patterns. An outline is drawn first in pure white (rice flour) and then filled with colours or left on its own to depict auspicious symbols or objects of beauty. A rangoli is conventionally painted on festive days alone, but some communities like the Parsees and the people in the entire southern belt of India perform this art as a daily ritual.

At times, a fine powder sold as rangoli powder in the markets, is used to draw the rangoli pattern. It requires great dexterity and swiftness of the fingers while working with them. Finely ground rice powder is carefully drizzled through the fingers to outline robust, vivid descriptions. In other words, a pinch of white powder gathered between the thumb and the index finger and then fine sharp designs are traced out on plain dark floor. To achieve better results, dots are made first, depending upon the kind of design or pattern, and then they are joined with lines to get the desired pattern. In some cases, lines are drawn in different directions and then joined to form patterns.

While making rangoli, the most important factor is the floor. It should be clean, smooth and even. In villages, the floor is given a cow dung finish before making a rangoli. Thus, these floors provide a natural background for making rangoli designs. In the cities, to achieve this effect, many a times, a small section of the floor is specially given a mud and cow dung wash that emulates a typical village-like floor, so that a good contrast of dark and white effect is achieved.

Rangoli and Festivals of India

Rangoli is an integral part of the rich heritage of Indian culture and it infuses colour, joy and happiness. It is practised all over the country with some variations and has different names in different parts of India. Rangoli is thus credited to be one of the several threads which unites our country of diverse religions, languages and traditions.

Rangoli is specially associated with Diwali, a festival, which is celebrated all over the country, without much difference. On Diwali day, every house is decorated with flowers, thorans – Festoons (hanging patterns with flowers and leaves), a floor pattern with coloured powders and diyas (lights on mud bowls) outside the front side of the house, as a gesture of celebrations and meetings of friends and relatives. The other major festivals of India, where rangoli is ritualised are:

- Pongal/Sankranthi: January - Tamil Nadu, Karnataka, Andhra Pradesh, Maharashtra
- Basant Panchami: February - all over India
- Gangaur: March/April - Rajasthan
- Holi: March - Mainly in Northern India
- New Year Day: April - Tamil Nadu
- Baisakhi: April - Punjab
- Pooram: May - Kerala
- Naag Panchami: July-August - Mainly in Northern India
- Teej: July-August - Rajasthan
- Onam: September - Kerala
- Raksha Bandhan (Rakhi): August - Northern and Western India
- Ganesh Chaturthi: August-September - Mainly Maharashtra, Tamil Nadu, Karnataka
- Janmashtami: August - Maharashtra, Uttar Pradesh, Tamilnadu
- Dusshera/Ram Lila/Durga Puja/Navarathri: September-October - All over India
- Kartik Poornima: November - Tamil-Nadu.
- Onset of the month "Margazhi": December-January - Tamil Nadu.

Thus, it can be observed that rangoli is an art, which is ritualised in each and every festival/function/deliberations of India, throughout the year. There is no single month in a year, when celebrations are held without doing "rangoli" in the households. Rangoli is thus credited to be one of the several threads which unites our country of diverse religions, languages and traditions.

Besides religious festivals, rangoli is also an integral part of many of the cultural festivals held in India. The Kajuraho Dance festival (Madhya Pradesh) Thyagaraja Music Festival (Tamilnadu) and the Ellora Festival (Mumbai) are the examples of such cultural festivals where rangoli is done elaborately. Besides these festivals and cultural celebrations, a large number of fairs are also held in India. During these fairs, rangoli is not only practised, but also becomes a sale point for the various designs and materials that can be used to create and make rangoli patterns. In all the village fairs, one comes cross the real faces of India – simple folk, domestic animals, melodious songs, colourful dances, exquisite arts and crafts, including rangoli – in all its pulsating vitality.

Significance

The freehand floor drawings of India – popularly known as rangoli – needs to be mentioned as a form of decoration made on the floors for social gatherings, festivals, fairs and festivals, and the other family celebrations. A "toran" (festoon) or bunting of gold foils or mango leaves tied to the top of the main entrance doors of the house, a pair of cut banana trees with their flowing raw banana bunches, tiny clay oil lamps and an elaborate floral rangoli in the frontyard of the house, announces an auspicious event of the household to any passersby. No celebration is thus, devoid of a rangoli on the floors.

Painting the floor of the entrance to a house, seems to be a common practice in many other countries too. In Yorkshire, for example, there are reports to indicate that there used to be such a practice. Similar practices are also seen in countries like Srilanka, Malaysia, Thailand, China, Nepal, Bangla Desh etc. Some places in South Africa also, seem to follow similar practices on special occasions.

Rangoli, apart from being a medium of decoration, of the house, has also some religious significance. It is considered as being very auspicious and is made during festivals and the other social occasions and celebrations like marriages, visits of special guests, ceremonies, muhurat of new office or business, and house warming ceremonies. Any kind of celebrations seems to be incomplete without a pattern of rangoli on the floors of the entrance hall or courtyard.

Rangoli, thus, enhances the beauty of the surroundings and spreads joy and happiness all around. Semi-religious, or social occasions like those associated with the changing of the seasons, the return of peace and plenty after the dark days of the monsoon or the dry days of the summer month, prayers for successful harvests and thanksgiving, birthdays and weddings, prayers for the welfare of the family and community and for celebration and festival time generally, even the birth of the young shoots at the close of winter or the beginning of the spring season, the ripening of the paddy or the golden wheat or such family events such as the naming ceremony of the child, the first entry into the new house, or even the dawning of a new day, have all inspired the Indian womenfolk to indulge in a luxurious ornamentation of the floors and walls of the house and its courtyards with patterns of unique charm and graceful variety. These women pour their very gleeful souls into this art and transform the odd bare surfaces of the walls and floors into a glowing mass of decorative designs and patterns with mere flowers, powders and pastes.

So well done are these decorations, that although drawn free-hand, they look like decorative floor mats created by the fine handiwork of a true and sincere artist. Decorating the floor in this manner is symbolic of a picturesque embroidery or a traditional weaving network on the fabric of the earthern floor. This art work is as unique and exemplary as any other work of arts and crafts of India.

HISTORICAL PERSPECTIVE

Rangoli as an ancient art form is a traditional folk art with its origin difficult to trace. Centuries old cave paintings have sketches and designs which could be termed as some form of Rangoli. In epics like Mahabharata and Ramayana, and Jain ancient literature, there is mention of this art in various contexts.

As any other folk art, Rangoli has been handed down from generation to generation. Rabindranath Tagore, one of the great sons of India, has played a great part in bringing this art to the forefront and Shantiniketan propagated this art not only in its traditional form but also adapted it to the modern ideas. Here it is a subject of study in its art curriculum.

With the passage of time, Rangoli as an art has been making lot of progress. In the present modern era, Rangoli is being made with old as well as new materials such as grains, flowers, fruits and vegetables. The modern market is more versatile – by providing easy-to-use/practice stencils and plastic ready-made stickers. As a rural art form, it is passed down from mother to daughter but in urban mileu, there are various art workshops, hobby classes, organisations like All India kitchen Garden Association etc. and Polytechnics have made this as part of their curriculum. Besides these, in national and international summit and conferences, it is elaborately made for decoration purposes. Moreover, in some non-governmental organisations set up with the objective of bringing awareness and for propagation at our rich cultural heritage, Rangoli occupies as an important popular activity.

Rangoli is also taught at the university level in all the Home Science Colleges and Departments as a part of their Interior Decoration Programme.

Rangoli on the floors is a part of a daily ritual in Southern parts of India, whereas it is not so common in urban North. However, Rangoli has made a revival and it is done commercially by decorators to beautify pandals in marriages and parties, conference halls, etc. all over India.

RANGOLI IN DIFFERENT STATES OF INDIA

In India, though there is generally a common culture, there is a lot of diversity in the way they are practised. Every State has some kind of its own uniqueness in its traditional practices which also distinguishes its folk arts from those of the other States. Similar to the other folk arts, Rangoli too has been widely practised, but in different forms with different names and materials. The unwritten rules and the techniques are passed on from one generation to the other as a treasure. Rangoli can be seen in various forms and patterns, and in a variety of methods and materials, which are in harmony with the physical environment, social and cultural ethos of the particular part of the country (Figure 15.1A, B, C and D – see colour plate).

The rangoli patterns are drawn rhythmically on the mud floors and walls which are cleaned and mopped with a cowdung paste. This practice seems to be common, all over the country, though characteristics, materials, forms and the corresponding names may differ from state to state. However, it is found in its most developed form in the states of West Bengal, Orissa, Andhra Pradesh, Tamil Nadu, Maharashtra, Karnataka and Gujarat – practically in all the regions lying along the coastal lines of Southern India (Figure 15.2).

Table 15.1: Rangoli in different parts of India

State	Zone	Name
Kerala, Tamil Nadu	South	Kolam
Andhra Pradesh	South	Muggu
Karnataka	South	Rangavalli/Rangoli
Gujarat	West	Sathia
Rajasthan	West	Mandana
Maharashtra	West	Rangavalli/Rangoli
West Bengal	East	Alpona
Orissa	East	Jhunti
Uttar Pradesh	North	Chowk-pujan, Sona-rakhna, Sanghi
Bihar	North	Aripana, Appan
Garhwal (Uttaranchal)	North	Aapana

This art of decorating the floors is known by various names in the different parts of the country. It is called Alpana in Bengal, Kolam in Tamilnadu, Muggu in Andhra Pradesh, Jhunti in Orissa, Rangoli in Maharashtra, Rangavalli or Rangoli in Karnataka, Sathia in Gujarat, Mandana in Rajasthan, Aripana in Bihar, Sangli, Chowkpujan and Sohna-rakhna in Uttar Pradesh.

In the Southern States of India, like Andhra Pradesh, Karnataka, Kerala and Tamil Nadu, Rangoli as an art is visually seen everyday in front of every home and it fills one's heart with happiness. In these southern states, women make a Shiva Peeth Kolam on mondays, a 'Kalipeeth'

Pulli (Dot) Kolam – Tamil Nadu

Central pattern with lines

"Step" pattern with dots

"Star" pattern with lines

"Rath" (chariot) pattern with dots

"Snake" pattern with lines

Figure 15.2. Rangoli in different states of India (Contd.)

"Light" effect with Rangoli

Centre "Star" pattern

"Peacock" Motif

Muggulu – Andhra Pradesh

"Muggulu" (dots) pattern

"Tulsi" (Herb) pattern with dots and lines

Figure 15.2. (Contd.)

Aniyal – Kerala (To be filled with flowers)

Chowk – Himachal Pradesh and Uttar Pradesh

Mandana – Rajasthan

Mandana – Madhya Pradesh

Pakhambha – Manipur

Figure 15.2. (Contd.)

Alpona – West Bengal
(pattern dominated by fish and feet)

Figure 15.2. Rangoli in different states of India

on Tuesdays, a 'Swastik' on wednesdays and 'Lakshmi Kolam' on Fridays. Here kitchen (cooking area) and farms are also decorated with kolam designs. Conch shells 'gada' 'gopads' (cow's foot-marks) are important motifs of the ceremonial aspect of rangoli. On Gokulashtami day also known as Janamashtami, Krishna Jayanthi, (Birthday of Lord Krishna), tiny footprints using wet rice paste are drawn in every South Indian home, from the entrance to the worship area, depicting symbolically the entry of Krishna into the homes on the day.

Besides the frontyard, kolam is also made in the Puja room where idols and pictures of god and goddesses are kept. This place of worship is cleaned everyday and different patterns are drawn every day. Sometimes the pattern are done according to the day of the week, which is based on "navagrahas" (nine gods of the universe). Though Kolam is prepared just by using white rice powder everyday, it is done with wet rice paste on special occasssions/festivals. Its outer lines and patterns are highlighted by the use of red coloured must paste.

In South Indian villages, towns and cities, Rangoli is a 'must' during the Dhanush month of the Hindu calender, corresponding to mid-December-mid-January. In Tamil Nadu, the entire frontyard of every house, thereby covering the whole stretch of the street, is filled with patterns of "Kolam" during the month, "Margazhi" falling between 14th December and 13th January. Not only there is a variety in patterns, but also in the colours and the materials used in the making of it. The patterns are made using either dots or lines to achieve symmetry in the final design. After finishing the kolam pattern all along the frontyard near the entrance, a handful quantity (a tennis-ball size) of cowdung with a pumpkin flower on top, is placed. It is believed that this will ward off all evils. Each and every family competes with one another in producing a variety and large sized kolams early in the morning, i.e. before the sunrise. Later the cowdung ball and the pumpkin flower are flattened together in the form of a round cake and dried. The entire month's collection of these cowdung cakes are used as a fuel in the preparation of "Pongal", a sweet rice dish on the first day of the next month that makes the beginning for the harvest season. On this day, a bigger kolam is made not only in the frontyard, but also near the cooking and worshipping places.

Kerala state, in the southern tip of the country, is identified with lush greenery and swaying coconut trees. Rangoli is done here popularly with fresh flowers. Kerala people celebrate a special occasion on "Onam" festival to draw rangoli, only with colourful fresh flowers. When flowers are not available, dyed coconut husks are then used as fillers in rangoli patterns.

As we come to western India, colour gets added to rangoli. Whereas in *Maharashtra*, only little colour is used, in *Gujarat* and *Rajasthan*, the whole rangoli is very colourful. In Maharashtra, women after cleaning the house, choose a place and write 'Shri Ganpathi Prasna' in a decorative manner everyday. In Gujarat, and Rajasthan, it has taken a colourful character. The people of Rajasthan, perhaps to make up for their dry and drab environment of the desert, use bright and stunning colours in their dress and furniture and it applies to rangoli designs also. In the States of Rajasthan and Gujarat, it is a feast for the eyes to see beautiful designs and colourful patterns sparkling in front of every house during the diwali and Dussehera festival. During the Deepavali week the *Gujarati girls* make a thousand and one Rangolis with a variety of designs and colour schemes, both inside the house and on the streets, in the night time for the town folk to see and rejoice in the morning.

The designs are more of geometrical patterns. Similar to South Indian practices, rangoli with dots is very prevalent. Here dots are made at equidistance from each other without the help of a measuring scale or compass, and the designs are drawn with ease by joining these equidistant dots forming geometrical and other floral patterns – circle, square, triangle, hexagon, rhombus, quad-rangle, slanting lines, etc. are prepared. 'Boondo ka Chowk' is very commonly made. In Rajasthan dyed quartz powder is used whereas in Gujurat, where locally it is called Sathiya, glass mirrors may be used. In Rajasthan the *Mandana* patterns impart a picturesque beauty to the open grounds which are cleaned up and finished with cow-dung in crimson red, obtained by mixing Rati (red earth found locally there). The background of the Mandana motifs is prepared in black, chocolate, blue, or green. As white is the colour much favoured for drawing the patterns, chalk dissolved in water or rice paste mixed with water are generally used. 'Chokas' or squares, single and interwoven, have greater importance on ceremonial occasions, while polygons and circles abound on festival days.

Among the *Keralites,* there is an interesting form of decoration for festive occasions. A model of gay flower bed with white powder and coconut husks is laid out on the ground of the outer courtyard by the main doorway. It is called *Phookkolam.* The Malayali women fill it up with beautiful flowers of as many varieties and colours as possible and make designs of rare charm. So on the occasion of Onam, every house in *Kerala* wears a gay and festive appearance. Beautiful Kolam designs filled with richly coloured flowers are then seen everywhere in abundance. In *Uttar Pradesh,* rangoli is known by the name 'Sanghi'. The rangoli is drawn on the ground or on the wall and different patterns can be seen for various festivals like Raksha Bandhan, Diwali, Dusshera, etc. The patterns are a mixture of geometrical and floral designs.

Floor decorations made in *Uttar Pradesh* are also called 'Chowk'. There are different designs for different occasions and it is seen that every family has its own ancestral designs for different festivities. The design called 'Chawkpurna' is popular in Uttar Pradesh. Originally rice paste was used but now several substitutes like maida (wheat flour extract) are used. In *Uttar Pradesh* and *Rajasthan,* it is commonly seen that women paint their mud walls in natural colours and draw patterns with chalk. In Bihar, rangoli goes by the name 'Ahpan' whereas in Garhwal it is 'Aapana'. The scorpion, supposed to symbolise human suffering figures in the Chowks of *Uttar Pradesh* and *Bihar* to ward off evil influences. The figure of a fish, regarded as an auspicious object by the Hindus, is a very favourite motif and appears in almost all cases where water is shown or where different kinds of animals are depicted.

In the greenbelt of *Punjab* and *Haryana,* rangoli is made around grain stores, sanjhachullah (community cookstoves/Langar) etc. Decorative rangoli patterns around doors and windows are also commonly seen in houses in the villages there. In Punjab, the women in rural areas make perforated tiles in sun-baked mud. These are used as walls and curtains in the verandahs and courtyards. To give them the semblance of mirror, women use bright foils which resemble 'Sheesh Mahal' of the Mughal period.

Even in the extreme north, in the State of *Kashmir,* rangoli is made on wooden floors and walls. In Kashmir, it is the cultural practice even today to paint the front walls of the house with beautiful patterns of leaves, fruits and flowers to welcome the newly wedded bride. This perhaps depicts the wish that the bride should bring into her new home – prosperity, happiness and fertility. The unique feature of rangoli in Kashmir is that the designs are mainly drawn on the walls and not on the floor. This could be explained by the fact that the Kashmir valley is snowbound for a considerable period and under these circumstances floor-drawn rangoli may not be feasible.

Until recently the *Urban North* had not shown much interest in either retaining this art and its development. Now in the past few years, rangoli in flowers has made a revival and it is executed commercially by decorators to beautify pandals in marriages and parties and conference halls. But the real art is still kept alive in villages. The 'Swastika', an ancient auspicious sign (not to be confused with the Nazi symbol) is almost the only Rangoli pattern drawn in these homes. Even this is done by the pandit who comes to do the puja and very rarely by the lady of the house. Swastika is a harbinger of good luck and it also symbolises infinite movement. The meaning and significance of the swastika has been variously interpreted by different scholars.

In *Madhya Pradesh,* where rangoli is different designs are made on various important festivals. For example, on 'Hariyali Amavasya' the doors are decorated with unique patterns.

PLATE 33

Figure 15.1. (A) Patterns for Rangoli dry powders for filling

Figure 15.1 (B) Rangoli Pattern made with red mud and rice paste on a theme—"Women's Literacy"

PLATE 34

Figure 15.1 (C) Rangoli pattern filled with flowers and leaves

Figure 15.1 (D) Rangoli pattern made with dyed saw dust with a theme-helping the physically challenged

West Bengal, the traditional seat of all arts, has its special rangoli form in 'Alpona'. Rangoli is done by using wet colours in white and red. Sometimes flowers may also be used to beautify the floral patterns. Earlier milk extracted from wheat grains was used to make rangoli but now-a-days ground wet rice paste is used. Alpona is drawn mainly free hand yet it has the flow and rhythm in its lines. Some of the patterns employ an axis of symmetry to make them look uniform and attractive. The Alpona may be made in the kitchen and around chaukies (low stools) used for dining purposes. It may also be seen at the entrance of the house and at the place of worship. The flat peerhi (low plank) on which a bride and bridegroom stand during their wedding is also very beautifully decorated. For 'Lakshmi Pooja', footprints are also marked with the same paste from the entrance to the worship centre. This kind of rangoli is also prevalent in Southern States of India to prepare quick designs easily along with uniformity.

In Bengal, figures of the vermilion casket, 'baju' (an ornament for the upper arm), 'nath' (an ornament for the nose) bangles and ear-tops are objects of Alpona decorations connected with the worship by married women. Some Alponas involve as many as forty different objects, before each of which a flower is offered, and a song and a rhyme recited on important occasions. These designs, though sometimes markedly crude and stiff, are interesting and surprisingly refreshing.

Since time immemorial man has an unconscious desire to beautify his possessions and surroundings and rangoli as an art form is one of the most economical and beautiful ways of decorating the floors, walls, courtyards and puja rooms of houses all over the India in their own unique ways depending upon the local resources and prevalent culture. It can be said that the Alpana's symbols are universal yet very typical of the region.

The tribals of *Orissa* make rangoli patterns on their walls to appease the ghosts of their ancestors. Their belief in doing so is that these ghosts will find a new home and leave the rest of the tribe in peace.

The *Parsis* use perforated designs marked out on small tin trays into which powdered chalk is put. These are then stamped repeatedly on wet floors and beautiful designs are produced.

TYPES OF RANGOLI

Each piece of decoration, consists of two classes of design, *the ceremonial and ornamental design.* *Ceremonial designs* are traditional and are always in keeping with the occasion that is being celebrated. In such designs figures of birds, fish and animals, both real and legendary, are used. Sometimes with riders and sometimes without them, sun, moon, stars and chariots combined into many pleasing arrangements, have their place of honour in rangoli patterns.

But the purely '*ornamental*' part of these decorations is most fascinating as it consists a very real means for the expression of the hearts yearning for beauty on the occasion of reception of a distinguished guest. The place where the guest sits and dines and at the weddings, the place where the bride and the bridegroom sit for the ceremonies or at the religious worship places where the idol is to be worshipped, traditional patterns are drawn. On the coming of a new dawn, the courtyards are decorated with beautiful patterns and designs which are not traditional but fresh creation of artists' mind unfettered by any convention/rules of ritual decoration. There is an endless variety of geometrical designs which has most probably oriented in the simple enjoyment of lines. Some of the designs are very simple, just a network of lines but very cleverly and artistically done. Sometimes colour effects are introduced to enhance their beauty on auspicious occasions clay lamps glow in the centre of them. In the South they are filled with brightly coloured flowers and leaves. It is in the execution of floral designs that the richness and glory of this art is to be seen at its best. Motifs taken from the plant world are handled with exquisite taste and great skill and proportions and details are carefully portrayed.

Patterns of Rangoli

The hand-made traditional patterns in rangoli are generally done in white with multi-coloured insets and may take the shape of stylised flowers of all kinds, fruits such as mango and almond, animals such as elephant, and fish, birds like the parrot, peacock and swan, and some beautiful and intricate geometrical designs with floral insets, all symbolic in form and meaning. Although free hand, patterns created form beautiful rangoli as of a reputed artist.

To get a thorough understanding of the endless variety of patterns, keeping in mind the traditional and the present and the futuristic aspects of art of rangoli, various patterns of rangoli are classified into six categories. The patterns used in rangoli can be ritualistic, naturalistic, geometrical, stylised/decorative or a combination of any of these. The variety of patterns are therefore unlimited. It is basically one's imagination which is transformed into a beautiful pattern for Rangoli.

I. Ritualistic patterns are based on the various rituals, festivals and cultural practices for which our country is famous. Our cultural heritage is deeply associated with various religious sentiments and it appears to have found an expression in the form of Rangoli. Various festivals like Dusshera, Diwali, Karvachauth, Ahoi Raksha Bandhan and the other traditional ceremonies like births and marriages, house warming (grahapravesh), religious practices like fasting and special pujas and havans, are all depicted through artistically sketched Rangolis. In Rajasthan, 'Solah Deepak', (16 Diyas) is made on Diwali. 'Chowk Paglay on Holi, 'Chang' and 'Khera' are made on marriage occasions, whereas 'Kalash' is for the entrance of the house for happiness and prosperity.

On the occasion of 'havan', purohit (pandit) may make a platform of mud and then a Rangoli on this platform with different colours. It is a common practice to make colourful Rangoli depicting leaves and flowers and other delicately made abstract design paintings based on folk art. Lord Ganesh is considered auspicious and his figure is a very popular Rangoli subject (whether in Abstract or traditional form) on the occasion of Ganesh Chaturthi Puja and other ceremonies. On the occasion of Luxmi Pujan, women make 'Luxmi Tika Chowk' and 'Swastik' Rangolis are made. In Maharashtra, women after cleaning every nook and corner of the house, make a Rangoli and write Shri Ganpati Prasaan in the centre. It is an everyday activity. In Madhya Pradesh '50 Boondo Ka Chowk', (Rangoli with 50 dots) is made. 'Aapna' in Garwal and 'Addhan' in Bihar has religious association and folk culture, where a mixture of cowdung and multani mitti is first plastered on walls and then Rangoli designs are made with terracota powder (Gheru) and white colour to please the gods. The colourful peacock design in the Rangoli depicts desire for happiness and pleasure. After plastering with mud and cowdung if certain areas are not covered with designs of Rangoli, it is considered inauspicious. Designs based on objects used in rituals, household objects and ceremonies in tune with other designs depicting plant vines, flowers, leaves, conch shell, birds, fish, footprints, plants, trees, and lotus leaf, lotus plant, (Palki) Palanquin, elephant, Paan leaf, sun and moon are the common ones.

For example, on the occasion of Nagpanchami day, rangoli designs and patterns must include figures of snakes, and on Ganesh Chaturthi day, figures of Lord Ganesh is drawn. Some ritualistic patterns are also drawn everyday to please ``navagrahas'' – the nine planets. The effects of

Monday (Planet-Moon)

Tuesday (Planet-Mars)

(Good health & Happiness)

(Happiness)

(Destruction of enemies and
success—also for 'Rahu")

Wednesday (Planet-Mercury)

Thursday (Planet-Jupiter)

Friday (Planet-Venus)

(Knowledge & Power of speech)

(Worthy Progenies, Removal
of obstacles)

(Blessings of navagrahas &
all kinds of wealth)

Saturday (Planet-Saturn)

Sunday (Planet-Sun)

(Clearance of debts & wealth)

(Destruction of difficulties and
progress—also for 'Kethu")

(Blessings of Goddess Lakshmi)

Figure 15.3. Rangoli Patterns to please "Nava grahas" (Nine Planets) and their significance.
Source: Navagraha Kshetra Puranam, LIFCO, Madras.

drawings near Puja place seems to vary from attaining health, wealth, knowledge, to the destruction of enemies, evils etc. These patterns and their significance are well illustrated in figure 15.3.

II. Naturalistic: When patterns are naturalistic in nature, they are based on designs of objects naturally occuring in nature like plants and animals and even human beings. Rangoli based on foot-prints of a human being is very popular for expressing welcome to guests. Peacock, plants, trees, elephant, fish, Kalash, candle, diyas (garvi) etc. are some of examples of patterns of objects and living beings occurring in nature. In a traditional "tree of life" pattern, a flurry of multicoloured parrots may look vying for honeyed blossoms in green and brown colour.

III. Geometrical: Rangoli as an ancient art form largely employs geometrical designs. Most of these designs are made by dots and lines forming decorative patterns. The patterns may be based on line, dot, dash, area, triangle, square, hexagon and circle.

Moreover, geometrical patterns give clarity in the range of design, as well as an idea of clearcut boundaries. All these enhances the beauty of the Rangoli pattern by balancing the use of the available space and colours. For example, On the floor, an area of 20" × 30" can be marked with dots in a gap of about one inch each both horizontally and vertically to get uniformly made dots. When these dots are joined in various ways, unlimited designs can be made.

IV. Plain-line and filled rangoli: This kind of rangoli pattern can also be looked upon as the plain line rangoli as compared to the filled designs. The Alpona in Bengal and Kolam in South are some of the plain line rangoli, which could be geometrical, naturalistic, abstract or ornamental in nature. The filled designs may be filled with rice powder, coloured saw dust, flower petals and grass, mirrors, pebbles etc. so that rangoli design is filled and the floor surface is not exposed at all.

V. Ornamental or Decorative: This is a kind of pattern which is not primarily based on traditional ritual or connected with religion but rather the emphasis is for purely ornamental or decorative purposes.

Floral rangolis especially made for welcoming delegates to conferences or other official functions in hotel lobbies, etc., lend a very beautiful ambience in addition to being very welcoming.

Using bright colours and interesting harmonious accessories, such as statues, mirrors, beads, puja thalis etc. give an ornamental look to this rangoli.

VI. Abstract: Like any other art, rangoli is also practised in abstract form. However, the basic significance is still retained and a symbolic representation of ideas takes place. Although symbols are dependent upon one's views and frame of reference but the ability to symbolize is cultivated in all cultures. Not only artifacts, but also nature's works have always been the objects of man's interpretation. For instance, tree has always been a symbol of life since time immemorial.

In abstract forms, Rangoli has donned a modern outlook and like modern art with its typical characteristics of having no boundaries and no-barred looks Rangoli with modern abstract patterns may reach new horizons of intimate imaginary artistic expression of talents. For instance, a Rangoli may show an abstract pigeon with leaves in its beak as a symbol of peace.

It can be said that the abstract designs are concerned with a personal interpretive expression. Since these kinds of patterns have no fixed rules or formulas for correct proportions, their only test is the effect of the rangoli on a sensitive and knowing viewer's eye.

Although patterns may be classified as ritualistic, naturalistic, geometrical, ornamental and abstract, it may be kept in mind that these various categories are not mutually exclusive but may overlap each other and thus the resulting patterns may be a combination of two or more of these types of patterns, which are popular or in fashion at that time.

Thus, we see that although some of the patterns are very conventional the spirit behind the art of rangoli is capable of manifesting itself into infinite variations. This can be easily be observed in the way this art is practised at present.

Techniques and materials used in the Making of Rangoli

The various ways and materials used in the making of rangoli, are listed below:

- Dry rangoli
- Wet rangoli
- Floral rangoli
- Rangoli on thalis
- Rangoli with etching
- Accessorised rangoli

Dry Rangoli

Dry rangoli was traditionally done with rice powder. However, dry powders of lime white stone or other coloured powders are also used to make rangoli. It is made mostly in white colour and also with several dry colours. At some places, it may also be done with marble powder or quartz powder. However, locally available materials are also used. For adding colour to the rangoli, either the standard medium i.e., rice, is dyed or some naturally occuring mediums are used. Some of the traditional ones are geru-mitti (red mud), coffee powder, neel (blue), roli and haldi (tumeric), charcoal powder, Kalonji seeds, sindoor. Mica chips or mica powder may also be used to give a sparkling effect.

Among the other dry coloured materials, dyed saw dust, holi colours, dyed broken rice, pulses (whole and broken), suji (plain and dyed), henna/mehandi, dyed small marble chips and stones, fruits etc. are popular in the making and filling rangoli designs. Geometric lines are most common and are usually made by making dots and lines resulting in attractive patterns. At first an appropriate pattern is drawn on the floor with dots and lines, and then its colour scheme is decided. The colours are always filled from the centre of the pattern to the outside. When the colours are filled, the various parts of the pattern may be highlighted with a fresh thin line of white rice powder as the outer line. It is also important to keep in mind that there should not be too many colours in one pattern though the number and choice of colours may also be influenced by size, design and purpose of rangoli. After finishing the rangoli, the surroundings may be cleaned to enhance the neat and attractive look of the rangoli.

A beginner can, in the first instance, practise drawing the pattern in a note-book as it gives the idea of joining the dots on the floor. The floor may be wiped with a wet cloth so that the powder will stick to the floor easily. In the rangoli using grains and pulses, it should be kept in mind to retain the sizes of grains fairly uniform to provide a smooth and good effect. When whole pulses are used, then no broken pulses should be combined with the whole ones. For making big carpet patterns in dry rangoli pattern, it is better to make use of sawdust, both in natural colour and dyed ones. Any dye used to colour cotton fabrics can be used to dye them using spirit. Animal figures, geometric designs and images of worship made with bright coloured saw dust look very attractive. Sawdust provide a course texture, making it easy to work with.

Sometimes, dry rangoli may be made on damp floor so that the rangoli design stays for some time. Either earth is moistened or mud layer is spread and then moistened and finally rangoli design is made for getting such effects.

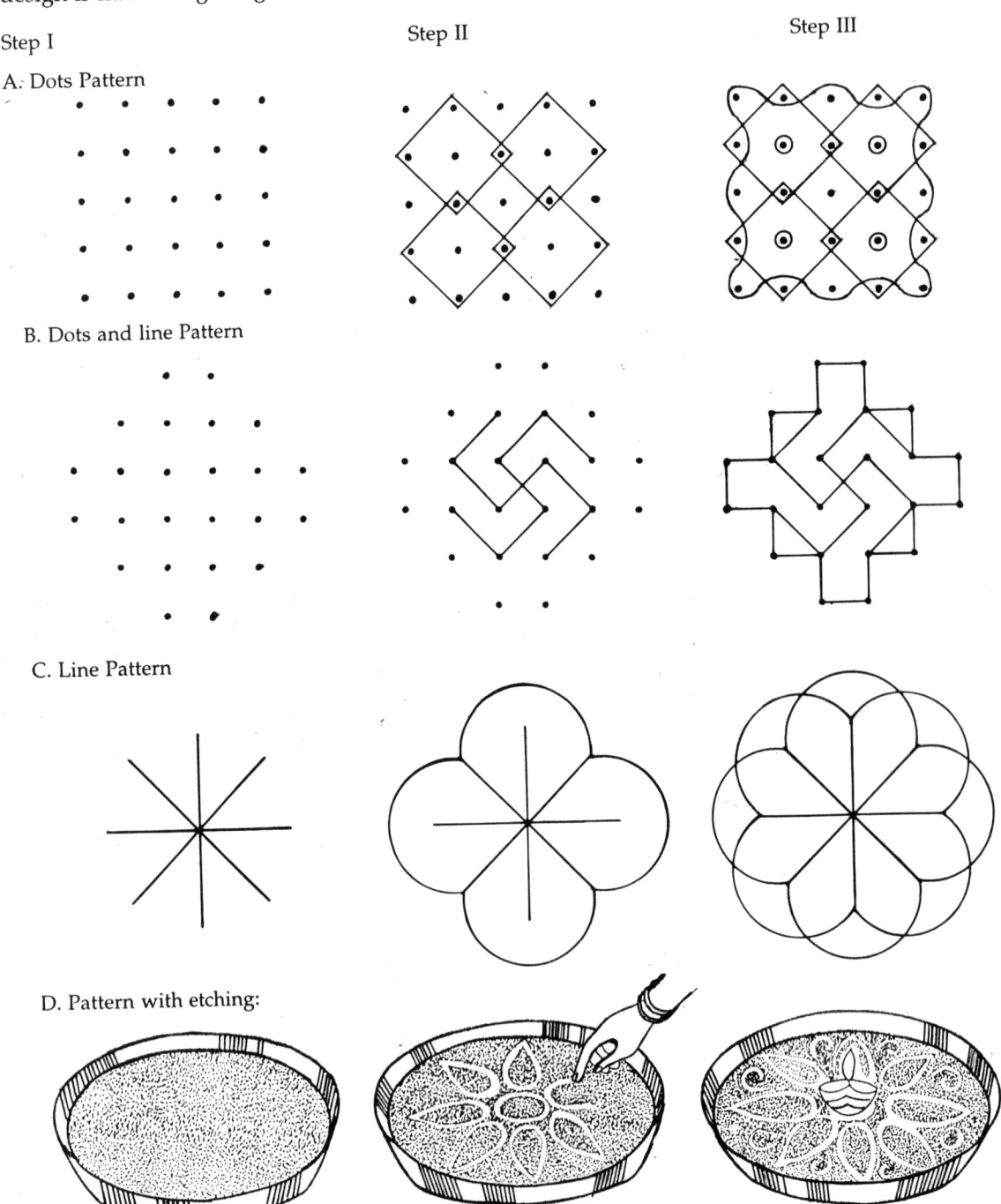

Figure 15.4. Stepwise Patterns for Rangoli

For making dry rangoli, the medium should be free flowing. Thus fine granular material is preferred to very fine powder which generally sticks to the fingers. It does not have a good fall on to the surface and it may fly around with minimum wind.

Dyeing of materials in dry rangoli

Different colours may be made by mixing colourful dyes to a number of base materials like dry rice powder, powdered quartz or powdered marble or marble chips. There are various methods for colouring such as mixing the dry colours with base materials and getting a number of shades of the same hue by mixing more or less of the dye colour. The resultant colour so made may not be very bright.

Another method involves mixing of sand with coloured 'kumkum' (vermilion). But this method gives good results only in cases of dark colours.

But the best colour effects are obtained by the method of mixing organic dyes (kuccha dyes). For quick and easy dyeing results, any organic dye is mixed with a little spirit and in this medium, the material gets easily dyed. If dark shades are needed, then more colour is added and for light shades less colour is used in the same amount of spirit. Spirit evaporates quickly and so it is preferable to use it instead of water. The material may be spread in old newspapers under the shade to get dryed for some time. If needed, it can be passed through a sieve to get an even texture. Evenly textured material is needed to draw especially neat and thin lines or to spread them evenly. It is important to keep the mixture away from moisture to avoid spoilage. The material may be preserved in an air-tight container, and re-used. While dyeing materials, one should wear Plastic/ rubber gloves to prevent the hands from getting the dye colour and to avoid skin allergies.

Most of the dry colours are made from dyes and chemicals. Some of them are organic dyes which are harmless. But those made from oxide, such as red oxide or lead oxide, are not desirable. Constant exposure and use of these colours may harm the lungs. There may not be any health hazard when the use is limited to festive occasions. But even then it is better to use organic dyes and prepare the colours at home. This is safe as well as economical. The colours can be made according to the predetermined pattern and thus much saving can be effected. One must draw the design on paper well in advance, colour it, with felt pens and after ascertaining the best effect, the patterns can be colored accordingly.

Wet Rangoli

For preparing a wet rangoli, traditionally rice grains are soaked in water for an hour or so, and then ground to get a fine paste. Rice powder in wet form is thus used to make a rangoli. In the place of rice powder paste, chalk powder, paste is used for making inexpensive wet rangoli. In modern times, wet poster colours are used increasingly to make this kind of rangoli. These are not only available easily in a large number of shades, but are also easy to paint using a painting brush, for which no extra skill is required.

Normally floral figures or geometrical patterns are drawn using wet rangoli paste. No brush is used, but the patterns are drawn with the finger tips. Though these are generally made in white colour (rice powder), usage of geru-mitti (red mud) paste along with the white lines enhances its attractiveness. Usually the rangoli lines are drawn with the ring finger, which comes with practice. However, a fine brush can be used by those people who have not mastered this art. Though dry rice powder is used everyday, wet rangoli is made on special occasions like festivals, marriages, house-warming etc. in south indian homes. Special patterns are drawn for "Navgrah" and "tantrik" pujas for ritualistic purposes.

In West Bengal, the combination of red and white colours in rangoli is common. Alpona, the floor decoration of Bengal, is also done mainly with two colours – red and white – is prepared with wet colours. A little bit of lime powder (Chuna) is added to rice paste to make it bright. It is also done with freehand drawings. The pattern is built around an axis to create symmetry and uniformity, which enhances its beauty.

Alpona is considered auspicious and its generally non-geometrical element represents prosperity and contentment. The major features of Alpona are paddy stalk, plough sickle, fish etc.

In a nutshell, the wet rangoli is made by mixing water and any material in a powder like rice powder, chalk powder, red mud etc. These days, even white and coloured paints may be used with fingers or brush to make rangoli especially at the entrance of the house, which also lasts for a long time. Wet rangoli is common in Southern Indian States, West Bengal and Uttar Pradesh They are made during festivals like Dushara, Janmashtami, Bhai dooj, Diwali, Holi etc. and other celebrations.

For wet rangoli or Alpona, soaked rice is ground finely and a few pieces of chalk are added to provide a bright white colour. This mixture may stay good for a longer time if it is mixed with a little jaggery. The designs in rangoli are done by hand with ease. The liquid mixture is taken between the thumb and the forefinger along with a pad of white cotton or fine muslin cloth, and the lines are drawn. The paste should be smooth flowing. Many a times a woman with a lot of practice in making Alpona can draw two or three parallel lines simultaneously by gently allowing the mixture, held between the thumb, ring, fore and middle fingers.

Floral Rangoli

This kind of rangoli is perhaps the oldest one. Natural materials like flowers and leaves, dry twigs etc. are always the foremost sources of inspiration for human beings for all kinds of arts. It is, therefore, not surprising that natural materials were used initially to fill in the rangoli patterns. Natural materials like flowers not only provide a colourful appearance, but also a fresh and fragrant atmosphere. They are also easy to work with. Any person who is starting to learn to make rangoli, flowers are the best materials to work with. After getting some practice with flowers, one can switch over to other materials like holi colours, coloured saw dust, rice flour etc.

Floral rangoli gives very warm and welcoming look. It is easy to make on any type of pattern. Most of the patterns are made using one's own imagination.

Same size of flowers should be used or if different sizes are used then the difference in size should be used to create a pattern. The other option is to cut the petals of different flowers in equal size. The flowers should be clean and fresh. Roses, chrysanthemums, margret and marigolds are some of the commonly used varieties. Green colour in the form of grass or thuja gives a beautiful background and enhances the colours of the flowers.

A base of wet river sand is highly recommended to give a three dimensional look as well as to keep the flowers fresh for long. The design is drawn with a stick on sand and accordingly flowers are put in their places. Start from the centre and end with a green surrounding of the pattern. Wet gunny cloth can be used in place of sand and the pattern can be drawn with any powder on it. After finishing the pattern, the flowers should be gently sprayed with fresh water. In summers the rangoli could be covered with wet muslin cloth for sometime to keep it fresh for long.

Floral patterns are also made on glass table in which case the pattern is so designed the some of the glass in it is not covered by flowers thus giving it a transparent look. It is best to choose big patterns while making rangoli with flowers.

As mentioned, earlier, Kerala has a special occasion during 'Onam' festival to draw rangoli with colourful flowers. A model for a fresh flower bed with white powder and coconut husk is laid out on the ground of the outer courtyard by the main doorway. It is called *Phookkolam*. The women fill it with beautiful flowers of many varieties.

Rangoli on Thalis

When people want to make rangoli other than on floors, the best choice is to make it in a "thali" – a brass plate. Traditionally Indians are used to brass plates while keeping various materials like flowers, fruits etc. during pujas and rituals. Similarly, a portable miniature rangoli can also be made in a "thali".

Beautiful rangoli is made on thalis of different sizes for all occasions. Thalis for arti, puja or tikka can all be decorated with rangoli patterns. These have several advantages. Firstly, they can be made in advance, secondly, there is no problem of space, and thirdly, these can be taken from one place to another.

For making rangoli design on thali or tray it is first oiled or an adhesive may also be used. The thali is then dusted with the desired colour and then extra colour is removed. On this surface, a design is either etched or selected parts may be coloured differently i.e., various treatments mentioned may be given to get a final rangoli finish.

Smaller patterns and thinner mediums are used for thalis. Colourful dry rangoli especially etching work on thali looks very beautiful. Thalis are also decorated with thin petals for floral decoration. Other materials to decorate thalis are sabudana, suji, atta etc. Floating arrangements with flowers can also be made in a thali containing water. Any kind of arrangement in a brass "thali" will give a traditional look. Plates made with clay and then baked can also be used to make such rangolis.

Rangoli with etching

Apart from the traditional rangolis, there is etching work with dry powder. It can be done on the floor, thali or on a platform. The powder is first evenly spread on the surface and the pattern is then etched on this powdered surface. Etching may be done with dry felt pen, spoon, fork, comb, match-stick or with the tip of the fingers on white powder. (Refer to Figure 15.4).

Unlike for floral rangolis where bigger patterns are used, etched rangoli can be made for small intricate patterns delicately executed, although big sized etched rangolis may also be made as per the demands of the occasion. Etched rangoli may be employed on any kind of pattern from the various patterns discussed earlier.

Since dry powders are used for making etched rangoli patterns, it is wise to make them where there is not much wind. Otherwise a fine coating with a transparent gum can be done on the surface before applying dry powder to prevent this powder to fly away with the wind. Care should be taken to see that the gum does not come in the way of etching. A thick gum coat may make it difficult to flaunt the etched portion of the rangoli pattern. Though the pattern can be of any kind, in this case fine floral motifs or ritualistic patterns like Lord Krishna/Ganesha etc. would provide an attractive appearance.

Rangoli with accessories

We have seen how the activity of art in India, including rangoli, is closely related to her people. Even rituals have developed a highly decorative character. The Indian woman has responded to the instinct for artistic and creative embellishment with so much vigour in the field of rangoli, that beautiful rangolis are made with mere powder or pastes along with a host of accessories suitable to the design and significance of the occasion. The most common accessories used in rangoli patterns are lamps (earthern and brass), flowers, statues and incense. Innumerable brass or golden/silver lamp-stands throwing fantastic shadows in many colours all around, flowers and incense wafting all the fragrance of the Heaven make a spectacle of rangoli a 'masterpiece of refined showmanship'. Specially on auspicious occasions like Diwali, etc., clay lamps glow in the centre of them, and in the South of India, these lamps are accompanied with brightly coloured flowers and leaves.

Rangoli thalis along with coconut ornamented with yellow/red thread and mango leaves, designer diyas (lamps) flowers and garlands of flowers, objects of worship, sandalwood, shells, pulses, saboodana, saunf, fruits, etc. as accessories provide a wonderful background for rangoli patterns. The accessories are used to accentuate an occasion and as a part of ritualistic custom. But keeping in mind the ornamental aspects of rangoli, the accessories may be inspired by its modern perspective, creativity and desire of the artist to be unique in its art presentation. Therefore, the choice of accessories may be from an unlimited range of objects – naturalistic, modified, abstract or stylised.

GUIDELINES FOR MAKING A RANGOLI

Skill development can help in improving the quality of Rangoli. Knowledge of geometry and symmetry gives better results in the making of a rangoli. A circle, as big as one wants, can be made with a piece of chalk and a desired length of string. The chalk tied to one end of the string is taken round. The other end is the centre and the string makes the radius. Similarly, hexagons can be drawn by cutting arcs of equal radius at the circumference. A scale/inch-tape can be used to mark the dots either horizontally, vertically/diagonally to make the figure perfect. Perfect drawings in chalk give an added beauty when the colours are filled (Figure 15.4).

For free-hand rangoli, an axis of symmetry must be taken and this comes only with practice.

The selection of colour is very important in rangoli. Either allied colours/or contrasting colours must be used. Use of too much of dark or dull colours like grey, blue or green certainly makes a less attractive pattern. A cheerful atmosphere should always be created by the use of colours like red, yellow, orange or warm colours which psychologically cheer and boost one's morale. Conversely, dark and deep colours like grey, black and blue give depth and if used sparingly give a three-dimensional effect. The spreading of colours can be even if dusted through a metal tea-stainer. This gives a thin layer and imports the effect of a painting, particularly when one is doing a landscape.

Similar to the other forms of art, rangoli patterns can also be made more effective and attractive by judging them in terms of art principles like balance, rhythm, harmony, emphasis and proportion. There should be balance in the rangoli pattern in terms of its various parts and the colours used. All parts of a rangoli pattern should be proportionate to each other. The size of the rangoli pattern should also be in proportion to the room or small patterns in a big room will look dis-proportionate. The lines and designs of a rangoli pattern should harmonise with each other. The texture of the material used should be in harmony with the texture of the surface on which it is made. Some amount of contrast in colour can be used to emphasize a particular section like the centre of the rangoli pattern. Too much of contrast will only create confusion and the impact of the rangoli for decorative purpose will be lost. Emphasis can also be achieved in rangoli patterns through the use of accessories like lamps/diyas, kalash (brass vessel along with mango leaves and a whole coconut fruit), statues etc. Thus one can realise that it is not difficult to apply art principles as the guiding factors in making a rangoli to be effective and decorative.

Many of our traditional beliefs and practices have a scientific basis. This is true in the case of Rangoli too. First, the sprinkling of cowdung water in the front portion clears all dirt from place and makes it clean. Secondly, cowdong is a cheap, effective and easily available disinfectant which could be used without any side or harmful effects. Designs drawn in white on the moss-green cowdung background are striking and add beauty to the place. Rice powder is used as a medium as it also serves as food to ants and birds – an indication of the concern for other living beings.

Lime powder, quartz powder/marble powder all contain Calcium carbonate ($CaCO_3$). When this is applied over a wet surface, it gets mixed with H_2O. When this is exposed to atmosphere, Calcium bicarbonate [$CaH(CO_3)_2$] is formed by absorption of CO_2 from the atmosphere. Thus an excess of Co_2 in the atmosphere is cleaned. This was the way thought of by our forefathers to eliminate the harmful Co_2 created by the burning of a large number of lamps and candles and the bursting of crackers during Diwali. Thus as an anti-pollutant measure *Rangoli* can make its own valuable contribution besides being an ornamental one.

Floor decorations, though a traditional art, and mostly specific to a State, have now evolved into common and popular form because of its beauty and colourful effects. It is also now emerging as a commercial art as it provides variety and a festive look. Anyone interested in the art of interior decoration, cannot overlook this sparkling and inspiring floor decoration—rangoli.

Chapter 16

INTERIOR DECORATION : TRENDS IN INDIA

While the different aspects of art in India have attained great heights, it would appear that Interior Decoration has not received the kind of attention it deserves. Nevertheless, a study of Interior Decoration in India, through different periods, not only reveals interesting facets but emphasizes a conscious effort towards beautifying human dwellings to make it more attractive and conducive for a comfortable living environment. Being an integral part of architecture and closely bonded with the daily life of a man, Interior Decoration was consistent with the evolving pattern of Indian society in a civilized manner.

Indians have been firm believers in things of enduring value, art of the highest order and a way of life that is both simple and economic. The varied strands of our culture are woven into the fabric of our national life. It is no wonder then that life in our country was moulded in an aesthetic environment. The beauty of architectural features, enhanced by the fine and final touches of a few functional objects that served more purposes than one, thus formed an important element in Interior Decoration. For example, a single gorgeous carpet, a couple of attractive bolsters and an exquisite lamp could well and truly complete the interior decor of an Indian home.

Interior decoration is really a setting appropriate for a congenial home. It aims at providing comfort, pleasant working conditions, relaxation, a warm and friendly atmosphere and aesthetic surroundings, factors that combine to make home what it should be, a place where the family finds satisfaction and cheer and friends feel an air of cordial welcome.

Is there such a thing as interior decoration in India? Since Interior Decoration conjures up in the mind rich furnishings, expensive furniture and rare curios tastefully arranged in a modern home the concept of interior decoration has to be understood in its proper context. Each country has its own tradition of interior decoration. Who would say that Japanese homes, known for their elegance and aesthetic charm, are devoid of interior decoration merely because furnishings and furniture are not much in evidence?

Man's habitation has to be a place where he can carry on his household activities, relax after his day's hard work and participate in a social life with his family and friends. Towards that end he aspires, works and uses all his energies.

Different countries have different types of homes, built in keeping with their climatic conditions, the particular ways of life of the people of the country and their traditional architectural trends. They are also furnished with a view to his work, leisure and comfort and as aesthetically possible within the means available (Figure 16.1 – see colour plate).

India has had her own style of architecture, mode of town planning and homes with provisions for amenities according to the prevailing conditions through the ages. India's vast treasures of cave dwellings and archaeological findings and architectural splendour found all over the country, provide sufficient evidence that the Indian men have always attempted to make their environment congenial to their living styles, artistic, inspiring and above all, functional. A glimpse of the earliest artistic expressions can be had from the fascinating drawings of prehistoric man who scribbled them on the walls of caves in Madhya Pradesh and Uttar Pradesh in the Kaimur and Vindhya ranges and the Raigarh and Mirzapur caves respectively. The once prosperous, buried cities of Lothal, Mohenjo-Daro and Harappa that have now been unearthed and are treated as prehistoric sites are proofs of the way people lived, the types of houses built and the things they constantly used. Next in chronological order are the rock-cut architectural splendours of Ajanta and Ellora, dated between 2nd century B.C. and the 7th century A.D., magnificent, awe-inspiring and breathtaking in their beauty with minute details. They are not only the pride of India, but are also the landmarks in the history of art in India. They were followed by the monasteries and universities that were the centres of learning and scholarship and the glorious medieval temples that were community centres in themselves, built between the 5th and the 13th century A.D. and even later. The medieval period of India's history is also not lacking in its superb creations of art.

Interiors of the Golden age

This glorious age starts about the 6th century B.C.,—the birth of Mahavira and Buddha, and passing through the Gupta age, which is known as India's Golden Age of Art and Literature, on account of great heights attained by the country in both these fields, trace the progress of Interior Decoration upto the 7th century A.D. or the period of early medieval India.

This was an era of what may be described as the great happenings. The literature, sculptures and paintings of this great era of material prosperity, intellectual vigour and artistic and literary genius, provide a vivid representation of the life and activities of the people. The frescos of Ajanta are illustrious examples of the art of interior decoration. A detailed study of the Ajanta caves and the murals that adorn their walls leads one to the conclusion that while public monuments, monastries and temples were built in rock, domestic architecture made use of wood for the construction of houses, and especially for the purpose of their Interior Decoration. Apart from the architectural beauty of the interiors of the private dwellings, with their sculptured columns, friezes and alcoves, rooms in those dwellings were furnished with couches, divans, charpoys, Chowkies, stools of different shapes, foot-rests, bolsters, cushions, trays, book-rests, carpets, curtains, pelmets and musical instruments and painted ceilings and murals. Even the stairs leading into the interior of the houses are ornamented with figures of animals or some natural design elements like trees, plants, flowers etc.

The high seats with backs and invariably with rests for the feet that one sees in the frescos of Ajanta are proof that high seats somewhat like chairs, were also in vogue at that time. There were also combined seats for two persons known as love seats or settees, or for more than two persons, and even benches. Writing tables, such as we know them today are not found though the rests for books existed. Perhaps low stools were used for writing when necessary. Many types of chairs, both plain and ornamental, were certainly in use at that time itself.

Round cushions for sitting on and bolsters as rests for the back and arms seems to have existed in the earlier times. Cane muras or stools, cushions with tassels are also noticeable.

Among the pots and pans for household, one finds vessels of many beautiful shapes, pots for storing water, containers for toilet requisites, decorative pots set in niches, vessels with spouts and handles like flagons and flasks. There are bowls and tumbler-like receptacles for drinking purposes trays and thalis (plates) for carrying flowers and other things are also included. A pan with a handle, used by bhikkus or Buddhist monks, is a beautiful ornamental piece to be seen and admired.

That curtain valances and divan covers or bed-spreads with gathered or plain hems were also in vogue along with decorative chair covers and carpets as furnishings. Carved boxes for keeping jewels and decorative lamps were included among the household articles of utility combined with beauty that formed items of Interior Decoration. A bedroom scene shows the lavish Interior Decoration prevailing at that time. Apart from the bedstead, bolster and a decorative column, a niche in the wall is adorned with curios. The intricate and exquisitely painted designs that adorn the ceilings of the Ajanta caves combine with the sculptures and inimitable frescos to heighten the richness of the interior of the caves and create a sense of beauty that have been the wonder and envy of the succeeding generations.

Interior Decor in Medieval India

Under the influence of Hindu renaissance which was making its influence felt, there was a spate of temple architecture which influenced domestic architecture as well. It is the ancient Jain literature which gives us a glimpse of the use of wood and stone sculptures in the medieval India. This period was rich in the exquisite miniature paintings of Rajasthan and the later pictures from Kangra and the Hill States of the Punjab, which illustrate in detail the daily life and the interior decoration of homes of medieval India.

What could be gathered from public monuments is now represented in the paintings of medieval India. There was a distinct aesthetic development. In the interiors of Rajasthan, homes have a rare style and beauty in the elegance of the architecture, decorative balconies, exquisite screens and furnishings.

With the advent of Muslims into India especially during the reign of three great Mughals, Akbar, Jehangir and Shah Jehan the architecture and the industrial arts greatly flourished. Greater patronage was given to ceramics, metalware, enamelling, glass-ware, inlay work and mosaic work. Gujarat has some beautiful toranas or gateways of medieval period. The main doors of houses in Gujarat, Rajasthan as well as temple and palace doors had rare wood carving, even exquisitely inlaid with ivory.

The Western Interlude

The introduction of chairs and tables for general use appears to have caused a radical change in interior decoration especially during the four hundred years of Western colonisation in the country, first by the Portuguese and lastly by the British. During this time, living rooms came to be flooded with period furniture of French and British styles. Interior decoration with mirrors, gilded picture-frames, clocks for the mantel-shelf, porcelain vases, and tapestry for chairs, came into fashion. All beautiful things like Indian carpets, brass and bronze ware, phulkaris, gold and silver embroideries, figured muslins and carved woodwork found their way into the international markets.

The exquisite art-crafts of India ceased to be attractive to the Indian eye and elaborate carvings and sculpture were worked upon as gaudy and outdated. But foreigners were fascinated by the ornamental furniture made in India. Carved wooden furniture was being continually produced in Surat, Ahmedabad, Mysore, Travancore, Moradabad, Saharanpur and Harpanhalli. Some lovely articles of furniture were also produced at Madras, Monghyr and Goa.

THE INDIAN RENAISSANCE

A consciousness of the aesthetics of the traditional arts and crafts of India was revived among the masses after Independence. The freedom movement, five-year plans by the Government of India led to spurt in the development of crafts. Opening of craft centres like craft Museum, Dilli Haat, Craft Resource centres, Government emporiums, in the national and State capital cities of India have also brought a spurt in the activities of the artisans of India. Regular annual art fairs like Trade Fairs, Suraj Kund Mela etc. had also opened up the opportunities for the craft persons to exhibit their talents and products in the important towns and cities, thereby bringing in the revival of the arts and crafts of India. All these circumstances brought a remarkable change in interior decoration. Many crafts which were on the brink of extinction have now been revived and homes are being tastefully furnished with articles of indigenous manufacture, for example, the traditional Sankheda or lacquered furniture of Gujarat. Newer forms of handicrafts such as sculpture from natural twigs and wood and art objects made of waste and indigenous materials are coming into vogue. Many handicrafts have become large cottage industries in India though the mode of production has not undergone any change.

Vastu Shastra

The concept of five elements (air, water, fire, earth and space) have been woven into the fabric of India. Indian beliefs have always been animist—dedicated to the worship of nature. Although ancient Western beliefs recognize the first four elements, the fifth—space is a more abstract principle. To many Indians, the notion of space represents the vast vacuum beyond ordinary perception, and symbolises the huge power of an unseen force. In common with these beliefs, ancient sciences such as Vastu Shastra (the science of construction and decorating) were directly or subtly linked with nature. Buildings and homes were built to balance the elements observed in nature, to encourage the flow of positive energy and thereby assure domestic health, wealth and happiness. As Indians happily embrace the cycle which links humankind with nature and the universe, Vastu Shastra is still influential today and the five elements are widely celebrated in the decorative schemes of many Indian homes.

Air (Vayu): The Vastu Shastra emphasizes the importance of fresh air (Vayu) in homes and buildings. The rooms in a home, the ancient text advises, must have efficient air circulation to ensure good health for the inhabitants. Because most of India is intensely hot, people try to circulate as much cool air through their home as they possibly can, be it with air conditioning, fans and ventilators. Some old homes still have the remnants of the hand-drawn 'punkhas' (fans) which servants (punkha-wallahs) would operate. The element of air, often symbolised by white and silver, is also ushered into Indian interiors by using materials that emphasize its qualities—long, light fabrics that dance in the summer breezes.

Jal (Water): Indians celebrate water (Jal), symbolised by colour blue. **Fire** is still sacred today and is symbolised by colours red, orange and gold. Sun is regarded as a special motif throughout India especially in Gujarat, Rajasthan and Orissa where it embellishes furniture and fabric.

Bhoomy (earth): According to vastu Shastra, 'Bhoomi Poojan'—earth worship is necessary before any house construction begins. Earthy colours and textures feature profusely in Indian homes and are used to stain floors, walls, furniture and fabric, sometimes to create a backdrop for painted images. In the hottest regions, huts are made out of mud to keep the blistering heat at bay, and objects crafted out of clay abound in homes throughout the country.

Akash (space): In India the vast landscape stretches beyond the realm of the naked eye. Yet in the cities, houses are small and clustered together, and in a land where social structures are based on the extended family, the western concepts of personal space and privacy are irrelevant in most rooms, even bedrooms are communal. It is the allocation of space which is more important to Indians.

The expansion of the visual space is an art that has been well developed by Indian architects for hundreds of years. Walls are quite commonly punctuated with small windows, niches and simple doorways. Open arches and filigreed screens between rooms demarcate areas that do not demand privacy.

Feng Shui based decoration

Similar to Vastu Sastra, Feng Shui is fast becoming popular among Indians. Some Feng Shui tips, having its roots in China, are given here. Let us discuss how we can try placing different decorative objects keeping some fundamental Feng Shui techniques in mind.

North: It is the career area. Here we could use wrought iron furniture, which is in vogue these days. This sector is ruled by the metal element, so metal trophies, and decorative statues made of metal could be placed here. However, while placing furniture or other decorative objects in this sector or anywhere in the house, we need to make sure that the furniture has round edges. Objects with sharp pointed corners are to be avoided. Paintings should not be abstract or those depicting unhappiness, struggle and battles. Paintings of meandering rivers and fishes framed in wood are ideal for this sector. An aquarium could be placed in the North of the living room. All paintings depicting water or waterlife should never be placed in the bedrooms. The colours that are needed in this sector are blue, silver, grey and gold.

East: This sector is associated with health and family happiness. Here we could place wooden furniture, especially tall wooden cupboards and cabinets. Handicrafts made of wood, paper mashes could be used to decorate this sector. Family photographs, especially those taken during a happy occasion, could be displayed in wooden frames here. Metal frames or objects should not be placed here as this cuts away the energies of wood, which is the governing element of the East. Introducing the color green in the upholstery here would also be ideal. Green healthy plants could be placed in the East of the living room. Paintings of landscapes, rising sun or meandering rivers could also be placed in wooden frames in this sector.

South: This area symbolizes name, fame and recognition. We need to introduce the fire element here. Introducing the colors red in the upholstery would energize this sector. Candles, pictures of the sun, fire in wooden frames could be hung here. Having plants in the South of the living area is very important as plants represent wood, which provides nourishment and energy to Fire, which is the ruling element. We could place wooden tables and other handicrafts here. A wooden windchime with nine rods could be hung here. However, do remember to avoid the colors

PLATE 35

Figure 16.1. Traditional Styles in Home Interiors

Figure 16.2. Contemporary Trends in Home Interiors

PLATE 36

Figure 16.3. Eclectic Styles in Home interiors

Figure 16.4. Simple, functional and attractive arrangement in the present day homes

blue and black here as these symbolize the water element, which extinguishes the Fire element of the South.

West: This is the area associated with children and creativity. This again is ruled by the metal element. You could also do up the West of your study by placing pens, photographs of children, a cluster of seven crystals. Keep in drawers in this area projects you intend to work on in the near future.

What is Indian Style

India is a land of glorious flamboyance and excess. This vast sub-continent is suffused with a riot of vibrant colours, earth-rich smells, tingling tastes and a tapestry of textures and sounds. Celebrations of festivals, wedding processions and election campaigns are all larger than life events that call for public participation and collective self-expression. India is an enormous melting point of lifestyles and social and cultural customs. Its vast population some 1000 million, the majority being Hindus and others being Muslims, Christians, Parsees, Sikhs, Jains and the Buddhist comprises the nation with it 28 states, fourteen official languages, hundreds of dialects and over thirty recognized political parties, India is a diverse nation but one of the characteristics that binds the whole nation together is the people's love of ornamentation and there is a profusion of sights, sounds, smells and textures. Yet, despite the vast range of form and influence, in every design, there will always be a distinctive motif, a particular flourish or a telling use of colour that fixes the style firmly as Indian.

However, the tranquil and at times awe-inspiring scenery belies the bloody history of the subcontinent, constantly ravaged by marauding foreign powers and divided by internal factions. After each battle for independence from foreign rule, India has survived and assimilated some of the culture that was introduced with external conquest. Islamic, Portuguese, British and other conquerors left nuances of their language, way of life, cuisine and architecture, which took roots and flowered along with indigenous styles, giving rise to a strange but beautiful hybrid form that is now so comfortably settled in India that the original styles are no longer easily discernable.

Each region of India has a definitive style influenced by climate and other geographical conditions, which is echoed throughout the aesthetics of its homes. For example, Kerala, on the south west coast, is known for its wooden houses which derive the strong lines with the natural wealth of woods and Plantations found abundantly in the state.

Until recently, Indian style was considered by many foreigners to be synonymous with Rajasthan, the northern desert State famous for its bright colours. However, the delicacy of north Indian style is also embodied in cities like Lucknow and Banaras where homes reflect an understated pride and orderliness and where cool white curtains and smooth marble walls are more likely to feature.

In Calcutta, Bombay and Chennai, colonial buildings abound, yet each city has its own identity. Bombay is a biggest meeting pot, the heart of India's movie industry, where homes reflect influences from the entire sub-continent as well as abroad and a variety of architectural styles. The folk arts from the region have amalgamated with urban sensibilities to create an aesthetic impression that enfolds European, Indian rural and city tastes.

Today, the Indian interior decorator can choose from a rich repertoire of arts and crafts that will fit into the changing pattern of life (Figure 16.1 – see colour plate 35).

Contemporary Indian Decor

The sophisticated home-maker who has the money power to spend, seeks the advice of professional architects or interior decorators for beautifying the home. But their number is negligible in this country, and they are, besides, to be found only in large cities. Big firms, business houses and the hospitality industry, make use of professional persons, but the average family has to plan their home with such ingenuity and imagination as they may be endowed with in order to keep within the limitations of their budget (Figure 16.2 – see colour plate).

Interior decoration is concerned with the selection and arrangement of things required for a home so that while providing the greatest comfort they blend harmoniously in size, form, design and colour to make an artistic unit. The concept of what is artistic has undergone radical changes in recent times. Today, any form, colour or design is looked upon as beautiful provided it is well composed and vigorous and where comfort and pleasure reign supreme. While comfort is a physical necessity, and easily understood, pleasure is more difficult to define. A home, in order to be ideal, has to offer much more than mere pleasure. It must give aesthetic and mental satisfaction, contain cherished objects, have a tranquil atmosphere, relieve boredom and monotony, provide scope for an occasional new look and above all, its maintenance must be easy and within one's means. Further, families have unlimited opportunities of making an artistic and radiant home with the wealth of beautiful things that the country offers as her traditional decorative arts.

The traditions of designing Interiors in India relate chiefly to architecture especially with built-in sculpture, mosaics, murals, wall niches, torans (or gateways) with traditional motifs. Fortunately, Indian craftsmen provide an answer to this perplexing question and enable the family to introduce all these features in a home, may be in a slightly different way, but relatively similar to choose and design an expensive interior or a modest one.

While in the ancient home architectural features like pillars, wood or stone carving, ornamental doors and windows, murals, mosaic patterns were prominent and the furniture and furnishings played a comparatively minor role, the Indian home of today attracts pieces of Indian carved furniture, well-designed carpets, ornamental screens and antiques. These now constitute an important element in interior decoration.

A well-arranged Indian home expresses not only its own personality or distinctive character, but reflects the personality of its members, their culture, tastes, hobbies, special interests and outlook. Further, a home should have an air of cosiness and should radiate warmth. In modern living conditions in India with old and the new jostling with each other, living space becoming more difficult to obtain and economic conditions worsening from day to day, the requirements of a good, cheerful home have to be considered from many angles. Besides interior decoration from the point of view of aesthetics, a home has to provide equipment for vocational use, education, social affairs, recreation and comfort. The idea is growing all over the world that every object in a house should be both functional and aesthetic, an idea that has been deeply rooted in our country and which has provided a basic pattern of living, simple yet artistic.

Moreover, moving and circulating space, and the working and relaxing areas are more important for comfort and pleasure in a home than being hedged in by unnecessary furniture. With the breaking up of large families and ancestral homes, small units are inevitable and families face problems peculiar to western countries today.

Limited space has necessitated furniture that can be put to more uses than one and which can be split up into separate units, whenever necessary, has become popular.

In this modern age of science and technology, when utility, function, simplicity and ease of maintenance are the hinges on which life revolves, and mass production, being more economical, is the order of the day. Though the Industrial Revolution and the Machine Age have dominated all spheres of life, the hand still retains its supremacy in creative art.

Modern materials that are cheap, colourful, eco-friendly, colourful and have ease of maintenance such as plastics, chromium, stainless steel, glass etc., have caught the fancy of the present age along with modern tastes and home conditions. A new concept in interior decoration, in keeping with these trends, is surely well on its way.

ECLECTIC DECORATING

The present generation is a mix of old and new, so are its tastes. This generation would like to continue with the old, but would also like to adopt new, innovative styles. This is specially true in the case of decorating the home interiors. Such a mix of the old and new can be called Eclectic Decorating (Figure 16.3 – see colour plate 36).

Eclecticism: The term eclectic comes from the Greek Word Eclectics, originated by a school of ancient Greek philosophers who believed in borrowing from several systems of thought and rearranging them to form a new philosophic composite. Today eclecticism is the marriage of the old with the new, the past with the present, the native with the foreign. It combines art forms and furnishings from various periods and adapts them to today's living. It is a challenging approach that immediately expresses the owner's personality.

It was not too long ago that styles from different periods were not used together in a room. The dictates of good taste and good design determined that an acceptable presentation of a certain period (be it modern, or one of a specific period) had to maintain a positive consistency. Only items from the same period could be used together for floor, walls, furniture coverings, and assorted elements, such as lighting fixtures and accessories. Even the fabric had to be derived from a single period—either past or present.

For example, to achieve the provincial look, the rules dictated the use of natural finishes of chestnut, walnut, oak, or fruitwood for furniture and floors; curved furniture lines; rough-textured materials; peasant patterns; copper, brass, or pewter accessories; and bright yellows, blues, and greens. If the chairs were upholstered in crimson antique satin, then the carpet had to be in the same colour. A chair of another colour that did not conform did not belong. Today we are living in a casual, dissenant world, and many regulations have been abandoned. We no longer say "you cannot do that" or "you must do this". Strict adherence to rules is no longer the fashion and we are expressing this freedom in eclectic decorating.

Possibly one of the most exciting things to happen in the field of decoration in recent years is the acceptance of eclecticism for a good sense of colour, proportion, and consistency, not expressive of period taste, but of a personal taste that can prevail to make our life styles more suitable to our daily circumstances. Eclecticism allows home owners to furnish their houses by combining the tastes and personalities of individual family members. The concept makes possible the visual excitement of combining the old world furniture with the contemporary sofa and oriental rug recently purchased for home (Figure 16.4 – see colour plate 36).

Suitability is a key word in eclectic decorating—the suitability of an object or a design to the style and purpose of a room, or to a person's lifestyle. It is not the custom today to maintain the same period furnishings within a house or even within the same room. Some professional decorators urge their clients "to let go"—to satisfy their own standards of beauty and taste, instead of being confined by strict adherence to the guidelines of decorating. Many famous decorators even frown on a room decorated with a suite of matching period or contemporary furniture because it may be boring and lacking in creativity. Because it is possible to mix freely (within the constraints of good taste) items from any period, one must not get the impression that electricism is organised clutter. It is not. To be successful an eclectic room should have common elements throughout. One style may dominate, with others used for accent. Antiques are mixed with modern, and there are no set rules except that pieces should be compatible in scale and degree of formality. A well planned colour scheme will tie the different styles together.

Different fabrics such as florals, geometrics, and solids may also be combined in a room. If two or three patterns are combined, it is recommended that they all have something in common like colour, flowers or being derived from the same historical period. Fabrics can be coordinated if the same sized (scale) patterns are used together.

Other examples are the selection of different furniture styles of the same wood, or the combination of the sleek lines of glass and steel combined with a chair (wooden chair with no carvings) The choice of style depends on many factors. The final goal is to make the home functional as well as satisfying to all the people who live in it.

Previously the purchase of a particular item might mean the expense and problem of total redecoration in order to accommodate it; today the consumer can collect individual pieces and use them together to convey a personal statement. In order to achieve an eclectic approach, follow these guidelines:

- Do not separate contemporary pieces from traditional. Freely group them together to produce a complementary blend.
- Combine straight and curved lines together, rather than all straight-line items in one area and all curved-line items in another.
- Use a variety of different materials such as wicker, glass, wood, brass, chrome, and fabric. Be sure to distribute them carefully to avoid the predominance of one material in any given area.
- Achieve a visually pleasing balance of size and scale. Do not use all large, heavy pieces or small, delicate items in one area.
- When these guidelines are used, the result is an exciting mixture of old and new.

Today, it is possible to pick and choose to suit one's life style and personality, mix or match interesting combinations of period and accessories, if you are so inclined. It represents a growing sophistication of our own times, and it is not only acceptable but inevitable as the world grows smaller.

BIBLIOGRAPHY

Anoop Parikh, (1995) The Book of Home Design Weidenfeld and Niolson, London.

Barker, Deborah, (1999), Home Inspirations – Over 60 decorative projects for every room in the home, Lorenz Books, London.

Barty Phillips, (1990) Fabrics and Wall Papers, Bulfinch Press Book, Little, Brown & Co. London.

Beotter (en el Jane) and Lockhart Bill (1969), Design for you, Octopus Books Ltd. London.

Bhavani Enakshi (1982), Decorative Designs and Craftsmanship of India, D.B. Taraporevala Sons & Co. Pvt. Ltd. Bombay.

Black Maggie (ed.) (1970) Design For Living, Octopus Books Ltd. London.

Blake, Jill and Fisher, Joan (edited) (1975), The compete Book of Handicrafts, Octopus Books Ltd. London.

Boehm, Peggy and Matsuda, Shizu, (1961), Flower arranging by number, The Oak Tree Press London, Melbourne and Cape Town.

Boehm P. and Matsuda S. (1961), Flower Arranging; The Oak Tree Press, London.

Chandler, B. (1983), Home Decorating and Design, Hennerwood Publications Ltd.

Chatterjee T.M. (1965), Alpona, Orient Longmans Ltd. Calcutta.

Chattopadhaya K. (1969), Carpets and Floor Coverings of India. D.B. Taraporevala, Bombay.

Cerver F.A. (1997) Decoration—Arco colour, Roto Vision, Hove, England.

Clare E. Horne, (2000) Geometric Symmetry in Patterns and Tilings. Woodhead Pub. Ltd. Cambridge, England.

Colvin, Brenda (1970), Land and Landscape – Evolution, Design and Control, John Murray, Albermarle Street, London.

Deutch, Yuonne (1977) Flower Craft, Marshall Cavendish Books Ltd. London.

Dommelen, David B. Van (1965) Designing and Decorating Interiors.

Dongerkery, Kamala S. (1973) Interior Decoration in India-Past and Present, D.B. Taraporewala Sons & Company Pvt Ltd. Bombay India.

Dorthy Stepat-De van, Introduction to Interior Design.

Faulkner (1965), Inside Todays Home Hot, Rinehart and Winston, Inc.

Female Magazine's Beautiful Homes – Malaysia and Singapore, MPH Magazines Put. Ltd., Singapore.

Fisher K. (1972) Living for Today, Thames & Hudson, London.

Francis D.K. Ching, (1987) Interior Design Illustrated, Van Nostrand Reinhold, New York.

Georgie Davidson, (1970) Classical Ikebana, W.H. Allen & Co. Ltd. London.

Gerritsen, Frans (1975), Theory and Practice of Colour, A Colour Theory Based on Laws of Perception, Studio Vista Publishers, London.

Goldstein and Goldstein, Art in Everyday Life, Prentice Hall Inc. New Jersey.

Hannerbaum L.G. (1994) Land Scape Design Prentice-Hall Inc. New Jersey.

Hatje Gerd and Kaspar Peter (1975), Design for Modern Living: A practical Guide to Home Furnishing and Interior Design.

Healey, Deryck (1986) The New Art of Flower Design, Villard Books, New York.

Kelvin McCloud (2000) The Outdoor Decorator, Ebury Press, London.

Knot Larson, (1962) Rugs & Carpets of the Orient, Fredrick Warne & Co. Ltd. London.

Lesley Taylor, Design & Interirors, New Holland, U.S.A.

Meintjes A. and Karen Ross, Fast Decor, New Holland Pub. London.

Merelle Soutar, (1967) The Driftwood Flower Arrangement Book, Arlington Books, London.

Michael, Michele (1966), The New Apartment Book – Inspiring Ideas and Practical Projects for Decorating Your Home, Clarkson Potter, New York.

Monisha Bharadwaj, (1998) Inside India, Kyle Cathie Ltd. London.

Moubray, Amicia de and Black David (1999), Carpets for the Home, Rizzoli International Publications, New York.

Ohi, Minobu and Ikenobo, Senei (1972), Flower Arrangement – The Ikebana Way, Shufunotomo Co., Ltd. Tokyo, Japan.

Oki, Morihiro (1989), India-Fairs and Festivals, Gakken Co., Ltd., Tokyo, Japan.

Pearsall, Ronald (1977), A Connoisseur's Guide to Antique Pottery and Porcelain, Tiger Books International, London.

Phillips B. (1998) Hamlyn Book of Decorating, Octopus Pub. Group Ltd. London.

Quartermaine, Ruth (1966) Better Homes, Flower Arranging; Collins, London and Glasgow.

Rayworth, Jenny and Berry, Susan (1966) Reader's Digest Fresh Flowers For All Seasons – A complete guide to selecting and arranging fresh flowers through out the year, The Reader's Digest Association, Inc. Pleasantville, New York/Montreal, Collins and Brown Ltd.

Robert de Calatchi, (1967) Oriental Carpets, Konecky & Konecky, New York.

Roger Banks-Pye, (1997) Inspirational Interiors, Ryland Peters and Small, London.

Rutt, A.H. (1969), Home Furnishing, Wiley Eastern Pvt. Ltd. New Delhi.

Sahni Nidhi (2002) Floor Covering Industries of Panipat. Unpublished Master's Dissertation, Institute of Home Economics, University of Delhi, Delhi.

Shibukawa, Ikuyoshi and Takahashi, Yumi (1984) Designer's Guide to colour, Kawade Shobo Shinsha Publishers, Japan.

Sparnon Norman (1970) A Guide to Japanese Flower Arrangement Shufunotoma Co., Ltd. Tokyo.

Strewart & Watton S. (1999) Decorating Ideas, Anness Pub. Ltd., Singapore.

Steward & Sally Walton (1997) The complete Home Decorator, Anness Pub. Ltd. London.

Swarup S., (1968) 5000 Years of Arts & Crafts in India & Pakistan, D.B. Taraporevala Sons & Co. Pvt. Ltd. Bombay.

Walton S. (2000) Colourful Home, Lorenz Books, Anness Pub. Inc. New York.

GLOSSARY

Abstract: usually used in designing art objects/paintings, separated from particular examples, and not concrete.

Applique: A design which is cut out and attached to the surface of another material

Architecture: The art of designing buildings, an art that combines beauty with strength and stability.

Architectural Lighting: Glamorous effect for general lighting achieved by the mounting of simple fixtures in valences, cornices or cover or behind translucent panels. To light a window area, wall ceiling, floor or any other special area for emphasis.

Art: Art is a word used specially to describe the useful, decorative and fine arts, particularly, the latter.

Artifacts: The relics left by primitive people. They provide the archaeologists with clues to ancient cultures and the decorators with ideas.

Atmosphere: A surrounding environment that evokes a feeling, such as that per vades a work of art, and can be created in any room through decoration.

Backgrounds: The floors, walls, woodwork and ceiling of a room supply a backdrop for other furnishings, which can be subtle or bold.

Bakelite: A brand name for vinyl resins made by Union Carbide Corporation. It is made from phenol and formaldehyde, and therefore, known as phenolic resin. The name is derived from its inventor, Leo, H. Background, an American Chemist. Today bakelite is used mainly for insulation, radio cabinets, handles etc.

Basket Weave: A style to weave, produced by interlacing of two or more warp ends with the filling yarns.

Batik: An ancient method of decorating fabric by waxing and dyeing technique.

Beech: A tough, strong, easily worked hard wood that is widely used for flooring, furniture and general lumber. It is close-grained and reddish-brown heartwood often used for veneers.

Birch: A very strong, heavy, close-grained hard wood of any of the several species of birch trees of Europe, Asia and North America, popularly used for doors, woodwork, panelling and furnishings of many kinds.

Bone China: An English porcelain that contains calcined bone. Bone China is softer than true porcelain, but is easier to make. It is considerably more durable than soft-paste porcelain.

Chafing Dish: A portable warming disher heated from beneath by a lamp. Originated from France in the seventeenth century, this device was originally developed to solve the problem of re-heating foods or keeping them warm at the dining table. Early chafing dishes were made of ornamental silver and fitted to small, three-legged stands. Modern versions are available in many styles and materials, and are primarily used for serving hot food at the dining table for large gatherings and parties.

Chair: A single seating unit with a back. The usual parts of chair are feet, legs, stretchers, apron, brackets, rails, seats frame, arms, splat and top rail.

Chair table: A convertible chair, also known as a "monk's chair" with a back that can be folded forward to transform the piece into a small table. Although it is known to have existed in the mid-fifteenth Century, the chair table was produced mainly in the seventeenth century.

Chair Web: A network of narrow woven fabric used to support upholstery springs in furniture. The web, made in widths from 2 inches to 3 1/2 inches, is woven of varying combinations of cotton, jute, hemp and flax.

Chamois leather: A soft leather originally obtained from an alpine antelope. Present-day chamois is taken from the underside of sheep skins. It is extensively used as a cleaning and polishing cloth.

Chandelier: A method of lighting which is used to refer a lighting fixture suspended from a ceiling or roof.

Chasing: A method of ornamenting a metal surface. The pattern is produced by embossing or cutting away parts of the metal, using a burin or graver on furniture items.

Cherry: A hard, compact, fine-grained wood, highly suitable for cabinet making because of its strength and beauty. Red-brown in colour, it takes a fine polish and resists warping.

Chestnut: A fairly soft wood with a coarse open grain. It is greyish brown in colour and is used for veneer cores.

Chests: Originally, containers or boxes with hinged lids. Drawers were added to the chest and the chest of drawers evolved there after.

Chinaware: A broad generic term for the many types of ceramic utensils and pottery.

Chroma: The degree of intensity or brilliance of a colour. Chroma does not apply to black, grey or white.

Collage: A picture or composition constructed of pieces of paper, cloth, and other materials glued or fastened to a surface. Collage is a French world for gluing.

Cornice lighting: concealed behind a cornice at the wall to give dramatic effect on a mural, pictures or wall textures.

Cove lighting: placed in a trough, near the ceiling, gives soft and uniform light directed towards the ceiling.

Design: Organisation of elements for a special object of art. Also refers to the act of designing, selecting and arranging of the elements for a particular purpose or to create beautiful objects.

Decorative Design: Any line, colour/material applied to structural design to add a richer quality as surface enrichment.

Eclecticism the term comes from the Greek Word Eclectics, meaning the combining art forms and furnishings from various periods and adopts them to today's living.

French Walnut: Fine walnut native to France prized for its clearly defined figures and deep brown colour. Walnut, along with ebony, was a favourite wood of the French, since English walnut was considered inferior.

French Windows: : Large rectangular casement windows that open from vertical halves in the same manner as a door. French windows are flush with the floor, and they swing inward to open. Such windows usually permit passage to porches, patios, front lawns or gardens.

Ferene Wood: The hard wood of the ash tree, tends to a white colour marked by brown streaks.

Frill: An ornamental edging often used on curtains. A frill, if gathered, crimped or pleated, produces the effect of a simple flounce.

Fringe: An ornamental edging often used on curtains, draperies or upholstery. They are composed of a variety of materials, often showy and luxurious.

Frontal: A hanging, richly embroidered curtain or an elaborately carved movable panel used in front of an altar. They are made of richest material and finest craftsmanship.

Fuller's earth: In textiles, a clay used as a filler in finishing cotton goods. Fuller's earth is also used as a cleaning agent because it can absorb common staining agents.

Furnishings: A term used in general sense to refer to the total equipment of any interior.

Furniture hardware: Structural or ornamental metals used in the construction of furniture and includes items such as locks, hinges, handles, leg mounts etc.

Furniture Veneer: The thin sheets or strips of wood used in furniture making to decorate or simply cover the flat furniture surfaces. These are usually cut from wood notable for a distinctive or appealing grain.

Gable: In architecture, the end portion of a triangular roof. The term is also used to refer to windows constructed in a triangular form. In furniture design, a gable is a triangular shaped member, usually a lid or a cornice, similar in shape to a gable roof.

Gallery: In architecture, a long rectangular hall used to display pictures, as in a museum. A gallery may also be a raised structure, such as a mezzanine, open on one side to a larger area.

Garages: In contemporary domestic architecture, structures designed to protect automobiles from inclement weather. They are also used as versatile units for carrying activities like laundry centres, storage, home workshops etc.

Glassware: A general term for glass objects used at the table.

Glaze: The glossy liquid coating applied to ceramics. The glaze effect can be either transparent or opaque and applied to decorate as well as to protect the object.

Gold leaf: Very thin sheets of gold foil applied to surfaces for decorative effect.

Golden coblong: The greek proportioning system to create pleasing rectangles having a ratio of 2:3 for flat surfaces and 5:7:11 for solids.

Guilding: The practice of decorating objects or parts of objects either by means of powder or gold leaf (a thin sheet of gold). This kind of decoration, prized for its sumptuous effect, has been practised since antiquity but was a favourite in Renaissance mode of decorating furniture, ceramics and metal works.

Hall marks: In gold and silver, a symbol or set or symbols used in the manner of a trademark to indicate the origin or maker.

Hardware: Metal fixtures, such as mounts, hinges and drawer pulls, used in the construction of furniture and other furnishings.

Hemp: A textile fibre obtained from the outer bark of the Heinz plant. The fibres are commonly used to make ropes, cloth and inexpensive and durable floor coverings.

Head boards: The head section of a bedstead made in different styles and decorative treatments.

Heika or Nageire are the arrangements of flowers which is made in a tall container, a bottle-shaped or cylindrical vase, or any vessel having depth.

High Chair: A general term for a child's armchair with unusually long legs set at an angle to the seat in order to ensure the stability of the structure.

Hooked rugs: Colourful and durable rugs made by hooking narrow strips of cloth or yarns through a canvas foundation.

House plants: A general term for all plants and shrubs capable of flourishing indoors. They provide variety and a decorative effect in a room.

Hue: The term used to refer the colour itself such as red hue or blue hue.

Ikebana means literally the bringing to life to plants, and its basis is the recreation of plants and flowers in their own natural wild settings. Also a popular kind of flower arrangement.

Insulation: The materials and techniques used in building to prevent the transfer of electric current, heat or sound.

Intensity or tone: A term used to refer to the brightness or dullness of a hue.

Intermediate (Tertiary) Colour: A hue formed by mixing two colours and falling midway between a primary and a nearly secondary colours.

Ivory: The white, bone-like tusks of elephants and the horns or tusks of other animals, such as rhinoceros, Whale, and Walrus. Ivory carving is an ancient, universal and highly sophisticated art, and used since ancient times in furniture construction and decoration.

Jacquard: A complex weaving technique named after the nineteenth-century French inventor, Joseph Marie Jacquard. This kind of weaving allowed greater freedom in patterns and figured fabrics.

Kapok: The name for a filler resembling silk and cottons, made from the fibres found in the seeds of the kapok tree used as filling for mattresses, pillows and sleeping bags and as insulation.

Kneehole: (desk): An opening in the facade (usually in the centre) of a desk. The kneehole provides space for the knees and thus allows the sitter to draw the desk chair closer to the writing surface.

Knife pleats: Sharp-edged peats, all folded in the same direction in a drapery, curtain etc. to give a neat and vertical line effect.

Lace wood: A hardwood native to Australia and similar to Mahogany in colour. The lace like grain of the wood has made it a favourite for Veneers.

Laminate: A plastic material built up in layers and then bonded together into a single unit. Because of its glossy surface, it does not hold dust and grease, and therefore easy to clean and maintain.

Lampshade: A variety of coverings use to shield, direct, or diffuse a light.

Lighting: A term used to refer to any lighting equipment and its effects.

Linoleum: A durable floor covering material made from linseed oil and cork.

Lounge: The term refers to a room or large area intended for occasional resting or relaxation.

Love seat: A settee, upholstered with open or closed arms, designed to seat two persons comfortably.

Macreme: Decorative knotting used originally in the Near East as a fringe in lace making.

Mahogany: As one of the principal woods used in the making of furniture pieces, Mahogany has a red colour, displays a variety of appealing grains and can be obtained in unusually large dimensions.

Malabar Carpet: A knotted carpet of coarse wool decorated with colourful motifs of Indian origin.

Mantel (mantelpiece): The wide ledge placed in the manner of a cornice above the opening of a fireplace. The term mantel piece is used to refer not only to the mantel, but also to the elements that surround the fireplace opening.

Maple: A light brown or yellow wood displaying a variety of grains used to make fine textured furniture.

Modern: A term used broadly to describe contemporary styles in furniture and architecture.

Mono-chromatic: A colour scheme that relies upon the effective blending of shades and tints of the same colour.

Moribana is a style of arranging flowers practiced by Japanese, means "piled-up flowers in a flat basin", (low bowls) and represents the first serious attempt to utilize western flowers in their arrangements.

Mortise and Tenon Joint: An ancient device for joining two pieces of wood inserting a tenon (a projecting ledge) of one piece into a mortise (a fitted cavity) of the other piece. The joint may be re-inforced by glue or a pin inserted so as to pass through both pieces of wood.

Mosaic: An extensive surface decorated with a portrait, scene, or pattern composed of small inlaid pieces of coloured stone, glass, ceramic tile or other suitable material.

Mural: A painting or a decorative surface (such as wall paper) applied to a wall, as an applied surface.

Muslin: A light weight cotton cloth often used for curtains and bed-spreads.

Myraid: infinitely great number, sometimes referring to ten thousand (in greek).

Net: A light-weight mesh fabric made from a variety of fibres and used for curtains and similar items.

Neo-catholic Style: A term used to describe the periodic re-awakening of interest in styles of medieval Europe. During nineteenth century, Gothic styles were a major contributing influence in both architecture and church furniture, though started in early and middle Georgian periods.

Nylon: A versatile synthetic material often used in place of silk and rayon. It is durable and elastic and can be manufactured in many forms.

Oak: A hard and durable wood very suitable for carved furniture etc. and native to Europe, North America and many other parts of the world. Oak ranges in colour from brown to nearly white.

Olive Wood: The hard wood of the European olive tree. Olive is yellow with brown streaks, and it is used principally for inlays and veneers.

Oriental rugs: Hand woven rugs from Asia and Asia Minor. The rugs are made from a variety of materials and in a variety of designs and colours, and divided into six major categories as Turkoman, Cancasiam, Turkish, Persian, Chinese and Indian.

Oriental Styles: The furniture styles of Far East, which have exerted an unique influence upon traditional and contemporary design.

Outdoor lighting: The application of electric light to the illumination of gardens and pathways for purposes of safety and decoration.

Outdoor Planting: Flowers and small shrubs suitable for cultivation in beds or in planting boxes used to decorate patios and porches.

Overstuffed: A term used in a general way to refer to fully and puffily upholstered sofas and Armchairs.

Padding: In textiles, the process often used in extensive production to apply dyestuffs to fabrics.

Paisley: The name give to a distinctive pattern appearing on woven or printed fabrics, originally made in Paisley, Scotland. Large, brightly coloured scroll forms of patterns, derived from Persian motifs, are popular for clothing, draperies and decorative items.

Pantries: A small area associated with kitchen or a separate area intended for the storage of foods, food articles and rarely used kitchen wares.

Papier-mache: Paper, cut into small pieces, mixed with glue or other ingredients to give colour or texture, then mashed to a pulpy consistency suitable for working like clay or plaster for making accessories like boxes, clock cases, pen holders etc.

Partition: An interior divider used to separate one room or area from another.

Pastel: In decorating, the term is often used when referring to soft, sometimes pale colours or tints of colours.

Patio: A roofless area within the house, such as a courtyard.

Pedestal: A base designed to support a statue, vase or column, specially for a lighting fixture, fan etc.

Pile: The napped, or softly resilient surface, carpets and fabrics like velvet.

Pine: An unusually versatile wood found in both hard and soft varieties and ranging in colour from white to yellow.

Plaster: A surfacing material composed of water, sand and lime or gypsum.

Plastic: A great number of synthetic or partially synthetic compounds like nylon, vinly and styrofoam.

Prefabricated materials: Materials used in a building technique in which components of a house are pre-constructed, then quickly assembled at the site.

Primary Colours: Red, blue and yellow are the primary colours which are combined to form the secondary colours.

Quilt: Usually a coverlet composed of two layers of fabric with soft filling between them, the layers are held together by quilting stitches executed in a decorative pattern.

Rangoli is an important art originating from India and is used to decorate mainly the floor, and in some areas, walls too, with colourful designs and patterns.

Rayon: A synthetic fibre formed from cellulose filament, either used alone or in combination with other fibres in the manufacture of a great variety of fabrics.

Recessed lighting: a type of lighting that includes built-in panel fixtures on the walls or ceiling to provide accent lighting.

Resilient floor covering: Floor covering that has a degree of elasticity underfoot like vinyl, rubber, cork etc.

Rikka means "standing flowers" and is the oldest form of Japanese flower arrangement usually large, elegant and magnificent, at times being massive in their proportions.

Rose wood: A dark brown to red wood marked by strong brown to black streaks. It is popular with furniture designers.

Sandal wood: A fragrant wood of yellowish colour and fine even grain. Sandalwood is native to India and is used for carving, and because of its fragrance, for drawer linings.

Sateen: Sateen is a cotton fabric woven with satin weave, producing a material with one smooth side and one rough side.

Satin: A silk fabric is prized for its richness of surface and weight. Its distinctive lustrous surface is especially attractive when draped, and the material is often used for hangings as well as for upholstery.

Satin wood: A hard, light-coloured, straight grained wood from the southern regions of Asia. Capable of taking a high polish, it is popular for inlays and veneers.

Scale: The capacity to select objects and arrange them so that they look well-proportioned.

Screen: A comprehensive term for a separating device, to define space, provide privacy, or shield an area against drafts.

Settee: A small upholstered seat intended to accommodate two or three people, and is distinguished from sofa by its smaller size.

Shade: A term used to refer to a darkened colour by the addition of black to a hue.

Shoka is derived from the compound meaning "living flowers" which may also be read Seika or Ikebana.

Sky lights: is a glazed opening in the roof of a house constructed to allow natural light inside the house.

Slip cover: A cover of fabric, plastic, or a similar material that can be fitted over an upholstered chair or sofa to protect the upholstery or provide added decoration.

Slate stone: A dark brittle stone easily cut into plates suitable for roofing and flooring.

Son et lumiere: entertainment given by night at historic buildings etc. with recorded sound and lighting effects to give dramatic narrative of history (sound and light).

Splint: A slender strip of hardwood, often woven to form chair seats and backs or boxes.

Stencil: A sheet of paper or metal or plastic into which patterns have been cut. The pattern is transferred to a piece of furniture or wall surface or fabric by brushing paint over the sheet.

Stone ware: A very heavy, hard and non-porous form of pottery related to porcelain.

Structural design: The design made by the size, form, colour and texture of the object in space or its drawing worked out on paper.

Swag: Cloth gathered and draped in a single curve or a series of similar curves. The term is also used to refer to the same effect in a carved motif in furniture.

Symmetrical: A term used to refer to any arrangement that has two equal halves, the one half a reverse of the other (a mirror image).

Taffeta: A tightly woven fabric with a lustrous surface and a graceful draping quality, woven from any of several different yarns.

Teak: A hard, durable, light-brown wood that has a strong black markings, popularly used in modern wood furniture.

Tenon: In wood working, the projecting portion of a mortise and tenon joint.

Terra-cotta: A coarse and porous baked clay varying in colour from red to dull tan, commonly used for making artistic as well as utilitarian objects.

Terrazo: A surface material composed of crushed marble blended into cement, then polished. Due to the colouring of the marble, terrazzo forms a highly decorative surface suitable for walls, floors and other walkways.

Tertiary Colour: A colour produced by mixing a primary colour with a neighbouring secondary colour on the colour wheel.

Texture: In general, the term is used for the tactile and visual qualities of a surface like satin with a smooth texture and concrete, a rough one.

Tile: A thin plate of varied size, shape and material used as the basic unit in covering surfaces like floors, walls and roofs.

Tint: A term used to refer to a paler colour produced by the addition of white to a brighter hue.

Tone: refers to a range of tints and shades of a colour, often obtained by the addition of grey to a colour.

Tongue and groove: A device used to join two pieces of wood by inserting a tongue (a projecting edge) of one piece into a groove (a fitted cavity) of the other piece, a joint similar to mortise and tenon.

Upholstery: The methods and materials used in stuffing an article of furniture and covering it with a suitable material.

Urn: A large vase usually of pottery or metal, traditionally used as a decorative article.

Valance: The fabric used like a cornice to hide the tops of draperies. Another common type of valance is the decorating border of fabric around a bed from mattress to floor that hides the space below.

Valence lighting: is the lighting source which is mounted to direct the light towards the ceiling with some amount of light coming down on the draperies or the wall as accents.

Value: This refers to lightness or darkness of a hue.

Varnish: A liquid applied to wood surfaces to produce a glossy transparent finish, and is composed of oils like linseed oil, containing dissolved gum or resins.

Veneering: The art of covering furniture surfaces with thin sheets of wood for a decorative effect.

Venetian blind: A panel or window hanging made of slats of wood or metal or plastic that rotates outward or inward to control visibility and air circulation.

Walnut: A light brown to golden wood found suitable for both solid and veneered furniture.

Wardrobe: A term used to refer large vertical cabinet intended for hanging clothing.

Welting: A length of fabric, usually tubular in shape, used to reinforce and disguise seams on upholstery, slipcovers and bed spreads.

Wormy Chestnut: Pock-marked chestnut wood. A soft wood, wormy chestnut is sometimes used for decorative wall paneling.

Wrought iron: Iron that has been worked into decorative shapes, used for structural as well as decorative parts of furniture.

X-shaped chair: A portable folding chair or stool with an X-shaped base in place of the customary legs.

Yew: A hard reddish brown wood often used for utilitarian furniture.

Zebra wood: A heavy hardwood with characteristic dark stripes making a light brown colour, and is used mostly for decorative effects.

Zigzag-motif: A continuous lined formed of equal angles and often used as an ornamental border.